CHARGED AND REACTIVE POLYMERS

VOLUME 2

POLYELECTROLYTES AND THEIR APPLICATIONS

CHARGED AND REACTIVE POLYMERS

A SERIES EDITED BY ERIC SÉLÉGNY

Volume 1

POLYELECTROLYTES

*Papers initiated by a NATO Advanced Study Institute
on 'Charged and Reactive Polymers', held in France, June 1972*

Edited by Eric Sélégny

Co-edited by Michel Mandel and Ulrich P. Strauss

VOLUME 2

POLYELECTROLYTES AND THEIR APPLICATIONS

Edited by

ALAN REMBAUM

Jet Propulsion Laboratory, California Institute of Technology, Calif., U.S.A.

and

ERIC SÉLÉGNY

Université de Rouen, France

D. REIDEL PUBLISHING COMPANY

DORDRECHT-HOLLAND / BOSTON-U.S.A.

Library of Congress Cataloging in Publication Data

Rembaum, Alan.
    Polyelectrolytes and their applications.

    (Charged and reactive polymers ; v. 2)
    Includes bibliographical references and index.
        1.   Polyelectrolytes.   I.   Sélégny, Éric, joint author.
    II.   Title.   III.   Series.
    QD565.R52   541'.372   74-34151
    ISBN 90-277-0561-5

Published by D. Reidel Publishing Company,
P.O. Box 17, Dordrecht, Holland

Sold and distributed in the U.S.A., Canada, and Mexico
by D. Reidel Publishing Company, Inc.
306 Dartmouth Street, Boston,
Mass. 02116, U.S.A.

Printed in The Netherlands by D. Reidel, Dordrecht

TABLE OF CONTENTS

INTRODUCTION

Polyelectrolytes and their Applications is the second volume in the series 'Charged and Reactive Polymers'. The important areas of polyelectrolyte applications, i.e., biomedicine, water purification, petroleum recovery and drag reduction, are presented along with discussions of the fundamental principles of polyelectrolyte chemistry and physics. This book should be of interest to scientists such as physicians, biochemists, polymer chemists and chemical engineers involved in applications of these materials.

The first part of the book is devoted to the basic properties of polyelectrolytes in general, namely to the factors influencing the chain conformation of charged polymers in solution and to their counterion selectivity. It also contains methods of synthesis and new concepts of charge stabilized polymer colloids and of polyelectrolyte catalysis.

The second part describes recent information on the properties and biological effects of already well-known natural polyelectrolytes such as heparin and DNA and recently developed polymers such as pyran and polyionenes. The effects of polyanions and polycations on normal and transformed cells as well as on acetylcholine receptors follow. This part is of particular interest to scientists involved in biological research.

The third part deals with problems of importance to our environment and energy sources, including a recent approach to water purification by means of charged hollow fibers, the formation of charged membranes for reverse osmosis and the removal of mineral acids by means of ion exchange resins. The reduction of frictional drag by means of high molecular weight polyelectrolytes is a phenomenon which offers practical potential in the design development of modern marine systems as well as in the improved extraction from oil wells. Similarly an understanding of the retention of polyelectrolytes in porous media is also required for an efficient oil recovery. Recent investigations in both these areas are discussed in the last two chapters.

The major part of this volume is based on lectures presented at a symposium held at the California Institute of Technology on May 23, 24 and 25, 1973. The symposium was sponsored by the Polymer Group of the Southern California Section of the American Chemical Society and was organized by a committee which included Dr. A. Rembaum (chairman), Dr. T. F. Yen (cochairman) of the University of Southern California, Dr. N. Tschoegl (cochairman) of the Chemical Engineering Department of the California Institute of Technology, Dr. D. Kaelble of the North American Science Center and Dr. J. Moacanin of the Jet Propulsion Laboratory. We are indebted to the members of the committee for making it possible to present such a stimulating program.

Some other subjects have been added during the editing period in an attempt to widen the spectrum of authors and topics. However, in view of the scope and vast number of problems in the polyelectrolyte field, the above topics could undoubtedly be further enlarged, but they do harmoniously complete the fundamental information contained in the first volume of the series and illustrate many of the applications. It is hoped that these volumes together will be of great help to all those interested in the study, teaching or use of charged polymers. In order to help the reader, an exhaustive index is included.

It is a pleasure for the editors to thank Miss Albane Sélégny and Drs. Braud, Fenyo and Muller for their valuable help in the preparation of the index and for proof readings.

A. REMBAUM
E. SÉLÉGNY

PART I

GENERAL PROPERTIES AND SYNTHESIS

HYPERCOILING IN HYDROPHOBIC POLYACIDS

P. L. DUBIN* and U. P. STRAUSS

*School of Chemistry, Rutgers University, The State University of New Jersey,
New Brunswick, N.J. 08903, U.S.A.*

Abstract. A series of alternating copolymers with the formula

$$\left(\begin{array}{cccc} ---CH--- & CH--- & CH_2--- & CH--- \\ | & | & & | \\ COOH & COOH & & OR \end{array}\right)_x,$$

where R is a normal alkyl group, was synthesized. The properties of these compounds in aqueous solution reflect a balance between cohesive hydrophobic interactions of non-polar side-chains and electrostatic repulsive forces of charged carboxyl groups. A similar accommodation of groups of widely differing hydrophilicity must be a major factor in determining the conformation of biopolymers in solution.

Intrinsic viscosity measurements in dilute NaCl and in THF-cyclohexane mixtures reveal that the copolymers with side-chains containing four or more methylene groups exhibit at a low degree of dissociation (α) dimensions inferior to the unperturbed values; hence the appellation 'hypercoiled'. With increasing charge density, the dimensions of the 'butyl' and 'hexyl' copolymers are observed to increase rapidly over a narrow range of α indicating cooperative destruction of the compact state. Solubilization measurements with these two polymers suggest that intramolecular micelles which exist at low pH are destroyed as the charge density increases over the same range of α. While 'butyl' and 'hexyl' copolymers undergo pH-induced conformation changes, 'methyl' and 'ethyl' compounds behave as typical weak polyelectrolytes. At the other extreme, 'octyl' and 'decyl' polymers appear to be hypercoiled even at high charge density and thus exhibit typical 'polysoap' behaviour.

Analysis of the proton titration curves of these polyacids according to a treatment first applied to ionic polypeptides leads to thermodynamic parameters for the (hypothetical) conformational transition at zero charge. The effect of side-chain length, temperature, ionic strength, and the presence of protein denaturants on these parameters has been studied.

Cohesive intra-molecular interactions among apolar moieties has long been recognized as a major factor determining the unique compact conformations of globular proteins in aqueous solution. Such hydrophobic interactions strongly influence the solution dimensions of a number of synthetic water-soluble polymers as well. The polymeric compounds of the studies described here [1, 2] represent a series of macromolecules whose hydrophobic content may be varied systematically. The presence of identical repeating ionizable groups makes possible experimental procedures which lead to thermodynamic parameters for the cooperative interactions of the apolar groups.

Alternating copolymers ('interpolymers') of maleic anhydride and alkyl vinyl ethers are prepared by free radical polymerization in solvent, non-solvent or bulk media. Products prepared from ethyl-, butyl-, hexyl-, and octyl-, and decyl- vinyl ethers are described in Table I. The decyl-copolymer was prepared by Dr A. W. Schultz. A methyl-copolymer ('Gantrez AN' General Aniline and Film Corp.), not described in Table I, was included in some studies. Molecular weight estimates were

---

* Current address: Dynapol, 1454 Page Mill Road, Palo Alto, Calif. 94304.

Alan Rembaum and Eric Sélégny (eds.), Polyelectrolytes and Their Applications, 3–13. All Rights Reserved.
Copyright © 1975 by D. Reidel Publishing Company, Dordrecht-Holland.

TABLE I

Synthesis of:

$$\{CH-CH-CH_2-CH\}_x$$

Polymerizations at 60 °C with 0.1 wt.% AIBN

with structure featuring:
$O=C$, O, $C=O$ groups and O — C_nH_{2n+1}

Code	n	Type of polymerization	$[\eta]_{THF}$, l.eq.$^{-1}$	Approx. DP
A	2	Heterogeneous (Benzene)	11.8	650
B-I	4	Homogeneous[a] (THF)	0.5	20[b]
B-II	4	Heterogeneous (Benzene)	9.9	500[b]
B-III	4	Bulk[c]	37.0	5000[b]
C	6	Homogeneous (Benzene)	20.4	1700[b]
D	8	Homogeneous (Benzene)	28.6	2800[b]
E	10	Homogeneous (Benzene)	7.0	340[d]

[a] No initiator necessary.
[b] From Reference 8.
[c] 0.02 wt.% AIBN.
[d] From Reference 3.

arrived at by interpolation between the viscosity molecular weight relationship obtained for the dodecyl-copolymer [4] and a relationship derived from data for the methyl copolymer [5] and must be regarded as rough approximations only.

In aqueous solutions the anhydride units of these copolymers hydrolyze to form dicarboxylic acids. In a manner common to all polycarboxylic acids, the physicochemical solution properties of these compounds manifest strong pH dependence because the polymeric linear charge density varies monotonically with the degree of dissociation of acid groups. It is convenient to define the degree of dissociation α as [COO^-]/[monomer residues] such that its value ranges from 0 to 2. The dependence of the intrinsic viscosity $[\eta]$ (in l eq.$^{-1}$) on α in 0.04 M NaCl at 30 °C is presented in Figure 1. The units of $[\eta]$ arise from the concentration units that were chosen to facilitate comparisons among the several polymers. Conversion to the more familiar units of dl gm^{-1} is accomplished by dividing by the factor (equivalent weight) $(10)^{-1}$ which typically leads to reduction by one order of magnitude.

Viscosity data for the ethyl-copolymer, A, over the range $\alpha=0$ to $\alpha=1$, reveal the typical electrostatic expansion behavior of a weak polyacid such as polyacrylic acid. Polymer B-II, on the other hand, exists in an abnormally compact state at low α. The curves for polymers B-I and B-III, excluded from Figure 1 for clarity, are identical in shape and differ only in relative magnitude. The five-fold increase in $[\eta]$ over the range $0.35 < \alpha < 0.65$ is indicative of a process in which an initially compact state, presumably stabilized by hydrophobic interactions, is destroyed by electrostatic repulsive forces. A similar region of expansion is observed for the hexyl copolymer, C, in the vicinity of $\alpha = 1.1$. The low values of $[\eta]$ for C at high α presumably reflect its relatively unfavorable solvent affinity. The octyl-copolymer is insoluble in 0.04 M NaCl at acid pH; hence the data for D of Figure 1 were obtained in 0.01 M NaCl. The

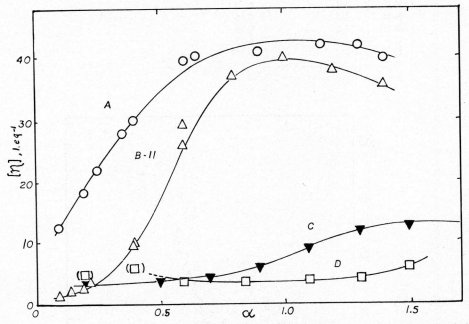

Fig. 1. Intrinsic viscosity as a function of degree of dissociation in 0.04 M NaCl at 30° for ethyl- (○), butyl- (△), and hexyl- (▼) copolymers; and for octyl- (□) copolymer in 0.01 M NaCl.

increase in $[\eta]$ for this polymer that occurs as the value of α falls below 0.5 is presumably related to aggregation. There is no compelling evidence from viscosity data for a transition to an expanded state at high values of α for the octyl-copolymer.

Meaningful comparisons of hydrodynamic behavior among these copolymers of varying molecular weights may be made by considering the ratio of the intrinsic viscosity in a given solvent to that observed in an ideal or θ-solvent. The parameter $[\eta]/[\eta]_\theta$ is the cube of the Flory expansion coefficient, and is relatively insensitive to molecular weight. Approximate values of $[\eta]_\theta$ were obtained from measurements in THF-cyclohexane mixtures in which the concentration of the second component was slightly less than that sufficient to produce phase separation. These ratios tabulated in Table II indicate that while dilute NaCl is a good solvent for the ethyl-copolymer at $\alpha = 0.1$, yielding an expansion coefficient similar to that in THF, the other co-polymers, under the same conditions, are more compact than under θ-solvent conditions. The anomalous values for polymer B-I in both solvents may reflect a discrepancy between a true θ-solvent and one chosen by the phase separation method for low

TABLE II

Viscosities in THF and dilute acid relative to
approximate θ-solvent values

	$([\eta]_{THF}/[\eta]_\theta)^{1/3}$	$([\eta]_{\alpha=0.1}/[\eta]_\theta)^{1/3}$
A	1.5	1.6
B-I	1.3	0.93
B-II	1.5	0.86
B-III	1.5	0.78
C	1.5	0.78
D	–	0.76[a]

[a] $\alpha=0.6$.

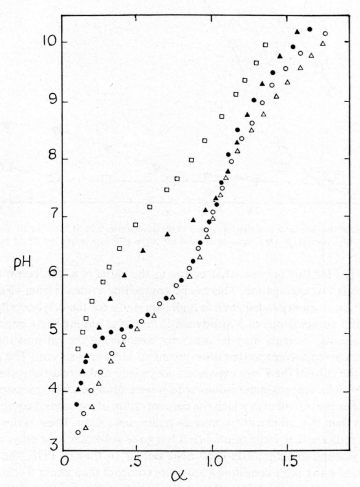

Fig. 2. Potentiometric titration data in pure water at 30 °C for octyl- (□), hexyl- (▲), butyl- (●), ethyl-
(○), and methyl- (△) (Gantrez AN) copolymers.

molecular weight polymers. However, the trend among the B polymers towards increasing compactness with increasing molecular weight is probably real and is in agreement with results for the styrene maleic acid copolymer reported elsewhere [6]. The appellation 'hypercoiling' [6] refers to the presence of a conformation more compact than the minimum theoretical value for the random coil.

When changes in pH result in a transition between one distinct conformation of a regular polyacid and another, the thermodynamic parameters of the transition may be evaluated from potentiometric titration data. This approach has been applied to numerous synthetic ionizable polypeptides. A characteristic feature of such data is a 'hump' in the titration curve in the region of the conformation change. The data of Figure 2 reveal that the butyl- and hexyl-copolymers exhibit such deviations from typical polyacid behavior as exemplified by the ethyl- and methyl-copolymer titration curves, over the same range of α in which viscosity data indicate the presence of a compact structure. Data for the octyl copolymer suggest that the compact structure is not fully disrupted over the pH range examined, corresponding to $\alpha < 1.3$, in harmony with the viscosity results of Figure 1.

Titration data for poly (monoprotic acids) are related to the thermodynamic parameters for the conformational transition by the expression [7]

$$(RT \ln 10)^{-1} (\partial G_{ion}/\partial \alpha) + pK_0 = pH + \log\left(\frac{1-\alpha}{\alpha}\right), \tag{1}$$

where pK_0 is the value of the apparent logarithmic dissociation constant $pK_a = pH + \log((1-\alpha)/\alpha)$ in the limit of $\alpha = 0$; and $(\partial G_{ion}/\partial \alpha)$ is the change in free energy accompanying an incremental change in ionization. The analogous expression for a poly (diprotic acid) is [2]

$$(RT \ln 10)^{-1}(\partial G_{ion}/\partial \alpha) + pK_1^0 =$$
$$= pH + \log\left\{\left(\frac{1-\alpha}{2\alpha}\right) + \left[\left(\frac{1-\alpha}{2\alpha}\right)^2 + \frac{K_2^0}{K_1^0}\left(\frac{2-\alpha}{\alpha}\right)\right]^{1/2}\right\}. \tag{2}$$

The left hand side of Equation (2) is plotted as a function of α in Figure 3. The regions of negative slope in the titration curves of B and C correspond to abrupt conformational transitions in the course of which the electrostatic free energy actually decreases with increasing *linear* charge density because of cooperative macromolecular expansion. Since the titration data of the ethyl copolymer correspond to the non-hypercoiling polyelectrolyte coil, the area between titration curve A and either B or C is proportional to the free energy change for the thermodynamic cycle:

$$(\text{hypercoil}_{\alpha=0}) \rightarrow (\text{expanded coil}_{\alpha=\alpha^*}) \rightarrow (\text{expanded coil})_{\alpha=0}.$$

This free energy change, ΔG_t^0, thus corresponds to the hypothetical transition from hypercoil to expanded coil at zero charge. Values for ΔG_t^0 are reported in Table III along with values of ΔH_t^0 and ΔS_t^0 obtained from temperature dependence studies. The behavior of the entropy and enthalpy terms is typical of processes controlled by

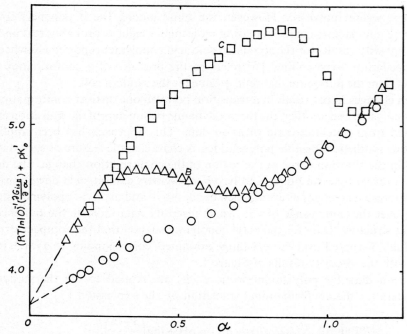

Fig. 3. Titration data for ethyl- (○), butyl- (△) and hexyl- (□) copolymers in pure water at 30 °C, plotted according to Equation (2).

TABLE III

Thermodynamic parameters of the transition from hypercoil to (hypothetical) extended coil at zero charge

		ΔG_t° (cal/mole)		ΔH_t° (cal/mole)		ΔS_t° (e.u.)	
B-II	25°	+308	±10	−700	±100	−3.4	±0.5
	45°	+323	±10	+500	±100	+0.6	±0.3
C	25°	+1130	±10	−1300	±200	+8.1	±1
	35°	+1120	±10	+2800	±400	+5.4	±1

hydrophobic interactions. The contribution of a methylene group to the stability of the uncharged hypercoil, $\Delta G_{CH_2}^0$, may be obtained from these results as 400 cal/mole [2]. It is of interest to examine the effect of changing solvent conditions on $\Delta G_{CH_2}^0$. Such an analysis has been undertaken in an attempt to elucidate the denaturant effect of concentrated urea solution [8].

The formation of a compact state stabilized by hydrophobic interactions in the copolymers containing side chains of four or more methylene groups suggests the presence of intramolecular micelles. Solubilization measurements [9] were made with a lipophilic water insoluble dye, yellow OB, by the method of Ito et al. [4]. In Figure 4, the solubilization parameter, S, expressed as moles of dye solubilized per

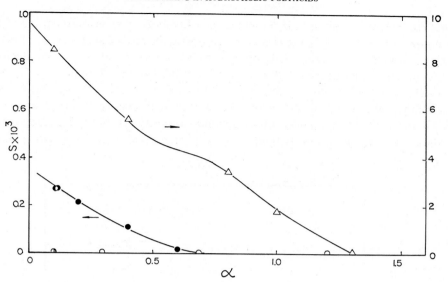

Fig. 4. The dependence of the solubilization parameter on the degree of dissociation in pure water at 22 °C for ethyl- (○), butyl- (B-I: ◑, B-II: ●, B-III: ◐) and hexyl- (△) copolymers. Polymer concentrations: 0.05 eq. l^{-1}, except for hexyl-, 0.01 eq. l^{-1}.

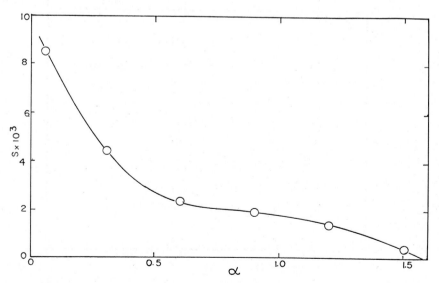

Fig. 5. Solubilization as a function of degree of dissociation for octyl- copolymer, 0.01 eq. l^{-1}, in 0.01 M NaCl at 30 °C.

mole equivalent of polymer, is presented as a function of α. Several aspects of the solubilization results of Figure 4 are in accord with the model suggested by viscosity and titration data. First, the ethyl-copolymer is seen to be incapable of solubilization while successive increases in S are observed for butyl- and hexyl-copolymers. Second, at values of α above those corresponding to the completion of the conformational transitions, determined from viscometric and potentiometric titration curves as α = 0.8 and 1.2 for butyl- and hexyl-copolymers respectively, the solubilizing power of these two polymers is observed to vanish. However, several features of these

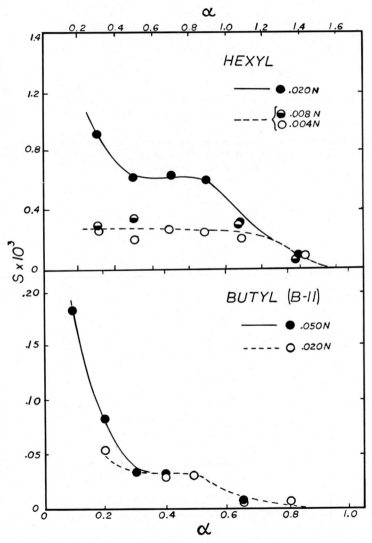

Fig. 6.   Dependence of solubilization on polymer concentration for butyl- and hexyl-copolymers in 0.04 M NaCl at 30 °C.

solubilization data are unexpected on the basis of preceding results. Polymer B-I appears to be incapable of dye uptake, indicating a lower limit on the number of successive hydrophobic groups required for the formation of a micellar site. Above this value, molecular weight independence is established by the coincidence of curves B-II and B-III. The initial negative slopes shown in Figure 4 are not fully in accord with viscosity and titration data, which suggest that the hypercoiled conformation is relatively stable with respect to α below the region of α identified with the conforma-

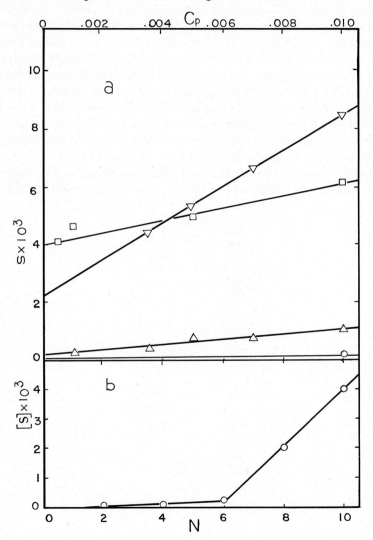

Fig. 7. (a) Concentration dependence of solubilization by butyl-, hexyl-, octyl- and decyl- copolyacids (stoichiometric degree of neutralization = 0) in pure water at 22 °C. Data for butyl-copolymer at higher concentrations not shown. (b) Dependence of the 'intrinsic solubilization' on alkyl side chain length (from data of Figure 7(a)).

tional transition. Similar behavior is observed for the octyl-copolymer (Figure 5); these results also support the earlier suggestion of a conformation charge for this copolymer in the vicinity of $\alpha = 1.5$.

The solubilization curves for hexyl- and octyl-copolymers (Figures 4 and 5) reveal regions of strongly negative α-dependence both at low pH and over the range of higher values of α corresponding to the conformational transition regions, separated by an intermediate plateau. One explanation for the behavior in the first region entails the successive disruption of aggregates with increasing α resulting in the disappearance of micellar sites formed from *inter*molecular associations. This hypothesis is supported by the concentration dependence of the solubilization curves as seen in Figure 6. The negative α-dependence at low α for the solubilization by the hexyl copolymer is observed to vanish at polymer concentrations below 0.008 N and the shape of the solubilization curve in the limit of high dilution is in close agreement with the existence of a stable hypercoiled conformation at low α. Because of the low solubilization by the butyl-copolymer measurements could not be carried out with adequate experimental accuracy at very low polymer concentrations; nontheless the available data for the butyl copolymer approach similar limiting behavior.

In view of the concentration dependence of the solubilization, it is appropriate to define the intrinsic solubilization $[S] = \lim_{c_p \to 0} S$ in analogy to the intrinsic viscosity.

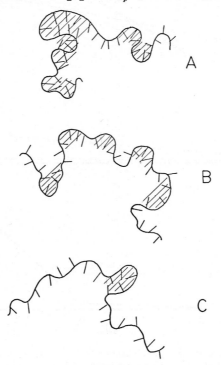

Fig. 8. Schematic illustration of intramolecular micellization of a hydrophobic polyacid at varying degrees of molecular expansion.

Extrapolations to zero concentration are illustrated in the upper part of Figure 7 for butyl-, hexyl-, octyl- and decyl- copolymers (B-II, C, D, E). The ordinate intercepts are plotted as a function of n, the number of methylene units in the side chains, in the lower part of this figure. Since total polymeric apolar content increases with n, the positive slope is not surprising. The break in the curve at $n = 6$ however suggests that increase in the length of the alkyl side chains beyond hexyl results in a more efficient solubilization process, possibly related to the formation of more densely hydrophobic intramolecular micellar regions.

A molecular model consistent with fundamental viscosity, titration, and solubilization behavior is illustrated schematically in Figure 8. In this model cohesive interactions occur among closely neighboring side chains only and a treatment analogous to those developed by Zimm and Rice [10] and by Ptitsyn and coworkers [11] may be applied to yield cooperativity parameters for the formation of micellar groups with values for σ of ca. 1×10^{-2} and 2.5×10^{-2} for hexyl- and butyl-copolymers respectively [9]. However, it is difficult to rigorously exclude the possibility of stable cohesive interactions among hydrophobic groups far apart on the polymer chain especially in view of the inordinately low viscosity of the hexyl copolymer at high α. Hence the abovementioned linear model should be viewed only as a first approximation to a molecular model.

Acknowledgements

This work was supported by Grants from the United States Public Health Service (Grant GM 12307) and from S. C. Johnson and Son. Inc.

References

1. Dubin, P. and Strauss, U. P.: *J. Phys. Chem.* **71**, 2757 (1967).
2. Dubin, P. and Strauss, U. P.: *ibid.* **74**, 2842 (1970).
3. Strauss, U. P. and Varoqui, R.: *ibid.* **72**, 2507 (1968).
4. Ito, K., Ono, H., and Yamashita, Y.: *J. Colloid Sci.* **19**, 28 (1964).
5. Woodruff, C. W., Peck, G. E., and Banker, G. S.: *J. Pharm. Sci.* **61**, 1916 (1972).
6. Dannhauser, W., Glaze, W. H., Dueltgen, R. L., and Ninomiya, K.: *J. Am. Chem. Soc.* **64**, 954 (1960).
7. Overbeek, J. Th. G.: *Bull. Soc. Chim. Belg.* **57**, 252 (1948).
8. Dubin, P. and Strauss, U. P.: *J. Phys. Chem.* **77**, 1427 (1973).
9. Dubin, P.: Dissertation, Rutgers, The State University, New Brunswick, N.J. 08903, 1970.
10. Zimm, B. H. and Rice, S. A.: *Mol. Phys.* **3**, 391 (1960).
11. Ptitsyn, O. B. and Birshtein, T. M.: *Biopolymers* **7**, 435 (1969).

COMPACT CONFORMATION OF POLYIONS STABILIZED
BY NON-ELECTROSTATIC
SHORT-RANGE INTERACTIONS

GUY MULLER, JEAN-CLAUDE FENYO, CHRISTIAN BRAUD,
and ERIC SÉLÉGNY

Laboratoire de Chimie Macromoléculaire de la Faculté des Sciences et des Techniques et de l'Institut Scientifique de Haute-Normandie, ERA 471, Université de Rouen, 76130 – Mont-Saint-Aignan, France

Abstract. The macromolecular behaviour of some hydrophobic polymers has been investigated by various methods, namely potentiometric, viscosimetric and optical measurements. All the reported results are interpreted by the existence of non-electrostatic short-range interactions leading to a partial or nearly complete screening of electrostatic repulsions. In water, the stabilization of compact conformations lies in a more or less wide range of ionization depending on the hydrophobic character of the polymeric chains and/or of the counter-ion. In the peculiar case of a polycondensate between L-lysine and 1,3-benzene-disulfonyl chloride (PLL), the observed transition from a very compact structure to a mean coil-like one is found to be strongly dependent on the length of apolar groups (and hydrophobicity) of N-tetraalkylammonium counter-ions. In the presence of bulky N-tetraalkylammonium (Bu_4N^+), the cohesive forces are predominant until α is around 0.7. By partial replacement of water by a good organic solvent of the polymeric chains, or in aqueous solution of urea, normal polyelectrolytic behaviour without any discontinuity in both potentiometric and viscosity changes is observed.

On the contrary, addition of dye (Acridine Orange) leads to a stabilization of compact conformation in a wider range of ionization as illustrated by the shift of the transition region to higher values of α as the concentration in dye increases. The optical data concerning PLL are in good agreement with the potentiometric and viscosimetric data. The observed features show that Bu_4N^+ acts in a similar way on the macromolecular behaviour of PLL to a molecule of dye.

Finally, the size of N-tetraalkylammonium counter-ion for PLL solutions is found to play a role similar to that of side-chain length for hydrophobic copolymers of maleic acid and alkylvinyl ethers.

Key Words

PLL : polycondensate between L-lysine and 1, 3-benzenedisulfonyl chloride
PLLMe : N,N'-dimethylated PLL
PLDAB : polycondensate between α,γ-diaminobutyric acid and 1,3-benzenedisulfonyl chloride
PAA : polyacrylic acid
PMA : polymethacrylic acid
COP_y : partially hydrolyzed N,N-disubstituted polyacrylamide.
AO : Acridine Orange
Potentiometric titrations
Viscosimetry
Dielectric dispersion
Optical activity
Dichroic spectra

Compact conformation
Non-electrostatic interactions
Hydrophobic interactions
Conformational transition
Extended form
Free energy change
Tetraalkylammonium counter-ions
Hydroorganic mixture
Urea solution

Alan Rembaum and Eric Sélégny (eds.), Polyelectrolytes and Their Applications, 15–30. All Rights Reserved.

Dye (Acridine Orange)
Stabilization of compact conformation

1. Introduction

Long-range interactions due to electric charges located on different units of macro-molecular chains are quite generally responsible for the physico-chemical behaviour of polyelectrolytes.

In fact, it seems to be a more realistic picture to admit that such interactions are in many cases a predominant factor controlling the average dimensions of chains, as pointed out by a good deal of work concerning polyelectrolytes.

In recent years, considerable interest has been shown in the effect of short-range interactions acting in an opposite way to the repulsive ones. Such cohesive inter-actions lead to the existence in aqueous solution of special conformations stabilized in spite of the increasing number of charges carried on the chains.

The role and the importance of non electrostatic forces was first pointed out in the case of biopolymers and then extended to synthetic polyelectrolytes.

In the case of globular proteins, it has been shown that the native conformation is stabilized through interactions of non polar groups and that denaturation occurs by disrupting them. A globular protein can be considered as an equilibrium state for three kinds of molecular forces: electrostatic repulsions, van der Waals cohesion of the uncharged side-groups and H-bonds between peptide groups [1].

In the field of synthetic polyelectrolytes, the special case of polysoaps (defined as synthetic polymers into which long chain soap-like groups have been introduced) was extensively studied [2]. Their compact conformation was shown to be primarily due to strong apolar-apolar interactions between the long aliphatic side groups carried by the main chain.

From viscosity and light scattering measurements made on a series of poly-4-vinyl-pyridine derivatives prepared by partial quaternization of the pyridine groups with dodecyl bromide, it was concluded that a transition from polyelectrolyte to polysoap is observed by increasing content of dodecyl groups on the flexible chain of the parent polymers. Also, results obtained on polysoaps derived from poly-2-vinylpyridine by partial quaternization of pyridine groups with alkyl bromide are in qualitative agreement with the above results and the polyelectrolyte-polysoap transition was also supported from solubilization of dye (Suddan III) [3].

Hydrolyzed copolymers of maleic anhydride and styrene [4] or alkylvinyl ethers [5, 6] were extensively investigated and in both cases the special behaviour of such polyions was attributed to the existence of strong cohesive forces between apolar side-groups.

Also the differences observed between polyacrylic acid (PAA) and polymethacrylic acid (PMA) have been intensively studied and correlated to a transition from a rather compact coil to an extended one, occuring during ionization in the case of PMA. All the differences between PAA and PMA are attributed to the presence of 'hydrophobic' methyl side groups in the PMA chain [7, 8, 9]. However, there is still some discrepancy concerning the exact nature of forces responsible for the compact structure of PMA at low ionization values.

In all these examples, the observed effects are related to a balance between 'cohesive forces' due to hydrophobic groups and 'repulsive coulombic interactions'.

For natural polyelectrolytes and polypeptides, the term 'transition' means the passage from an ordered state (e.g. helix for poly-L-glutamic acid) to a disordered one under the influence of different factors i.e. chemical agents, temperature, ionization....

It must be pointed out that for synthetic polyelectrolytes, the term 'conformational transition' as generally used is concerned with the passage from a rather compact state ('a' state) to an extended state ('b' state) occuring in a small range of ionization.

The transitions have been established by various experimental methods such as viscosity, optical rotation, potentiometric titrations or dielectric measurements.

TABLE I

Characteristics of polymer samples

Code	Formula	Reference
PAA	—CH—CH$_2$— $\mid$ COOH	[10]
PMA	—C(CH$_3$)—CH$_2$— $\mid$ COOH	[10]
PLL	—SO$_2$—ϕ—SO$_2$—NH—C*H—(CH$_2$)$_4$—NH— $\mid$ COOH	[11]
PLDAB	—SO$_2$—ϕ—SO$_2$—NH—C*H—(CH$_2$)$_2$—NH— $\mid$ COOH	[12]
PLLMe$_x$[a]	—SO$_2$—ϕ—SO$_2$—N(CH$_3$)—C*H—(CH$_2$)$_4$—N(CH$_3$)— $\mid$ COOH	[12]
COP$_y$[b]	$\left[\begin{array}{c}\text{—CH—CH}_2\text{—}\\ \mid \\ \text{COOH}\end{array}\right]_m \cdots \left[\begin{array}{c}\text{—CH—CH}_2\text{—}\\ \mid \\ \text{CO}\\ \mid \\ \text{N—C*H(CH}_3)\text{—C}_2\text{H}_5\\ \mid \\ \text{CH}_3\end{array}\right]_p \cdots$	[13]

[a] x is the percentage of N–N'-methylation.
[b] y is the molar fraction of acidic groups ($y = m/(m+p)$).
* asymmetrical carbon.
PAA: Poly(acrylic acid).
PMA: Poly(methacrylic acid).
PLL: Polycondensate between L-lysine and 1,3-benzenedisulphonyl chloride.
PLDAB: Polycondensate between α–γ-diaminobutyric acid and 1,3-benzene-disulphonyl chloride.
PLLMe: N–N'-dimethylated PLL.
COP: N-(sec-butyl)-N-methyl-acrylamide-co-acrylic acid.

In view of the increasing interest concerning the existence of non electrostatic interactions and their influence on the conformation of polyelectrolyte chains, we have extensively investigated in our laboratory some synthetic polyelectrolytes showing a pH-induced conformational transition.

The characteristics of samples examined here are reported in Table I.

2. pH-Induced Conformational Transition in Water

The interrelation between ionization free energy and polymer conformation makes potentiometric titration a good method of detection thus explaining its wide use.

The titration curves of polyacids which undergo a conformational transition in aqueous solution greatly differ from those of polyacids without transition. A discontinuous change of pK_a vs the degree of ionization (α) is observed in the former case which can be related to a transition from a rather compact state to a mean extended conformation; in the latter case, a gradual increase in the electric potential is observed in accordance with the absence of any special conformation. The cases of PMA and PAA clearly show such features although in our opinion PMA is a limiting case.

Anomalous titration curves for PLL in pure water are observed as illustrated by Figure 1 which shows the effect of various counterions on the potentiometric behaviour of PLL [12, 14]. A conformational change between a very compact form (stable at low α values) and an extended coil-like form clearly appears. Stabilization of the compact conformation is found to be more pronounced as the length of apolar chains of tetraalkylammonium counterion chains increases. The transition between the two forms, which is initiated at a critical α_c of 0.15–0.20 (in the presence of Na^+ or K^+), is shifted to higher values of α in the presence of tetraalkylammonium counterions attaining an α_c of about 0.7 with bulky (and hydrophobic) N-tetrabutylammonium ions (Bu_4N^+).

The compact conformation of PLL, stable at low α, is characterized by low and constant values of the viscosity (and therefore of mean dimensions of polyions) whereas a sharp increase in viscosity is initiated in the ionization range $0.2 < \alpha < 0.7$, depending on the nature of counter-ions. This is indicative of the destruction by electrostatic interactions of the compact structure of PLL stabilized at low α by non electrostatic cohesive forces.

Also electric permittivity measurements show rather abrupt variations of dielectric parameters of PLL–Na^+ in the ionization range corresponding to the transition which may be related to changes in mean dimensions of polyions [15] (Figure 2). The dielectric behaviour is also in agreement with that observed for PLGA during the helix-coil transition [16].

The very low values of $|\eta|^*$ for the system PLL–Bu_4N^+ in a wide range of ionization is significant of a specific effect of Bu_4N^+ on the average dimensions of PLL and shows that greater electric charge is needed to overcome the strong cohesive forces existing

---

* $|\eta|$: intrinsic viscosity.

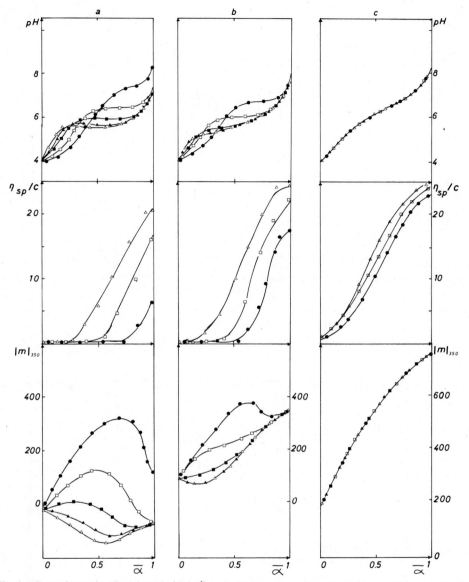

Fig. 1.   Potentiometric, viscosimetric (dl g^{-1}) and polarimetric titrations of PLL in water (a), and acetone-water mixtures: 21% acetone (b), 42% acetone (c). △△ KOH; ▲▲ Me$_4$NOH; ■■ Et$_4$NOH; □□ Pr$_4$NOH; ●● Bu$_4$NOH.

in such a system. In this regard, it is of interest to recall the peculiar behaviour reported by Barone *et al.* [17] for the potentiometric titration of PMA in the presence of Me$_4$NBr; a specific interaction of partially hydrophobic Me$_4$N$^+$ ions with PMA was then suggested. The herereported behaviour of PLL in the presence of Bu$_4$N$^+$ seems to agree with such a mechanism.

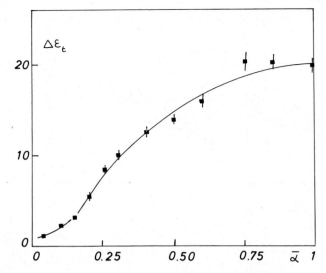

Fig. 2. Dependence of the dielectric increment $\Delta\varepsilon_t$ on the degree of neutralization $\bar{\alpha}$ for PLL solutions ($C_p = 2 \times 10^{-3}$ M) neutralized with NaOH. [15]

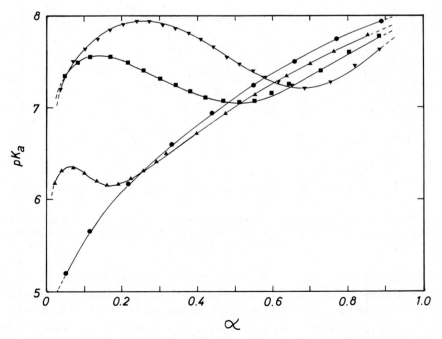

Fig. 3. Potentiometric titrations in water of COP_y by KOH, y: ● ● 0.96; ▲ ▲ 0.67; ■ ■ 0.49 ▼ ▼ 0.31.

It has been reported by Strauss *et al.* [5] that the transition region observed for alternating copolymers of maleic acid and alkylvinyl ethers was shifted to higher values of α with increased length of hydrophobic side groups.

A similar behaviour is also found here for hydrophilic-hydrophobic copolymers obtained by partial hydrolysis of a polyacrylamide (COP_y) [13]. By adjusting the percentage of hydrolysis, the hydrophobic character of chains can be varied and a behaviour from a normal polyelectrolyte to an 'abnormal' one can thus be observed. As illustrated in Figure 3, the compact conformation stable at low α is dependent on the relative proportion of statistically distributed hydrophobic and hydrophilic sequences [18]. The transition region is shifted to higher values of α as the hydrophobic character of the chain increases.

A major difference between PLL and the two above copolymers is due to the fact that in PLL, it is the nature of the counter-ions which is primarily responsible for the stabilization of compact conformation, whereas for both samples of Strauss and also COP_y, the stabilization is due to the presence of hydrophobic groups located on the chain.

For PLL, one may ask whether hydrogen bonds may be not held responsible for the compact form. The comparison between the potentiometric behaviour of PLL and PLDAB indicates that the latter behaves as a stronger acid than the former,

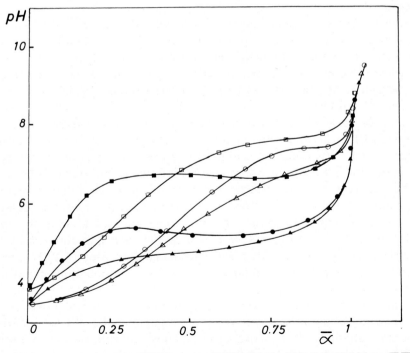

Fig. 4.   Potentiometric titrations by KOH in water of PLDAB ▲ ▲, PLL ● ●, $PLLMe_{100}$ ■ ■ and by Bu$_4$NOH in water of PLDAB △ △, PLL ○ ○ and $PLLMe_{100}$ □ □. [12]

whereas the reverse should be observed if H-bonds were a predominant factor for the stabilization. Moreover, from potentiometric data, PLLMe is found to behave as a weaker acid than PLL which is a further confirmation that the origin of the stabilization is not the result of H-bonds (Figure 4).

For PLLMe, the features found for PLL are even more pronounced. Due to its more hydrophobic character, the repulsive forces are nearly completely screened in the presence of Bu_4N^+ as indicated by its low and quasi constant viscosity over the whole ionization range (η_{sp}/C increases from 0.1 ($\alpha=0$) to 1 dl g^{-1} ($\alpha=1$)) as compared to the strong viscosity variations of PLL solutions between the same α limits.

3. Potentiometric and Viscosimetric Behaviour in Hydro-Organic Mixtures and in Urea Solutions

The preceeding data show that short-range interactions may be predominant over electrostatic repulsions as is obvious in the case of PLL neutralized with Bu_4NOH.

All the above-mentioned anomalies can be reasonably assumed to be due to the existence of 'hydrophobic interactions' and stress the importance of chain structure, the nature of side-groups and of the environment upon chain conformation.

Replacing water by a good organic solvent, or by urea, should lead to a decrease, and even a disappearance, of the compact conformation if hydrophobic interactions are destroyed.

As shown in Figure 5, in 40 and 50% methanol-water mixtures, the behaviour of PMA solutions becomes similar to that of PAA, i.e. there is no longer a detectable discontinuity in both potentiometric and viscosity titrations. It can thus be admitted that hydrophobic interactions are virtually destroyed and therefore the conformational behaviour of PMA follows directly the electrostatic repulsions between the carboxylic groups [10].

Figure 1 shows the effect of increasing percentage of acetone on the potentiometric titration of PLL. In acetone-water mixture (42%), the potentiometric titration of PLL becomes similar to that of a polyacid without any transition and a monotonous increase in both the viscosity and pK_a values vs α is found for all counter-ions.

Also in the presence of urea, which is known to act on non electrostatic interactions through alteration of water structure [6, 17], an increase in viscosity is found. Such increase is more important in the ionization range corresponding to the compact structure as illustrated in Figure 6. In 8 M urea, the anomalies observed in the viscosimetric behaviour of PLL disappear and the conformation of PLL is only dependent on the electrostatic repulsions as indicated by the monotonic increase of the viscosity vs α [19].

The effect of counterions and solvent composition on the optical properties is also found to be in good agreement with the above potentiometric and viscosity data. In Figure 1 the variation at 350 nm of the rotatory power of PLL $|m|_{350}$ as a function of ionization in various media is reported. In water and in 21% acetone-water mixture, $|m|_{350}$ does not vary monotonously with α and is dependent on the nature of the

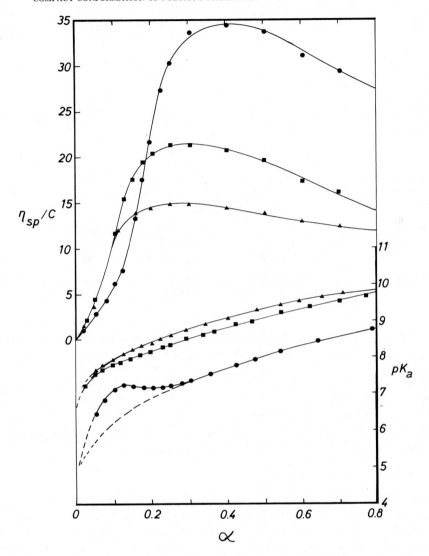

Fig. 5.   Potentiometric and viscosimetric titrations of PMA $(C_p = 2 \times 10^{-3}$ M$)$ in various methanol-water mixtures. % of methanol: ●● 0; ■■ 40; ▲▲ 50. [10]

counter-ion. At the end of neutralization, the values for $|m|_{350}$ become independent of the counter-ion; only the behaviour with Bu_4N^+ is peculiar.

In 42% acetone-water, a monotonous increase of $|m|_{350}$ vs α is observed. Moreover, in the whole ionization range, $|m|_{350}$ is found to be independent of the counter-ion. By disrupting the special conformation of PLL in water, acetone suppresses all the differences due to the nature of counter-ions and a single curve is obtained.

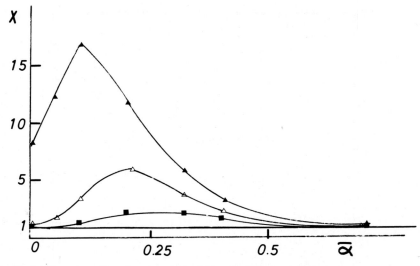

Fig. 6.   Ratio X $(\eta_{urea}/\eta_{water})$ as a function of $\bar{\alpha}$ for PLL neutralized by KOH in various urea-water mixtures.
Urea: ▲ ▲ 8 M; △ △ 6 M; ■ ■ 4 M. [19]

4. Free Energy Change for the Process Uncharged Compact Form – Hypothetical Uncharged Expanded Form

The free energy change ΔG_t^0 for the transition process at zero degree of ionization, which indicates the strength of cohesive forces, has been evaluated.

ΔG_t^0 is determined by measurements of the area (A) between the experimental curve and the hypothetical one for the 'b' state at zero ionization corresponding to a chain more open than the 'a' conformation, according to the relation [20]:

$$\Delta G_t^0/N = (RT/0.434)\,A,$$

where N is the number of ionizable groups per macromolecule.

The evaluation of ΔG_t^0 is directly related to the titration curve for the 'b' state.

For PMA, such a curve can be described when the macromolecule is unfolded $(0.4 < \alpha < 1)$ by different methods [21] and specifically by the use of the extended Henderson-Hasselbach Equation [17]:

$$\mathrm{pH} = pK_{(\alpha = 0.5)} + n\,\log\frac{\alpha}{1-\alpha}.$$

The influence of methanol on $\Delta G_t^0/N$ for PMA is reported in Table II. The negative or nil value obtained for $\Delta G_t^0/N$ in 50/50 methanol-water mixture is indicative of the stability of the unfolded state.

For PLL in water, even at $\alpha = 1$ the conformational transition seems to be not quite complete [12]. Therefore, contrary to the case of PMA, it is not possible to extrapolate the pK_a values at the end of neutralization towards $\alpha = 0$ by a mathematical

TABLE II

Free energy changes for the transition process 'a' state – 'b' state at 25 °C

Sample									
PMA	% in MeOH	0	5	10	15	20	25	30	35
	ΔG_t^0 cal/mole	185	170	155	130	110	70	50	10
PLLMe$_{37}$	% in acetone	6.3	16.8	33.6					
	ΔG_t^0 cal/mole	1040	730	200					
PLL	% in acetone	4	8.4	12.6	21	29.4	33.6		
	Δ_t^0 cal/mole	1035	920	790	470	252	170		
PLDAB	% in acetone	6.3	16.8						
	ΔG_t^0 cal/mole	570	350						

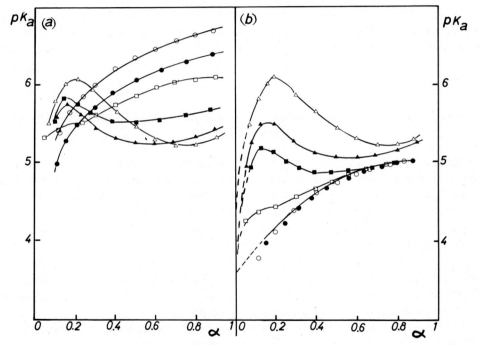

Fig. 7. pK_a values of PLL ($C_p = 2 \times 10^{-3}$ M) neutralized by KOH in water and acetone-water mixtures. (a) Without correction of pK_a. (b) After correction for the hydro-organic medium. Water △△; 12.5% acetone ▲▲; 21% ■■; 37% □□; 42% ●●; 50% ○○.[12]

method owing to the absence of an adequate model for PLL [21] or by means of Henderson-Hasselbach plots [7].

However, the following suggested method can be used: a reference curve can be obtained by measuring the pK_0 values of N-tosyl L-α-alanine, chosen as a model for PLL, in various acetone-water mixtures. The pK_a values of PLL in these mixtures can thus be corrected by taking into account the influence of acetone on the intrinsic dissociation constant of acidic groups of the model.

The different titrations curves are extrapolated to $pK = 3.6$. From Figure 7, it appears that for both acetone-water mixtures containing respectively 42% and 50% of acetone, the curves superimpose in the whole range of ionization; this already justifies their use as correct reference curves. Moreover, viscosity measurements show also that conformational transition no more exists in such mixtures. For high α, the curves also merge into a single one except for water and acetone-water (12%) where it can be admitted that the transition is not completed at $\alpha = 1$.

Each ΔG_t^0 was evaluated from the area between the corresponding titration curve corrected for the hydro-organic medium effect and the corrected reference curve. In Table II, the variation of $\Delta G_t^0/N$ as a function of % of acetone is reported. A linear decrease with increasing amount of acetone is observed until about 40% acetone.

As a result, by extrapolating to zero acetone content, a value of about 1200 cal/unit is obtained for PLL in water. The same method yields 750 cal for PLDAB and 1300 cal for PLLMe$_{37}$. Such order is in rather good agreement with the increasing hydrophobic character as the number of —CH$_2$— groups is increased. In this regard, a difference of 800 cal has been reported between butyl and hexylvinylethers copolymerized with maleic acid [5].

5. Effect of Dyes and N-Tetraalkylammonium Salts on the Stabilization of the Compact Structure of PLL

Some recent papers deal with the interaction between dyes and synthetic polyelectrolytes and the binding of dyes has been shown to strongly affect the macromolecular behaviour of polyions [22, 23, 24, 25]. For ionic dyes, electrostatic interactions may be involved in the binding of dyes to polyions. However, it has been reported that binding of crystal violet to PMA is possible for undissociated PMA [26], which can be an indication that ionic dyes are bound by non ionic interactions.

To try to explain qualitatively the effect of Bu$_4$N$^+$, the effect of dyes on the macromolecular behaviour of PLL has been investigated and compared to that of Bu$_4$N$^+$. The influence of bound dye (Acridine Orange: AO) on the conformational transition of PLL has been estimated by viscosity measurements.

In Figure 8 the changes in the viscosity vs α for PLL–AO solutions neutralized with KOH are plotted. In the whole ionization range, i.e. for both compact and extended mean conformations, addition of AO leads to lower values of viscosity than those corresponding to polymer solutions without dye.

As the concentration in dye increases, the transition region is shifted to higher values of α. Thus undoubted demonstrates that AO stabilizes the compact conformation of PLL and therefore unfolding of chains through electrostatic replusions is initiated at higher of α. Similar behaviour is thus observed in the presence of AO and Bu$_4$N$^+$.

As AO is not soluble in acetone-water mixtures, the effect of AO on the conformation of PLL has been investigated in 50% methanol-water mixtures.

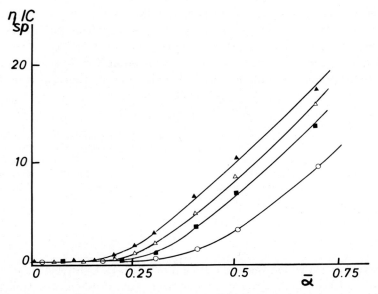

Fig. 8.   Dependence of reduced viscosity of PLL $(C_p = 10^{-3}$ M) neutralized with KOH on the AO concentration. C_{AO}: ▲ ▲ 0; △△ 2×10^{-5}; ■■ 5×10^{-5}; ○○ 10^{-4} M.[19]

Figure 9 shows that in such a medium, the transition which was initiated in water at $\bar{\alpha}_C = 0.2$ (with K^+) and 0.7 (with Bu_4N^+) is shifted to lower values of $\bar{\alpha}$ in both cases and occurs at $\bar{\alpha}$ 0.1 and 0.3 respectively. Contrary to what happens in 50% acetone-water, the transition does not disappear, which is indicative that non electrostatic interactions are not totally suppressed in a such medium, although diminished.

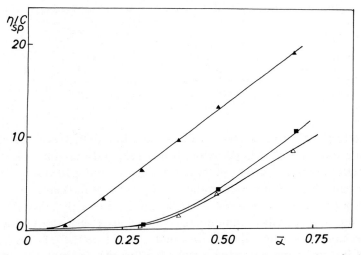

Fig. 9.   Ionization dependence on the viscosimetric behaviour (dl g^{-1}) of PLL $(C_p = 10^{-3}$ M) in methanol-water mixtures (50/50 v/v). ▲ ▲ KOH; ■■ Bu_4NOH; △△ KOH in presence of AO $(C_{AO} = 10^{-4}$ M). [19]

The close values found for the viscosity of PLL–K$^+$ (in the presence of 10^{-4} M AO) and of PLL–Bu$_4$N$^+$ (without dye) are in qualitative agreement with the assumption that Bu$_4$N$^+$ acts in a similar way on the stabilization as a molecule of dye.

The effect of dye on the macromolecular dimensions differs from a usual screening effect due to the presence of salts as indicated by the viscosity values given in Table III.

TABLE III

Effect of dye and salt on the viscosity of PMA and PLL solutions ($C_p = 10^{-3}$ M, degree of neutralization $\bar{\alpha} = 0.4$ by KOH)

η_{sp}/C_p (dl g^{-1})	$C_{salt} = 0$ $C_{AO} = 0$	$C_{KCl} = 10^{-4}$ M $C_{AO} = 0$	$C_{Bu_4NCl} = 10^{-4}$ M $C_{AO} = 0$	$C_{salt} = 0$ $C_{AO} = 10^{-4}$ M
PMA	39.5	29.3	32.0	11.0
PLL	7.4	5.4	3.2	1.1

The above data show that the viscosity of the system PLL–Bu$_4$NCl is smaller than that of PLL–KCl, while the reverse is found in the case of PMA, thus indicating the specific effect of Bu$_4$N$^+$ in the case of PLL.

A further evidence for the alikeness between the behaviour of AO and Bu$_4$N$^+$ is given by the viscosity and optical properties of PLL–AO and PLL-salt systems.

In salt-free water, no dependance of the rotatory power ($\bar{\alpha} = 1$) on the counter-ion has been found except for Bu$_4$N$^+$ (Figure 1). As shown in Table IV, in the presence of salts, the same counter-ion dependance on the optical properties as for viscosity is observed.

TABLE IV

Influence of salts MCl on the viscosity (dl g^{-1}) and optical properties of PLL at $\bar{\alpha} = 1$

Counter-ion	K^+	Me$_4$N$^+$	Et$_4$N$^+$	Pr$_4$N$^+$	Bu$_4$N$^+$
$|m|_{300}$ in water	-165	-165	-165	-165	$+200$
$|\eta|$ in water	~ 30	~ 30	~ 30	~ 30	~ 30
$|m|_{300}$ in salt ($C_{salt} = 10^{-2}$ M)	-530	-510	-460	-360	$+700$
$|\eta|$ in salt ($C_{salt} = 5 \times 10^{-3}$ M)	3.1	3.2	2.8	2.1	0.4

In Figure 10, the dichroïc spectra of aqueous solutions of PLL neutralized at $\bar{\alpha} = 0.8$ with alkaline and N-tetraalkylammonium hydroxides in the presence of AO show an abnormal induced optical activity in the region corresponding to absorption bands of AO. This means that AO is bound in the vicinity of the chiral centre of PLL and an 'extrinsic' optical activity is induced. These features indicate that the affinity of PLL chains for bulky N-tetraalkylammonium counter-ion is very strong and therefore that the mechanism by which Bu$_4$N$^+$ is bound is similar to that of dyes.

This fact justifies our assumption concerning the effect of bulky N-tetraalkylammonium on the macromolecular behaviour of PLL.

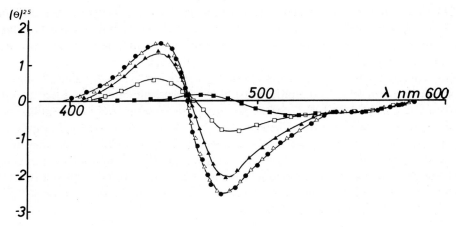

Fig. 10.   Dichroïc spectra ((θ) experimental data in millidegrees) of PLL ($C_p = 10^{-3}$ M, $\bar{\alpha} = 0.8$) in the presence of AO ($C_{AO} = 2 \times 10^{-5}$ M) for various counter-ions (1 = 0.2 dm). ● ● KOH; △△ Me$_4$NOH; ▲ ▲ Et$_4$NOH; □ □ Pr$_4$NOH; ■ ■ Bu$_4$NOH.[19]

6. Conclusions

The results reported above of various measuring methods used with different kinds of polyelectrolytes undergoing a pH-induced conformational transition may be combined to form a consistent picture on the main cause for stabilization of compact structure of polyions in water.

From our own data, together with other data found in the literature, it is obvious that non electrostatic cohesive interactions are predominant at low degree of ionization and are responsible for the maintenance of very compact structures. In such ionization ranges, the coulombic repulsive interactions are therefore more or less completely screened, depending on the strength of cohesive forces as clearly indicated by the respective behaviour of PMA, partially hydrolyzed N, N-disubstituted polyacrylamide (COP$_y$), PLL and related polyelectrolytes.

We have tried to elucidate the forces twisting the macromolecular chain into a compact structure. In this regard, the most important one seems to be hydrophobic interactions. A proof is given by the data obtained by changing the nature of solvent gradually. The results presented here show that changes in local solvent composition, by partially replacing water by an organic solvent (e.g. methanol or acetone) or by urea, are accompanied by unfolding of polyions. The anomalies characteristic of a transition from a compact form to a more extended one are then no longer detectable and a 'normal polyelectrolytic behaviour' is observed.

All the phenomena pointed out by Strauss et al. for alternating copolymers of maleic acid and alkylvinyl ethers are clearly related to the existence of hydrophobic interactions between alkyl side groups, such interactions strongly depending on the length of apolar groups. Our own results concerning COP$_y$ are in agreement with such behaviour. In the extensively studied case of PLL and its related compounds (PLLMe,

PLDAB), a somewhat different mechanism must be envisaged as a consistent explanation for the reported experimental facts. In the presence of bulky N-tetraalkylammonium counter-ions, the compact structure has been shown to be stabilized in a large range of ionization.

So compact a conformation, characterized by very low viscosity values, cannot be explained on a purely electrostatic basis. Non electrostatic interactions between the hydrophobic Bu_4N^+ and PLL chains can be envisaged, thus leading to stabilization of compact structure in water. The specific effect of such counter-ions disappears in hydro-organic or urea solutions as indicated by potentiometric, viscosimetric or optical results. In the case of PLL, Bu_4N^+ acts in a similar way as apolar side groups, but the phenomena is induced from 'outside'. The similarity between the effect of Bu_4N^+ and that of dye seems to be concordant with the stabilization of compact structure by non electrostatic interactions.

References

1. Kauzmann, W.: *Adv. Protein Chem.* **14**, 1 (1959).
2. Strauss, U. P. and Gershfeld, N. L.: *J. Phys. Chem.* **58**, 747 (1954).
3. Inoue, H.: *Kolloïd Z. Z. Polymere* **195**, 102 (1964).
4. Ferry, J. D., Udy, D. C., Wu, F. C., Heckler, G. E., and Fordyce, D. B.: *J. Colloïd Sci.* **6**, 429 (1952).
5. Dubin, P. L. and Strauss, U. P.: *J. Phys. Chem.* **74**, 2842 (1970).
6. Dubin, P. L. and Strauss, U. P.: *J. Phys. Chem.* **77**, 1427 (1973).
7. Mandel, M., Leyte, J. C., and Stadhouder, M. G.: *J. Phys. Chem.* **71**, 603 (1967).
8. Crescenzi, V.: *Adv. Polym. Sci.* **5**, 358 (1968).
9. Anufrieva, E. V., Birshtein, T. M., Nekrasova, T. N., Ptitsyn, O. B., and Sheveleva, T. V.: *J. Polym. Sci.* **C16**, 3519 (1968).
10. Braud, C., Muller, G., Fenyo, J. C., and Sélégny, E.: *J. Polym. Sci.* **12**, 2767 (1974).
11. Beaumais, J., Fenyo, J. C., and Sélégny, E.: *Europ. Polym. J.* **9**, 15 (1973).
12. Fenyo, J. C., Beaumais, J., and Sélégny, E.: *J. Polym. Sci.* **12**, 2659 (1974).
13. Braud, C., Vert, M., and Sélégny, E.: *Makromol. Chem.* **175**, 775 (1974).
14. Fenyo, J. C.: *Europ. Polym. J.* **10**, 233 (1974).
15. Muller, G.: *J. Polym. Sci., Polym. Lett. Ed.* **12**, 319 (1974).
16. Muller, G., Touw, F. van der, Zwolle, S., and Mandel, M.: *Biophys. Chem.* **2**, 242 (1974).
17. Barone, G., Crescenzi, V., and Quadrifoglio, F.: *Ric. Sci.* **36** (6), 482 (1966).
18. Braud, C. and Vert, M.: *Polymer* **16**, 115 (1975).
19. Muller, G., Fenyo, J. C., Beaumais, J., and Sélégny, E.: *J. Polymer. Sci.* **12**, 2671 (1974).
20. Nagasawa, M. and Holtzer, A.: *J. Am. Chem. Soc.* **86**, 538 (1964).
21. Nagasawa, M., Murase, T., and Kondo, K.: *J. Phys. Chem.* **69**, 4005 (1965).
22. Birshtein, T. M., Anufrieva, E. V., Nekrasova, T. M., Ptitsyn, O. B., and Sheveleva, T. V.: *Polym. Sci. SSSR* **7**, 412 (1965).
23. Stork, W. H. J., Hasseth, P. L. de, Schippers, W. B., Kormeling, C. M., and Mandel, M.: *J. Phys. Chem.* **77**, 1772 (1973).
24. Stork, W. H. J., Hasseth, P. L. de, Lippitz, G. J. M., and Mandel, M.: *J. Phys. Chem.* **77**, 1778 (1973).
25. Crescenzi, V. and Quadrifoglio, F.: *Europ. Polym. J.* **10**, 329 (1974).
26. Barone, G., Caramaza, R., and Vitagliano, V.: *Ric. sci.* **32** (6), 554 (1962).

IONIC SELECTIVITY OF POLYELECTROLYTES IN
SALT FREE SOLUTIONS

M. RINAUDO and M. MILAS

Centre de Recherches sur les Macromolécules Végétales (C.N.R.S.),
Domaine Universitaire, B.P. 53, 38041 Grenoble, France

Abstract. Experimental data concerning the ionic selectivity in carboxylic polyelectrolyte systems are given and discussed.

The selectivity is quantitatively expressed by the three thermodynamic parameters related to the nature of the counterions and to the charge density. The ionic selectivity must be attributed to site binding with reorganization of shells of hydration. The sequence of affinity for monovalent counterions is Li > Na > K > > Cs ~ TMA corresponding to the order of cristallographic radii. With divalent counterions the affinity is greater than with monovalent counterions but selectivities are negligible at 25 °C on C.M.C. The proposed site binding is based on experimental results and is in agreement with the ion condensation for $|Z| \lambda > 1$ predicted by Manning. The selectivity is found to increase linearly with the charge density over a critical value λ_c (for example $1 < \lambda_c < 1.6$ with monovalent counterions depending on the technique used).

1. Introduction

A polyelectrolyte can be utilized as a model to investigate the mechanism of ion selectivity of ion exchangers. The ion selectivity exhibited by polyelectrolytes is therefore an important property for practical applications. Our work deals specifically with polycarboxylic polyelectrolytes in salt free aqueous solutions. In the first part, the activities of monovalent and divalent counterions are determined and discussed in connection with the nature of the counterion and polyelectrolyte charge density. Then, experimental results are presented showing clearly the existence of selectivity and affinity sequences.

Finally the behaviour of polyelectrolytes as a function of their charge density is discussed in connection with ion pair formation. All the experimental results were obtained on carboxymethylcelluloses with various degrees of substitution; these polyelectrolytes are assumed to be rigid and their properties are interpreted as a function of the linear charge density.

2. Characterization of Polyelectrolytes

A polyelectrolyte is well characterized by its charge parameter λ introduced by Lifson and Katchalsky [1].

The parameter λ reflects the linear charge density of the polyion in a rigid rod model, λ is given by

$$\lambda = \frac{\nu \varepsilon^2}{hDkT},$$

(1)

ν = number of ionic sites on a polyion of length h,

   ε = electronic charge,
  kT = Boltzmann term,
   D = dielectric constant.

For carboxymethylcelluloses (CMC), the value of λ depends on the degree of substitution $\overline{DS}$ and on the degree of dissociation α of the polyacid; in this case, $\lambda = \alpha \cdot \overline{DS} \cdot \lambda_0$ with

$$\lambda_0 = \frac{\varepsilon^2}{bDkT} = 1.38 \quad \text{at} \quad 25\,°\text{C}.$$

The parameter b is the length of the monomeric unit. These CMC are good models to test a theoretical approach because they are rigid as proven by hydrodynamic measurements [2]. Theoretical models with cylindrical symetry are convenient to interpret the experimental data.

3. Free Fractions of Counterions

3.1. INFLUENCE OF CHARGE DENSITY

If the free fraction of counterions is determined, the number of counterions bound can be deduced. In fact the selectivity must be attributed to a certain number of these 'atmospheric' counterions.

The fraction of thermodynamically active counterions is obtained experimentally by conductimetry [3] or potentiometry [3, 4] with selective electrodes (Na^+ or Ca^{2+}).

The results are recorded in Figure 1 for monovalent and divalent counterions. The experimental results are compared to the free fractions of counterions calculated by the treatment of Oosawa (γ_0) [5] and to the activity of the counterions given by Manning (γ_M) [6]; these parameters are given for infinite dilution by:

$$
\left.
\begin{aligned}
\gamma_0 &= 1 \\
\ln \gamma_M &= -\frac{|z|\lambda}{2}
\end{aligned}
\right\}
\quad |z|\lambda \leqslant 1
$$

$$
\left.
\begin{aligned}
\gamma_0 &= \frac{1}{|z|\lambda} \\
\gamma_M &= \frac{0.606}{|z|\lambda}
\end{aligned}
\right\}
\quad |z|\lambda \geqslant 1
$$

$$(2)$$

with z equal to the valence of the counterion.

With monovalent counterions, in the theory of Manning, the effective charge parameter goes to a limit $\lambda_{\text{eff}} = 1$; as a consequence, for all the values of $\lambda > 1$, a fraction $(1 - 1/\lambda)$ of counterions is condensed; the fraction of bound counterions in the polyelectrolyte phase according to Oosawa is equal to $(1 - \gamma_0)$; the calculated values for the density of bound ions are given in Figure 2 for infinite dilution; in this representation, Oosawa and Manning's treatments give the same results.

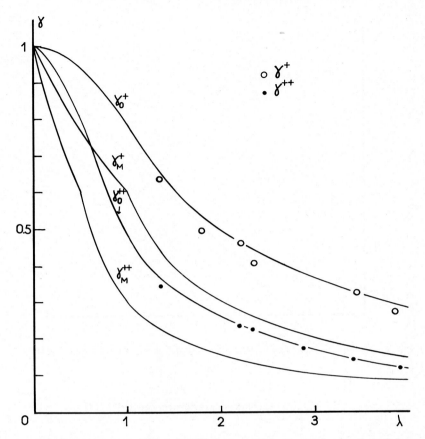

Fig. 1.   Activity coefficient of monovalent and divalent counterions on carboxymethylcelluloses. ○, ●
experimental values. ——— γ_O, γ_M theoretical values from Oosawa and Manning.

3.2. INFLUENCE OF THE NATURE OF THE COUNTERIONS

The free fraction of counterions has been obtained [3, 7] by different techniques; the
set of results given in Table I is for monovalent counterions and in Tables I and II
for divalent counterions. As shown, at equilibrium, the active fraction of counterions
is quasi independent of the methods used; especially, the osmotic coefficient Φ is very
close to the activity coefficient. The value is almost independent of the nature of the
counterions; nevertheless, it seems that there is a sequence related to the ionic selectivity $\gamma_{Li} < \gamma_{Na} < \gamma_K < \gamma_{Cs}$ (Table I).

This fact is more clearly shown in Table III; the values of free diffusion depend
primarily on the counterion and the ratio $(D/D_0)_{Cs}/(D/D_0)_{Na}$ increases with the charge
density [8].

A second point is that the coefficients of free diffusion are always higher than the
equilibrium parameter (Table I).

As shown in this part of our work, there is a good agreement between free fractions

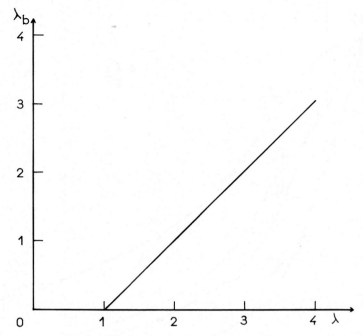

Fig. 2.   Density of bound counterions (λ_b) as a function of structural charge density λ.

TABLE I

Free fractions of different counterions [7]

$\overline{DS}$	X	Li	Na	K	Cs	Ca
	γ	–	0.640	0.650	0.640	0.300
1.1	Φ	–	0.600	–	–	–
	γ'	0.520	0.550	0.600	–	–
	γ	–	0.170	0.515	0.510	0.190
1.6	Φ	0.430	0.440	0.460	–	0.230
	γ'	0.490	0.470	0.520	–	–
	γ	–	0.350	0.400	0.400	0.145
2.5	Φ	–	0.300	–	–	0.140
	γ'	0.300	0.340	0.370	–	–

γ by potentiometry $\pm 8\%$.
Φ by osmometry $\pm 10\%$.
γ' by tonometry $\pm 0.05\%$.

of counterions and the theoretical values calculated by the Oosawa treatment; there is only a small influence of the nature of the counterions for equal valence and in fact the more important parameter is the charge parameter which controls directly the electrostatic binding. From the activity coefficient, the degree of binding obtained is certainly the sum of atmospheric binding and of ion pair formation. In this sense, the

TABLE II

Free fractions of divalent counterions by conductimetry and potentiometry

$\overline{DS}$		Mg^{+2}	Ca^{+2}	Sr^{+2}	Ba^{+2}	
2.9	Conductim	–	0.134	0.134	0.129	±0.005
	Potentiom	0.140	0.140	0.135	0.130	±0.010
2.49	Conductim	–	0.145	0.150	0.145	±0.005
	Potentiom	0.135	0.145	0.145	0.140	±0.010
2.1	Conductim	–	–	–	–	–
	Potentiom	0.165	0.170	0.180	0.180	±0.010
1.6	Conductim	–	0.205	0.210	0.210	±0.005
	Potentiom	0.190	0.190	0.190	0.190	±0.010
1	Conductim	–	0.323	0.334	0.323	±0.010
	Potentiom	0.300	0.305	0.280	0.297	±0.015

TABLE III

Autodiffusion of monovalent counterions [8]

$\overline{DS}$ CMC	$D_{Na^+} \times 10^5$	$D_{Cs^+} \times 10^5$	$\left(\frac{D}{D_0}\right)_{Na^+}$	$\left(\frac{D}{D_0}\right)_{Cs^+}$	γ_{Na^+}	$\left(\frac{D}{D_0}\right)_{Cs^+}\Big/\left(\frac{D}{D_0}\right)_{Na^+}$
1.25	0.820	1.260	0.630	0.665	0.520	1.05
1.7	0.786	1.230	0.604	0.650	0.410	1.07
2.4	0.610	0.992	0.470	0.523	0.290	1.11
2.9	0.630	1.072	0.485	0.565	0.275	1.17

$(D_0)_{Na^+} = 1.3 \times 10^{-5} \text{ cm}^2 \text{ s}^{-1}$; $(D_0)_{Cs^+} = 1.9 \times 10^{-5} \text{ cm}^2 \text{ s}^{-1}$.

tight binding with the cation ethidium [9] gives interesting results shown in Figure 3; this figure gives the density of association on ionic sites as a function of λ with a critical value λ_c for binding ($\lambda_c \sim 0.5$) confrontation of values taken from Table I with these last results gives for a CMC of $\overline{DS} = 2.5$:
– degree of ion pair binding (ethidium ion) 0.25,
– degree of total binding (sodium ion) 0.70.
In our mind, the ion pair formation is directly implicated in ion selectivity.

4. Experimental Evidence of Selectivity

Different techniques are convenient to show the evidence of ionic affinity and to test the selectivity sequences; in our work, conductimetry, potentiometry, microcalorimetry and ultrasonic absorption have been used.

4.1. CONDUCTIMETRY

In aqueous solution, a treatment has been given to relate the specific conductivity to the effective ionisation i, the ionic mobility of polyion λ_p and counterions λ_c [10]:

$$\chi = 10^{-3}c_p(i(\lambda_p + \lambda_c(1-k)) + \alpha'k\lambda_c),\tag{3}$$

c_p = concentration of the polyelectrolyte in equiv. 1^{-1},

k = conduction parameter ($k \sim 0{,}2$) independent of the counterion in first approxi-
    mation,

$i = \alpha'\Phi$ effective ionisation,

α' = degree of neutralization.

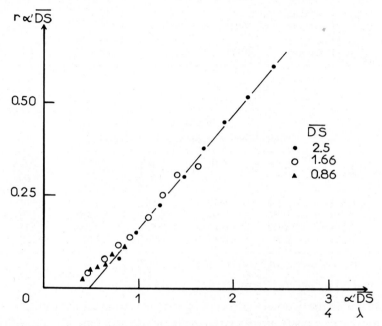

Fig. 3.   Density of Ethidium counterions bound as a function of charge
density on CMC [9].

From this relation, during neutralization, the value of effective ionisation is deduced; as a consequence, the degree of binding is obtained. If a competition between two monovalent counterions (index 1 and 2) in equal quantity is investigated, the theoretical value of the specific conductivity (χ_{th}) in absence of selectivity is equal to

$$\chi_{th} = \frac{\chi_1 + \chi_2}{2} = 10^{-3}c_p\left\{i\left(\lambda_p + (1-k)\left(\frac{\lambda_1 + \lambda_2}{2}\right)\right) + \alpha'k\left(\frac{\lambda_1 + \lambda_2}{2}\right)\right\}.$$

If there is an ion selectivity, the experimental specific conductivity χ_{exp} can be inter-
preted using the equivalent fractions X_1 of counterion 1 and $X_2 = 1 - X_1$ of counterion 2 free in solution and $\overline{X_1}$, $\overline{X_2}$ for equivalent fractions bound on the polyelectrolyte

[11]; χ_{exp} is written:

$$\chi_{exp} = 10^{-3} c_p \{ i \lambda_p + k(\alpha' - i) \lambda_1 \bar{X}_1 + [0.5 - (\alpha' - i) \bar{X}_1] \lambda_1 + \\ + k(\alpha' - i)(1 - \bar{X}_1) \lambda_2 + [0.5 - (\alpha' - i)(1 - \bar{X}_1)] \lambda_2 \}.$$

The difference between χ_{th} and the experimental value χ_{exp} gives the value of $\bar{X}_1$ through the relation:

$$\bar{X}_1 = \frac{\chi_{th} - \chi_{exp} + \dfrac{10^{-3} c_p (\lambda_1 - \lambda_2)(1 - k)(\alpha' - i)}{2}}{10^{-3} c_p (\lambda_1 - \lambda_2)(1 - k)(\alpha' - i)}. \tag{4}$$

We assume that λ_p, k and i are independent of the counterion.

For exact neutralization by an equimolecular solution (LiOH + KOH) of a CMC of $\overline{DS} = 2.9$, $c_p = 5 \times 10^{-3}$ N, the following results are obtained for different pairs of counterions:

1/2	Li/Na	Li/K	Na/K	K/Cs
$\overline{X}_1$	0.525	0.533	0.520	0.500

4.2. POTENTIOMETRY

Different determinations have been made. First of all, the *pH* curves were easily obtained during neutralization by different hydroxides.

With monovalent counterions a dissociation between the different curves appears on CMC for a critical value $\alpha' \sim 0.35$, $\overline{DS} = 2.49$ ($\lambda_c \sim 1.2$). This dissociation can be attributed to a difference of affinity between the different hydroxides tested [12]; an example is given in Figure 4 showing a sequence of affinity Li > Na > K > Cs > TEA corresponding to the sequence of crystallographic ionic radii.

With divalent-divalent counterions or divalent-monovalent counterions, the critical value is $\alpha' \sim 0.15$ for $\overline{DS}$ 2.49 or $\lambda_c \sim 0.5$.

Using specific electrodes for Na^+ and Ca^{2+} ions respectively, for exact neutralization by a mixture $(NaOH + Ca(OH)_2)$ with different compositions, the free fractions of both counterions have been determined at the same time for each equivalent fraction X_{Na} of the hydroxide. Results are given in Table IV [13]. The same experiment should be realized with two monovalent counterions. In fact, the displacement of Na^+ on Na form of the polyelectrolyte has been tested by addition of a salt XCl; for a ratio Na/$X = 1$, the fraction $\bar{X}_{Na}$ has been calculated [11]; the results on CMC $\overline{DS}$ 2.9 for different counterions are given in Table V.

These results are in agreement with the selectivity Li > Na > K > Cs; the independance of the free fraction of the counterions from its nature must be assumed.

The determination needs also the knowledge of the variation of the free fraction of counterions with the ionic strength; in addition, we assume that the activity of the neutral salt is unaffected by the polyion.

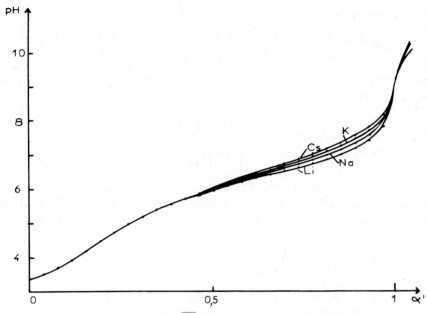

Fig. 4.   Neutralization of CMC $\overline{DS} = 2.9$ by monovalent hydroxides by pHmetry.

TABLE IV

Free fractions of monovalent and divalent counterions for different stoechiometric compositions [13]

		0	0.1	0.2	0.3	0.4	0.5	0.6	0.7	0.8	0.9	1
	X_{Na}											
$\overline{DS}$	γ_{Na^+}	–	1	0.980	0.900	0.870	0.852	0.840	0.800	0.730	0.680	0.630
1	$\gamma_{Ca^{2+}}$	0.320	0.310	0.280	0.237	0.174	0.126	0.075	0.033	0	0	–
	χ_{tot}	0.320	0.380	0.420	0.440	0.455	0.490	0.532	0.570	0.580	0.615	0.630
$\overline{DS}$	γ_{Na^+}	–	0.850	0.792	0.696	0.613	0.564	0.508	0.453	0.414	0.363	0.330
2.49	$\gamma_{Ca^{2+}}$	0.145	0.091	0.050	0.015	0	0	0	0	0	0	–
	χ_{tot}	0.145	0.166	0.199	0.220	0.245	0.282	0.305	0.317	0.330	0.330	0.330

X_{Na} is the equivalent fraction of Na^+ in the hydroxide solution.
X_{tot} is the total degree of dissociation of the polyelectrolyte.
$c_p = 2 \times 10^{-3}$ N.

TABLE V

Ion selectivity obtained by potentiometry

	Li/Na	K/Na	Cs/Na
X_{Na}	0.440	0.535	0.570

CMC $\overline{DS}$ 2.9; $c_p = 5 \times 10^{-3}$ N.

4.3. MICROCALORIMETRY

The determination of the enthalpy of exchange can be obtained directly by micro-calorimetry using two methods [14];

– adding HCl in two cells containing respectively the C_1–CMC and C_2–CMC forms

– adding C_1Cl in two cells containing respectively C_2–CMC and water $(C_2 = TEA^+)$.

In both cases the differential heat measurement corresponds to the enthalpy of exchange between C_1 and C_2.

In this work, the counterion tetraethylammonium (TEA) is chosen as a reference

TABLE VI

Enthalpy of exchange between monovalent counterions and the cation TEA as reference

		Li	Na	K	Cs
$\Delta H_{cal\ /site}$	(retrogradation 2 HCl/site)	+576	+134	−268	−378
$\Delta H_{cal\ /site}$	(exchange $X/TEA = 2$)	+580	+177	−257	−292

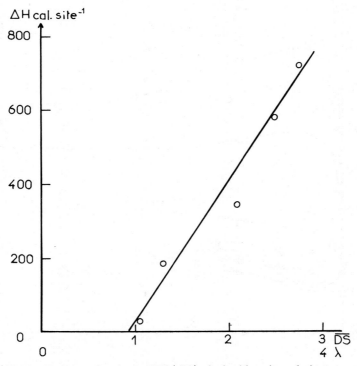

Fig. 5. Enthalpy of exchange TEA$^+$/Li$^+$ obtained by microcalorimetry as a function of the charge density.

for non specific binding (based on ultrasonic absorption); however, to-day, it seems that there is some indication of secondary effects with this ion and it would be better to take Cs^+ as reference.

The values obtained by both methods proposed are in good agreement (Table VI) [14].

The ΔH values obtained are a function of the counterion and of charge density (Figure 5). The heat effects are large enough to be measured experimentally for $\overline{DS}>1$ or $\lambda>1.4$. The ΔH values combined with the potentiometric data giving the ΔG fo exchange can be used to deduce the three thermodynamic parameters of exchange.

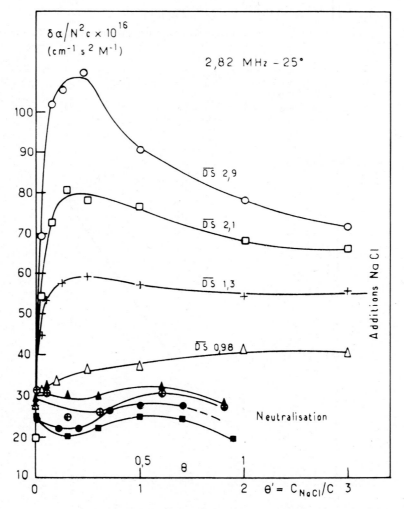

Fig. 6. Variation of the ultrasonic absorption ($\delta\alpha/N^2$) as a function of the degree of neutralization θ by TMAOH and of the excess of salt θ' for different $\overline{DS}$ (c_p=0.084 N to 0.095 N) [17].

4.4. Ultrasonic absorption

The application of ultrasonic methods to electrolytic solutions is convenient to study the mechanism of ion pair formation [15, 16]; the extension of polyelectrolytes gives information on the existence of site binding depending strongly on charge density and the nature of the counterions [17].

Solutions of CMC with various $\overline{DS}$ have been investigated; results are given on

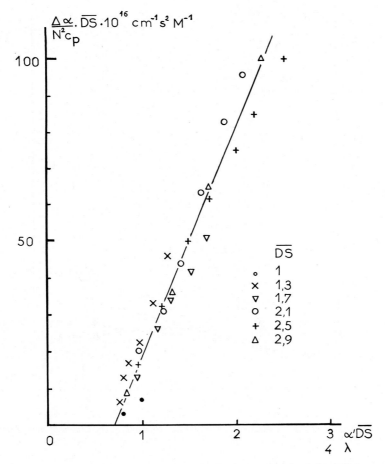

Fig. 7. Variation of the ultrasonic absorption between Na$^+$ and TMA$^+$ salt ($\Delta\alpha/N^2$) *by monomeric unit* as a function of the charge density (experimental results [17]).

Figures 6 and 7. The difference of absorption between solvent and solution $\delta\alpha$ at the frequency N, $\delta\alpha/N^2$, is related to the change of volume associated with the equilibrium between free and site bound counterions.

In Figure 6, different products are neutralized first by tetramethylammonium

(TMA) without perturbation; then an effect directly related to $\overline{DS}$ is obtained by formation of Na^+COO^- pairs during addition of NaCl.

In Figure 7, the difference $(\delta\alpha/N^2 C)_{Na} - (\delta\alpha/N^2 C)_{TMA}$ *by monomeric unit* is plotted as a function of *the charge density*; it shows well the existence of a critical charge density corresponding to the formation of ion pairs $(\lambda_c \sim 0.8)$.

On Figure 8 is given the influence of the nature of counterions; it reflects clearly the

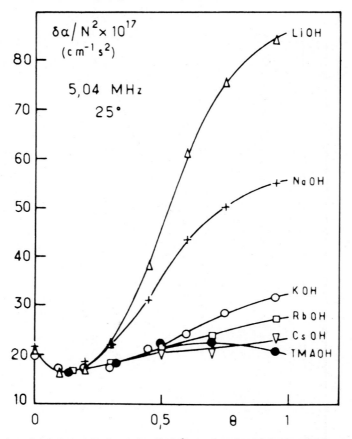

Fig. 8. Variation of the ultrasonic absorption $(\delta\alpha/N^2)$ as a function of the degree of neutralization θ by different monovalent hydroxides [17]; CMC $\overline{DS}$ 2.49 $c_p = 0.1$ N

sequence of affinity related to the volume of the perturbation: $Li > Na > K > Rb > Cs > TMA$.

The same sequences of perturbation are obtained if the specific volumes of polyelectrolytes are measured [18]. These ultrasonic experiments are the best proof of site binding and the importance of water perturbation at the origin of the ion selectivity; nevertheless it is a global measurement and only a spectroscopic experiment would be convenient to estimate the number of associated counterions.

5. Interpretation of Experimental Results

5.1. CONDUCTIMETRIC DATA

Relation (4) gives the expression of the fraction of counterions bound along the polyelectrolyte $(\bar{X}_1, \bar{X}_2 = 1 - \bar{X}_1)$ when two counterions are in competition. This relation can be used to follow the binding during neutralization for different pairs of cations.

This study was done with Li^+/K^+ on CMC with different $\overline{DS}$ and during neutralization; experimental results are given in Figure 9. The fraction $\bar{X}_{Li}$ bound depends only

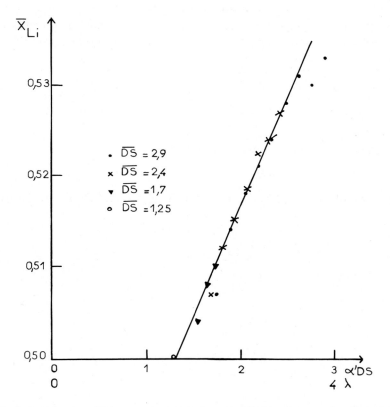

Fig. 9.   Equivalent fraction of Li counterions bound to the polyelectrolyte as a function of the charge density obtained by conductimetry for the pair Li/K.

on the charge density (proportional to $(\alpha' \cdot \overline{DS})$) and grows linearly with the charge on a single curve independant of the $\overline{DS}$ of CMC. From these results, we conclude that the selectivity is a purely electrostatically controlled phenomenon and begins for values of the charge parameter $\lambda > 1.6$ for the pair Li/K.

5.2. CHARACTERISTICS OF THE EXCHANGES

5.2.1. *Constants of Exchange*

The polyelectrolyte may be assimilated to an ion exchanger and the following equilibria are valid:

$$\bar{C}_1^+ + C_2^+ \leftrightarrows \bar{C}_2^+ + C_1^+$$

for a competition between two monovalent counterions

$$2C_1^+ + C_3^{2+} \leftrightarrows \bar{C}_3^{2+} + 2C_1^+$$

for a competition between a monovalent and a divalent counterion; $\bar{C}_i$ is related to the 'atmospherically' bound counterions; if $\bar{x}_i$ is the equivalent fraction of counterions i, the ion selectivity can be expressed by an exchange coefficient K_2^1 or K_1^3:

$$K_2^1 = \frac{x_2}{x_1} \cdot \frac{\bar{x}_1}{\bar{X}_2}$$

$$K_1^3 = \frac{x_1^2}{x_3} \cdot \frac{\bar{x}_3}{\bar{x}_1^2} \tag{5}$$

used to calculate the free energy of exchange by the relation

$$\Delta G_{\text{exch}} = -RT \ln K \tag{6}$$

The experimental values of $\bar{X}_i$ deduced from Tables V and IV give respectively

1/2	Na/Li	Na/K	Na/Cs
K_2^1	0.69	1.4	1.55
ΔG_{exch} cal/site	+220	−111	−260

$\overline{DS}$	$K_{\text{Na}}^{\text{Ca}}$	ΔG_{exch}
1	300	−3200
2.49	1100	−4200

These values of ΔG are related to the free energy of exchange for a competition between two counterions; the absolute value of ΔG is directly related to the difference of affinity of both counterions in competitions.

5.2.2. *Thermodynamic Parameters*

During titration of a polyacid, the free energy of exchange $\Delta G(T)$ is given by the potentiometric area between both neutralization curves; experimentally, these values are found to be independant of the polymer concentration. In fact, the pH values are related to the change of the electrostatic free energy ΔG_{el} by the following expression:

$$pH = pK_0 = \log((1-\alpha')/\alpha') + 0.434\Delta G_{el}/kT$$

in which K_0 is the intrinsic constant of ionization and α' the degree of neutralization. The electrostatic free energy per ionizable site at the degree of ionization α' is given by

$$G_{el}(\alpha') = 2.3kT \int_0^{\alpha'} (pH + \log((1-\alpha')/\alpha') - pK_0)\, d\alpha'$$

corresponding to the area under the graph $(pH + \log((1-\alpha')/\alpha')) - pK_0)$ vs α'. As a consequence, the area between two curves of neutralization gives directly the free energy (per charged carboxyl) of transfer of fully charged polymer from a solution in which it is associated with one counterion to a solution in which it is associated with the other counterion.

In Figure 10, the values of the ΔG of exchange for the pairs Na/Ca and TEA/Li are

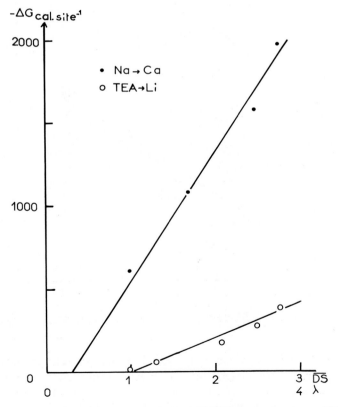

Fig. 10.   Free energy of exchange ΔG obtained by potentiometry at 25 °C as a function of the charge density.

plotted as a function of the charge density of CMC. From this linear representation $\Delta G_{exch} = 0$ when $\lambda \sim 1.3$ for the exchange TEA/Li but corresponds to $\lambda \sim 0.5$ for the exchange Na/Ca.

More recently, the neutralizations were followed at different temperatures [19]; these experiments gave $\Delta G(T)$ from which ΔH and ΔS are easily deduced for exchange between different counterions and a cation taken as reference (Figures 11 and 12).

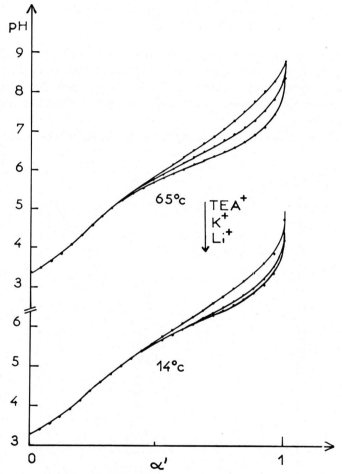

Fig. 11. Titration of H–CMC ($\overline{DS}$ 2.49; $c_p = 5 \times 10^{-3}$ N) by TEAOH, KOH and LiOH (10^{-1} N).

The experimental values are given in Table VII. At a given temperature, ΔG can be obtained by potentiometry and ΔH by calorimetry; from these values, ΔS is directly obtained; the set of values given by this procedure is in Table VIII; as shown, there is a good agreement between both sets of values (Tables VII, VIII).

The influence of the nature of the counterion is clearly shown. The particular

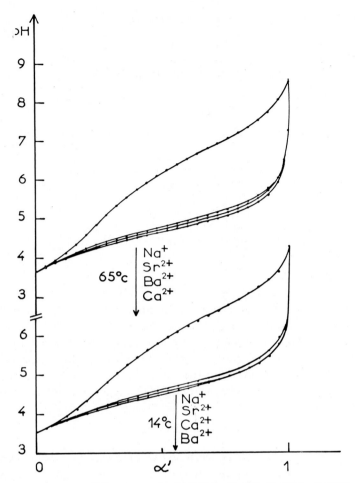

Fig. 12. Titration of H–CMC ($\overline{DS}$ 2.49; $c_p = 5 \times 10^{-3}$ N) by NaOH, Ca(OH)$_2$, Sr(OH)$_2$ and Ba(OH)$_2$ (all 4×10^{-2} N).

TABLE VII

Thermodynamic parameters of exchange
deduced from temperature
dependence of ΔG

Exchanges	ΔH^{a}	ΔS^{b}
TEA$^+$→Li$^+$	$+628$	$+3.10$
TEA$^+$→K$^+$	-231	-0.08
K$^+$→Li$^+$	$+860$	$+3.18$
Na$^+$→Ca^{2+}	$+901$	$+8.34$
Na$^+$→Sr^{2+}	$+694$	$+7.57$
Na$^+$→Ba^{2+}	$+330$	$+6.61$

^{a} Cal site^{-1}.
^{b} Cal deg^{-1} site^{-1}.

TABLE VIII

Thermodynamic parameters deduced by combination of
potentiometric and microcalorimetric data ($T=25\,°C$)

Exchanges	ΔG^a Potentiometry 25°	ΔH^a Calorimetry	ΔS^b
$TEA^+ \to Li^+$	-307	$+580$	$+2.97$
$TEA^+ \to Na^+$	-237	$+177$	$+1.24$
$TEA^+ \to K^+$	-220	-247	-0.10
$TEA^+ \to Cs^+$	-165	-292	-0.72
$K^+ \to Li^+$	-87	$+827$	$+3.07$
$Na^+ \to Ca^{2+}$	-1641	$+770$	$+8.1$
$Na^+ \to Sr^{2+}$	-1584	$+450$	$+6.85$
$Na^+ \to Ba^{2+}$	-1648	$+170$	$+6.1$

[a] Cal site^{-1}.
[b] Cal deg^{-1} site^{-1}.

entropic contribution can be attributed to a modification of hydration corresponding to site binding; this hypothesis has been proposed by Strauss [20].

These results confirm those obtained by ultrasonic absorption [17] or ΔV measurements [18]. The values of ΔG_{exch} obtained in Sections 5.2.1 and 5.2.2 for example for the pairs Na/Ca are different but they don't have the same meaning; in the first case, it is a competition and in the last one it is the difference of free energy between two ionic states.

6. Conclusion

In this paper, experimental data concerning the ionic selectivity have been given and discussed. The selectivity is quantitatively expressed by the three thermodynamic parameters related to the nature of the counterions and to the charge density. The ionic selectivity must be attributed to a site binding with reorganization of the shell of hydration. Nevertheless, it is always a global effect which is measured and experimentally to-day the difficulty is to calculate the number of ionic sites involved in the phenomena and to know exactly how each site is affected. On carboxylic polyelectrolyte, the sequence of affinity for monovalent counterions is $Li > Na > K > Cs \sim TMA$ corresponding to that of the crystallographic radii. With divalent counterions, the affinity is always greater but the difference between counterions is very small and negligible at 25 °C [19] in presence of CMC.

The theoretical prediction on condensation of Manning following which for $|z|\,\lambda > 1$ the density of bound sites should increase linearly with the charge density finds some confirmation in our work; in all experiments the selectivity appears with monovalent counterions for a critical value of λ ($1 < \lambda_c < 1.6$) depending on the techniques and grows linearly with charge density.

In fact, so far, no theory is really adequate to introduce characteristics of the counterions other than ionic radii in an electrostatic treatment.

A first approach had been given by Eisenman [21] to interpret the selectivity sequence as a function of the local anionic field and of the dimensions of counterions; a general discussion of our results in this way had been given elsewhere concluding to a qualitative agreement [22].

References

1. Lifson, S. and Katchalsky, A.: *J. Polym. Sci.* **13**, 43 (1954).
2. Hanss, M., Roux, B., Bernengo, J. C., Milas, M., and Rinaudo, M.: *Biopolymers* **12**, 1747 (1973).
3. Rinaudo, M. and Milas, M.: *Compt. Rend. Acad. Sci. Paris, Ser. C* **271**, 1170 (1970).
4. Rinaudo, M. and Milas, M.: *IUPAC International Symposium on Macromolecules*, Vol. V, com. V 24, Helsinki, 1972.
5. Oosawa, F.: *Polyelectrolytes*, Marcel Dekker, New York, 1971.
6. Manning, G. S.: *J. Chem. Phys.* **51**, 924 (1969).
7. Rinaudo, M., Mazet, J., and Milas, M.: *Compt. Rend. Acad. Sci. Paris, Ser. C* **276**, 1401 (1973).
8. Rinaudo, M., Loiseleur, B., Milas, M., and Varoqui, R.: *Compt. Rend. Acad. Sci. Paris, Ser. C* **272**, 1003 (1971).
9. Domard, M., Rinaudo, M., and Rinaldi, R.: *J. Chim. Phys.* **70**, 1410 (1973).
10. Rinaudo, M. and Daune, M.: *J. Chim. Phys.* **64**, 1753 (1967).
11. Rinaudo, M. and Milas, M.: *J. Chim. Phys.* **66**, 1489 (1969).
12. Gregor, H. P. and Frederick, M.: *J. Polym. Sci.* **23**, 451 (1957).
13. Rinaudo, M. and Milas, M.: *Eur. Polym. J.* **8**, 737 (1972).
14. Rinaudo, M., Milas, M., and Laffond, M.: *J. Chim. Phys.* **70**, 884 (1973).
15. Suehr, J. and Yeager, E.: in P. Warren (ed.), *Physical Acoustics*, Vol. II, Part A; Mason Academic Press, New York, 1965, p. 351.
16. Tamm, K.: *6th International Congress of Acoustic*, Tokyo, 1968, Communication GP 3-3.
17. Zana, R., Tondre, C., Rinaudo, M., and Milas, M.: *J. Chim. Phys.* **68**, 1258 (1971).
18. Rinaudo, M. and Pierre, C.: *Compt. Rend. Acad. Sci. Paris, Ser. C* **269**, 1280 (1969).
19. Rinaudo, M. and Milas, M.: *Macromolecules* **6**, 879 (1973).
20. Bagala, A. J., and Strauss, U. P.: *J. Phys. Chem.* **76**, 254 (1972).
21. Eisenman, G.: *Biophys. J.* **2**, 259 (1962).
22. Milas, M.: Thesis, Grenoble, 1974.

PREPARATION AND CHARACTERIZATION OF CHARGE-STABILIZED POLYMER COLLOIDS

R. M. FITCH

University of Connecticut, Storrs, Conn. 06268, U.S.A.

Abstract. Polymer colloids are readily prepared by 'emulsion' polymerization, and derive their stability as colloids from charged groups fixed at the polymer/water interface. These groups provide an electrostatic potential energy barrier to flocculation. We are particularly interested in the polymer surface, since this represents an 'interfacial polyelectrolyte'. Polymer colloids are uniquely suited for study since high specific surface areas, high charge densities, and variation in the kinds of groups introduced are attainable on well-characterized interfaces.

Polymer colloids are synthesized ordinarily by free radical polymerization of an olefinic monomer in water. Free radicals are generated by a water-soluble initiator which introduces charged end-groups onto the polymer chains. Particle formation occurs by homogeneous self-nucleation of growing oligomeric radicals in solution. Particle growth occurs by further imbibing and polymerization of monomer, and capture of un-nucleated oligomeric radicals from the aqueous medium. The consequences of this mechanism with regard to the formation of the interfacial polyelectrolyte will be discussed.

Variations in the nature and density of charged groups can be obtained by varying the initiator, by copolymerization of ionic monomers, by subsequent chemical transformations, or by chain transfer with ionic agents. Examples of each of these will be cited.

Characterization of the interfacial polyelectrolyte has been accomplished by converting all counterions to H^+ and conductometric titration of these with base. Some studies have also been made on specific ion selectivity in equilibrium ion exchange experiments.

When the polymer colloids possess monodisperse particle size distributions, they may exhibit true liquid crystallinity with Bragg diffraction in the visible region. This manifests itself in brilliant opalescence.

Implications with regard to heterogeneous catalysis, immunological research, double layer theory, and properties and adhesion of derived films will be considered.

1. Introduction

Polymer colloids are dispersions of small particles of polymeric substances in fluid media. The disperse phase may be organic, such as polystyrene or polymethylmethacrylate, or inorganic, such as silica or sulfur. The continuous phase may be aqueous or organic. This discussion is limited to synthetic lataxes of lyophobic, organic polymers primarily in aqueous media formed by emulsion polymerization. The polymer/water interface often contains ionizable groups, e.g. sulfonate or carboxylate, the electrostatic charge on which is largely responsible for the stability of the colloid. Particle sizes may be anywhere in the colloidal range from 10 to 1000 nm diameter; size distributions vary from monodisperse (Figure 1) to extremely broad or even sharply polymodal (Figure 2), depending upon the choice of experimental conditions during synthesis. At the lower end of this range, the size and electrophoretic behavior is similar to that of some globular proteins, although in synthetic latexes the polymer may be very hydrophobic. Because it is possible to obtain surface charge densities in the range of ca. 0.5 to 3 micro-Coulombs per square centimeter ($\mu C\ cm^{-2}$) and because of their high specific interfacial areas, these systems may be regarded as 'interfacial polyelectrolytes', and exhibit many fascinating properties of both theoretical

Alan Rembaum and Eric Sélégny (eds.), Polyelectrolytes and Their Applications, 51–69. All Rights Reserved.

and practical importance. For instance, colloids of polymethyl methacrylate (PMMA) have been made with particle diameters in the range of 10 nm and surface charge densities, σ, on the order of 0.65 μC cm^{-2}, which have a specific interfacial areas of ca. 60 m^2 g^{-1} and ion exchange capacities of approximately 0.2 meq. g^{-1} of polymer [1].

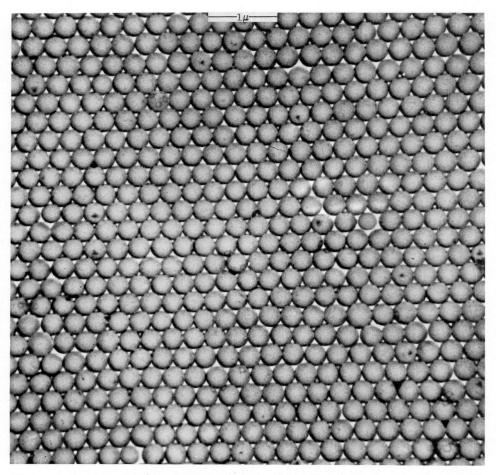

Fig. 1.   Monodisperse polyvinyl toluene latex. Surface replica electron micrograph of a dried film (E. B. Bradford, Dow Chemical Co., Midland, Mich.), 27, 830 × magnification.

In the succeeding sections the mechanism of the formation of polymer colloids is discussed, as it provides the ability and understanding required to exercise synthetic control of the properties of the system. Synthetic procedures are also discussed, and this is followed by brief treatments of methods of purification, surface characterization, some interesting interfacial chemistry, and finally, applications.

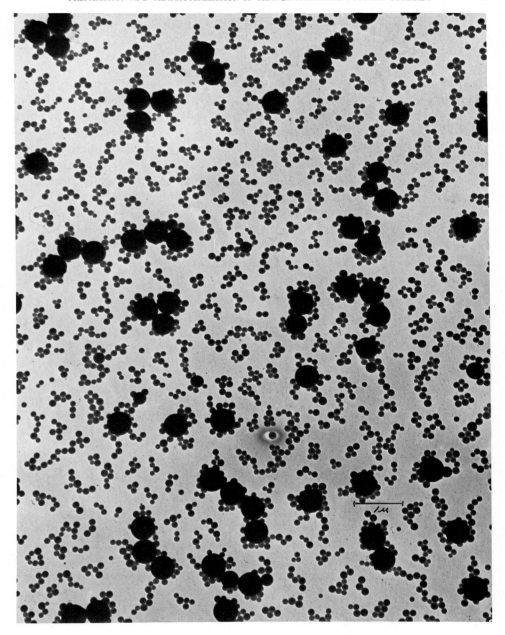

Fig. 2.   Didisperse acrylic latex. Electron micrograph, 18 200 magnification

2. Synthesis of Emulsifier-Free Latexes

2.1 MECHANISM OF PARTICLE FORMATION

In order to synthesize an interfacial polyelectrolyte the ionizable groups must be chemically bound to the interface and be an integral part of the polymer molecules comprising the colloidal particles. It is preferable to avoid adsorbed emulsifier, which is usually employed to stabilize these colloids, as it complicates the subsequent purification and quantitative surface characterization of the system. Thus the classical technique of emulsion polymerization, as described by Harkins [2], is not applicable. The systems are actually simpler, employing only monomer(s), water and initiator. It is most important that the initiator be very soluble in the water and that it form ionizable free radicals. The monomer must have a finite solubility in water, although this may be very small (even styrene satisfies this requirement at 0.038 wt.%), but may be present in amounts far in excess of this as an initially separate phase. Initiation of polymerization must then occur in the aqueous solution. For instance, if the initiator is potassium persulfate, $K_2S_2O_8$, free radical-ions are formed (along with some $\cdot OH$) by the thermal decomposition of the anion:

$$S_2O_8^= \rightarrow 2 \cdot SO_4^- . \tag{R-1}$$

These initiate polymerization by:

$$\cdot SO_4^- + CH_2 = C \overset{H}{\underset{X}{\diagdown}} \rightarrow {}^-OSO{-}CH_2{-}\overset{H}{\underset{X}{\overset{|}{C}}} \cdot \tag{R-2}$$

followed by propagation, until self-nucleation of the new polymer phase occurs [1]. Polymer particles are thus produced in great profusion – on the order of 10^{15} to 10^{18} per liter in a few seconds time.

Of great importance here is that the ionic polymer end-group, coming from the initiator, is held at the interface of the newly formed primary particle because of hydration forces. This is shown in Figure 3. The particles subsequently grow by imbibing monomer and polymerizing it within themselves by a process fully described by Smith and Ewart [3], Stockmayer [4], O'Toole [5] and Ugelstad and coworkers [6].

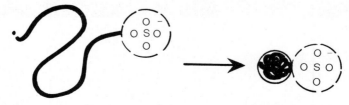

Fig. 3.   Self-nucleation of oligoradicals to form primary particles (schematic).

The free radicals are generated continuously throughout the entire reaction period. After all the particles have been formed, the ionic free radicals subsequently generated will be captured by polymer particles before they grow to the critical size for self-nucleation. The qualitative and quantitative aspects of these processes have been given by Fitch and Tsai [1] and Fitch and Shih [7]. The entry of a second oligoradical into a primary particle produces two results: (1) a second ionic end-group is placed at the interface and (2) mutual termination of the two radicals within the polymer/monomer particle ensues. This is shown schematically in Figure 4. The particle is temporarily

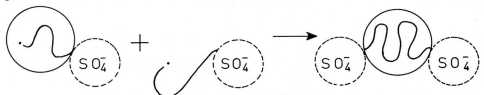

Fig. 4.   Entry of a second radical into a particle (schematic).

'dead' until the entry of a third radical. The process may be repeated tens to thousands of times, so that the particles finally possess a corresponding number of surface ions. The number of such groups on a single particle will depend upon the particle size and the rate of generation of free radicals. It has been experimentally shown by van den Hul and Vanderhoff that all the sulfate groups generated in a persulfate-initiated emulsion polymerization of polystyrene could be accounted for at the polymer/water interface [8]. When chain termination occurs by combination as depicted in Figure 4, there will be two ionic end-groups per polymer molecule; when termination is by disproportionation, then there will be one ionic end-group per molecule. In either case, the interfacial concentration of ionic groups (or more precisely, ionogenic groups) will be determined by just two factors, (1) the degree of polymerization (molecular weight) of the polymer, barring effects due to chain transfer and copoly merization of ionogenic monomers, and (2) the particle size:

(1) Degree of polymerization, $\overline{DP}$: In the simplest instance the $\overline{DP}$ is determined by the ratio of the rate of propagation to that of termination, which in emulsion polymerization is given by

$$\overline{DP} = \frac{R_p}{R_t} = \frac{k_p\left(\dfrac{\bar{n}N}{N_A}\right)[M]_p}{k_t\bar{n}\dfrac{(\bar{n}-1)}{v}\dfrac{N}{N_A^2}} = \frac{k_p[M]_p N_A}{k_t\dfrac{(\bar{n}-1)}{v}},$$
(1)

where R_p, R_t are rates of chain propagation and termination,

   k_p, k_t are the corresponding specific rate constants,

   $\bar{n}$ is the average number of free radicals per particle,

   N_A is Avogadro's number,

   N is the number concentration of particles,

   $[M]_p$ is the monomer concentration within the particles, and

   v is the volume of a particle.

The solutions to Equation (1) for various experimental conditions involve the statistical analysis of the distribution of free radicals among the particles to give values of $\bar{n}$, and the partition of monomer between the two phases to give $[M]_p$ [3, 4, 5, 6, 9].

(2) The Particle Size: Kinetic factors evidently determine the final particle concentration which in turn governs the particle size. Particles are formed by the self-nucleation of the oligomeric radicals. However as N increases, the rate of capture of oligoradicals by the particles increases, thereby decreasing the rate of primary particle nucleation:

$$\frac{dN}{dt} = R_i - R_c, \tag{2}$$

where R_i and R_c are the rates of radical generation and capture respectively [1]. The experimental evidence favors a diffusion-controlled rate of capture which may or may not be reduced by an electrostatic repulsion between oligoradical and polymer particle, depending on the size of the particle, its surface electrical potential, Ψ_0, and the ionic strength of the solution [7]. If the primary particles thus formed are colloidally 'stable', or more precisely, diuturnal [10], they will be produced in great numbers over a few seconds time, after which R_c rapidly approaches R_i, and dN/dt falls essentially to zero; a steady state is reached in which no new particles are formed. We may expect the time-dependence of N to be something like that shown in Figure 5.

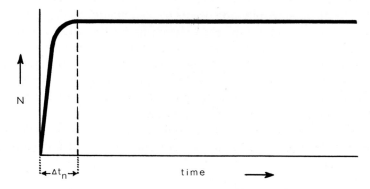

Fig. 5.   Kinetics of formation of colloidally stable primary polymer particles.
Nucleation time interval is given by Δt_n.

Most often, however, the primary particles are colloidally 'unstable', or caducous, and flocculate to a *limited* extent. In this process two particles upon collision adhere to each other and subsequently fuse, or coalesce, to form a single particle in which most, if not all of the surface groups remain at the interface. This is represented in Figure 6. The consequent reduction in surface-to-volume ratio leads to an increase In surface charge density, σ, and surface electrical potential, Ψ_0, which in turn produces a rapid decline in the rate of flocculation, R_f, as the process continues. [11]. Such behavior has been observed [1] experimentally (Figure 7) and probably occurs in most latex polymerizations. Equation (2) must therefore be modified to take into account the kinetics of flocculation:

$$\frac{dN}{dt} = R_i - R_c - R_f. \tag{3}$$

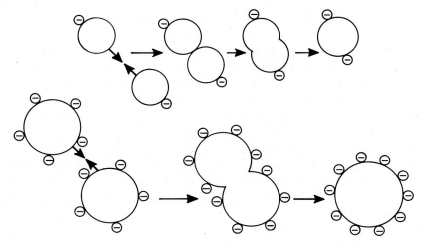

Fig. 6.   Limited flocculation: self stabilization through increased surface electrical potential (schematic).

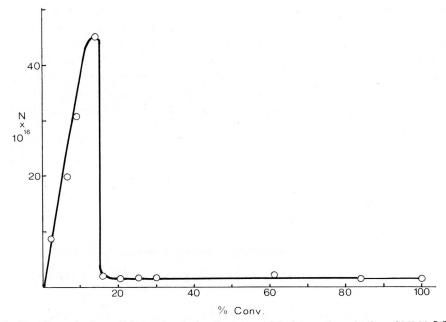

% Conv.

Fig. 7.   Change in the particle number during the course of the latex polymerization of MMA [1].

2.2. Experimental considerations: particle size and size distribution

The above considerations lead to some rather simple experimental procedures for controlling the particle size of the colloid. Since a high surface potential reduces R_f, introducing more ionic groups per unit area of interface will *in*crease the final number of particles. This can be accomplished through a reduction in the molecular weight of the polymer (more end-groups per gram), which in turn is brought about by an increase in the rate of initiation relative to the rate of propagation. An increase in the particle number, of course, produces a decrease in size.

An increase in R_f may easily be produced by an increase in the ionic strength of the aqueous medium. This reduces the thickness of the electrical double layer surrounding the particles and therefore the potential barrier to flocculation. If the number of particles is relatively small, and consequently R_c is small, it is possible that a steady state may be achieved, in which $dN/dt = 0$, governed principally by R_i and R_f in Equation (3). Under these conditions the DLVO theory [11] states that the primary particles disappear by flocculating into the large ones [12]. Very large particles can be produced by seeded polymerizations in which particles of a previously formed latex are present at the beginning of a polymerization. In order to ensure only growth and no nucleation of new particles, the size and concentration of the seed particles must be such that $R_c \geqslant R_i$ [7].

An especially attractive feature of polymer colloids is that they often can be synthesized with extremely narrow particle size distributions (see Figure 1) with ease. Monodisperse sols are formed when the nucleation of particles occurs over a time interval which is short compared to that used for growth [13]. This is evidently the case when the primary particles are stable and the nucleation kinetics are as shown in Figure 5, where Δt_n is short compared to the total polymerization time. Stabilization of the primary particles, however, ordinarily requires addition of relatively large amounts of emulsifier [1]. In emulsifier-free systems, where flocculation is likely to occur, the same principle may apply. In this case, however, any decrease in N due to flocculation must stop within a short time, Δt_n, as shown in Figure 8. This behavior

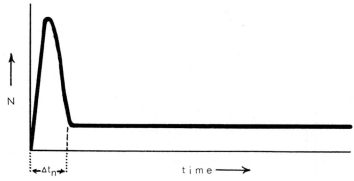

Fig. 8.   Formation of a monodisperse colloid under limited flocculationcontrol of the particle number, N.

will occur, according to Equation (3), when $R_i \gg (R_c + R_f)$ initially and immediately is followed by a short period when $R_f \gg (R_i - R_c)$ prior to the steady state when $R_i = (R_c + R_f)$. Experimentally this must be controlled by ensuring that R_f is never of the same order of magnitude as R_i during Δt_n, and that self-stabilization may occur through concentration of stabilizing groups at the polymer/water interface. The initial sharp rise in N probably always occurs since rates of initiation are chosen to be on the order of 10^{15} to 10^{18} radicals $l^{-1} s^{-1}$. The ionic strengths due to initiator concentrations of sufficient magnitude to bring about such rates of radical generation (e.g. 10^{-3} to 10^{-2} M persulfate) probably are adequate to give high rates of flocculation. Otherwise, small amounts of neutral electrolyte such as NaCl may be added [14]. To ensure some mobility of the surface groups to reposition themselves with each flocculating event (Figure 6) it may be necessary to add a good solvent for the polymer (volatile plasticizer) or to use a high concentration of monomer where the monomer is a good solvent for the polymer. That flocculation followed by coalescence may lead to rapid self stabilization under these conditions has been demonstrated by theory and experiment [15].

2.3. CONTROL OF THE NATURE AND DENSITY OF SURFACE GROUPS

There is apparently an almost endless variety of free radicals capable of initiating polymerization which can be readily formed in aqueous media.

Oxidation-reduction (redox) reactions, in which a one-electron transfer step produces a radical intermediate are probably the most numerous. Two of the most widely used are cited below.

(a) Persulfate-bisulfite (PBI) [16]:

$$S_2O_8^= + Fe^{+2} \rightarrow \cdot SO_4^- + SO_4^= + Fe^{+3} \qquad \text{(R-3)}$$
$$HSO_3^- + Fe^{+3} \rightarrow \cdot SO_3^- + H^+ + Fe^{+2} \qquad \text{(R-4)}$$

(b) Fenton's Reagent [17]:

$$H_2O_2 + Fe^{+2} \rightarrow \cdot OH + OH^- + Fe^{+3}. \qquad \text{(R-5)}$$

The thermal decomposition of peroxides and aliphatic azo-compounds is also widely employed [18]. The persulfate ion, cited above in Reaction R-1, is a typical peroxide in which the oxygen-oxygen bond is ruptured:

$$
\begin{array}{ccc}
O^- & O^- & O^- \\
| & | & | \\
OSO \sim OSO & \xrightarrow{\Delta} & 2\ OSO\cdot . \\
| & | & | \\
O & O & O
\end{array}
\qquad \text{(R-1)}
$$

weak bond

Perphosphates, percarbonates, peroxides of alkyl borons, hydrogen peroxide and organic peroxy acids, e.g. peracetic acid, are other examples of O—O bond-breaking compounds. Among azo compounds, the following two are of especial interest:

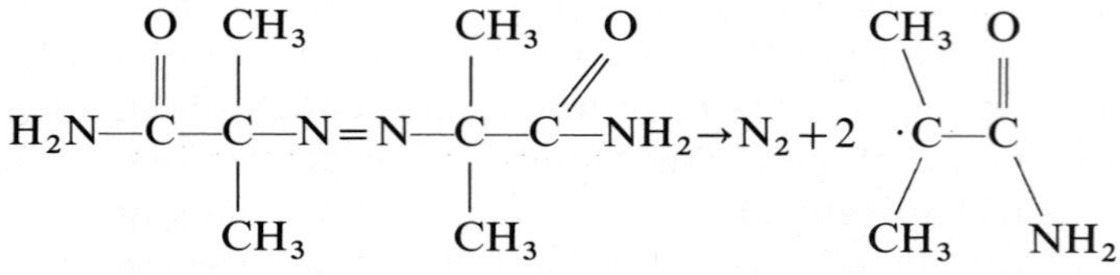

$$H_2N\text{—}\overset{\overset{\displaystyle O}{\|}}{C}\text{—}\overset{\overset{\displaystyle CH_3}{|}}{\underset{\underset{\displaystyle CH_3}{|}}{C}}\text{—}N=N\text{—}\overset{\overset{\displaystyle CH_3}{|}}{\underset{\underset{\displaystyle CH_3}{|}}{C}}\text{—}\overset{\overset{\displaystyle O}{/\!/}}{C}\text{—}NH_2 \rightarrow N_2 + 2 \ \cdot\overset{\overset{\displaystyle CH_3}{\diagup}}{C}\text{—}\overset{\overset{\displaystyle O}{\|}}{C}\underset{\diagdown \ NH_2}{\diagup \ CH_3} \quad , \quad (R\text{-}6)$$

and the corresponding amidine hydrochloride which produces positively charged radicals, giving cationic end-groups:

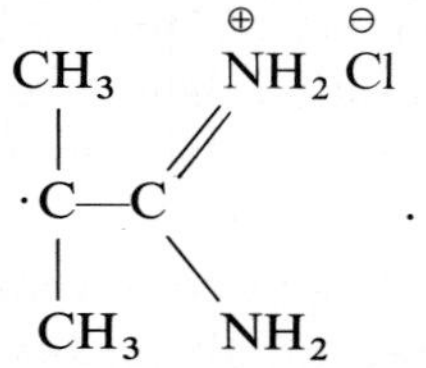

$$\cdot\overset{\overset{\displaystyle CH_3}{|}}{\underset{\underset{\displaystyle CH_3}{|}}{C}}\text{—}\overset{\overset{\displaystyle \overset{\oplus}{N}H_2 \ \overset{\ominus}{Cl}}{/\!/}}{\underset{\diagdown \ NH_2}{C}} \quad .$$

Photochemical reactions also often lead to the formation of free radicals, and a variety of different end-groups may be thus formed. However participation by water usually means that hydroxyl radicals will be formed [19, 20]. In emulsion polymerizations, the intense scattering of the incident light makes quantitative measurements of R_i as a function of light intensity extremely difficult, although some estimates have been made [20].

Copolymerization of small amounts of monomers containing ionogenic functional groups may also be used, especially to obtain a high concentration at the interface. To accomplish this, the monomer must either be adsorbed at the polymer/water interface prior to its copolymerization or it must be incorporated into the oligomeric radical in the aqueous phase prior to capture by a particle. These processes are illustrated in Figure 9. However, some of these groups may be found permanently in the aqueous

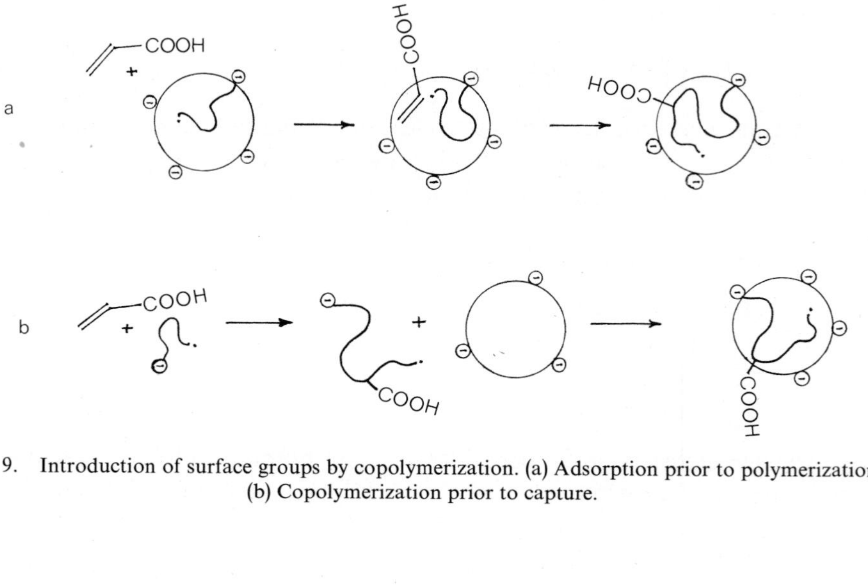

Fig. 9.   Introduction of surface groups by copolymerization. (a) Adsorption prior to polymerization. (b) Copolymerization prior to capture.

phase as a result of the formation of some water-soluble copolymer (containing relatively high percentages of the ionogenic monomer), and/or permanently 'buried' within the polymer particles because of the solubility of the monomer in the particle. For instance Greene has found that acrylic acid, when copolymerized with styrene/ butadiene in amounts from 1.25% to 5% of the polymer, was distributed among the aqueous phase, the polymer/water interface and the interior of the particles in the approximate ratio of 2:3:1; the amount in the aqueous solution could be greatly increased by raising the pH [21]. Presumably, by choosing an ionogenic monomer which strongly concentrates at the interface and is poorly soluble in both aqueous and polymer phases, one could obtain most efficient placement of the groups at the interface [22].

Besides acrylic and methacrylic acids, the following are commercially available ionogenic monomers:

Carboxylic Acids:
    Itaconic
    Maleic (used as anhydride)
    Crotonic
Sulfonic Acids:
    Sulfoethyl methacrylate
    p-Styrene sulfonic acid
    Vinyl sulfonic acid
Amines:
    Dimethylaminoethyl methacrylate
    Allyl amine
Alkyl Ammonium Halide:
    Methacryloxyethyl trimethyl ammonium chloride.

There are potentially many more functional monomers which could be synthesized, the possibilities of which are largely unexplored.

End-groups may also be introduced by chain transfer. There may be many transfer steps for each initiating free radical, so that the chemical nature of the initiator may become relatively unimportant under these circumstances. The problem here, as above, is to achieve efficient placement of the ionogenic groups derived from the chain transfer agent at the interface. The process is shown schematically in Figure 10, in which mercaptoacetic acid is used to introduce carboxyl groups to the interface. The ideal situation, as in the case of functional monomers, is probably one in which the transfer agent is strongly, preferentially adsorbed at the interface. The use of this technique is apparently almost totally unexplored.

2.4. EXAMPLES

(1) Goodwin and coworkers have made a rather extensive study of the experimental variables involved in the synthesis of emulsifier-free, monodisperse polystyrene colloids stabilized solely by end-groups derived from persulfate initiator [14]. They obtained particle sizes in the range of 100 to 1000 nm diameter by varying initiator

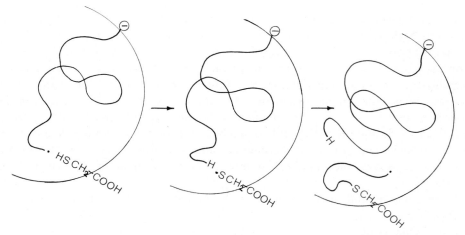

Fig. 10.   Placement of ionogenic polymer end-groups at the polymer particle
interface by chain transfer.

concentration, $[P]$, ionic strength, $[I]$, and the polymerization temperature, T. The dependence of the average particle diameter, D, upon these variables was found to obey the empirical equation

$$\log D = 0.238 \left[\log \frac{[I]\, M^{1.723}}{[P]} + \frac{4929}{T} \right] - 0.827. \tag{4}$$

The interfacial polyelectrolyte formed in this fashion evidently contains not only sulfate groups, but hydroxyl and carboxyl as well [8].

(2) Wright and coworkers describe a variety of different emulsifier-free polymer colloids, one of the most interesting of which contains exclusively alkyl quarternary ammonium chloride interfacial polyelectrolyte [23]. The particles consist of a copolymer of styrene and butyl acrylate along with the ionogenic conomer, methacryloxyethyl trimethyl ammonium chloride, while the initiator is α,α'-azobis-isobutyramidine hydrochloride. The polymerization is conducted at $80\,°C$ with the continuous addition of the monomers and part of the initiator in order to ensure compositional homogeneity of the polymer. These colloids are often monodisperse.

(3) Fitch and Tsai describe the preparation of monodisperse, emulsifier-free latexes made from methyl methacrylate and stabilized solely by end-groups [1]. The initiator system was persulfate-bisulfite-iron ((R-3) and (R-4)) and the temperature was always $30\,°C$. At a constant rate of initiation they found that the number of particles, N, was independent of the initial monomer concentration over the range 0.003 80 to 0.0951 molar, while the particle diameter increased from 35.4 to 105.4 nm. In another experiment at a lower R_i, they obtained a final particle diameter of 172 nm and $N = 1.1 \times 10^{15}\, l^{-1}$, compared to $D = 68.7$ nm and $N = 1.8 \times 10^{16}\, l^{-1}$ at the same monomer concentration (0.035 M) but higher R_i. They estimated in the former case that each ionic group occupied 24.7 nm^2 at the point when self-stabilization by

flocculation occurred. The final surface density of groups was approximately 2.5 nm^2 per group or 0.41 groups per nm^2.

2.4.1. *Purification*

Before a polymer colloid can be used for quantitative work, either for physical studies or as a chemical reagent, the residual salts and organic compounds, which might otherwise interfere, must be removed. In some cases emulsifier, added for particle size control during the polymerization, must also be removed. Dialysis would appear to be the method of choice, but it is extremely slow as the concentrations of impurities becomes low, and there is some evidence that even after exhaustive dialysis, appreciable quantitities of low molecular weight compounds may reamin associated with the colloid [8]. Greene claims that a combination of dialysis of ionic impurities against continuously running, deionized water for two days and absorption into, and adsorption onto nonionic bead polymer particles of organic impurities gave good results [21]. On the other hand, Vanderhoff and coworkers claim that use of a mixed-bed ion exchange resin, which has been previously exhaustively purified, is the only way of removing *all* impurities [24]. It is also likely that the ion exchange resin beads may remove non-ionic, organic impurities by ab- and adsorption. Because of the strong electrostatic interaction which may develop between the colloidal particles and ion exchange resin beads, it is possible that adsorption of the particles will also occur, in which case the concentration and size distribution of the latex may be altered upon treatment [25]. Of crucial importance in this method is the prior scrupulous purification of the ion exchange resin in order to remove any water-soluble polyelectrolytes [26].

3. Characterization

The measurement of particle size and size distribution will not be discussed here, although it is essential to the complete characterization of a polymer colloid. Collins has recently written a thorough review of available methods [27]. Of primary concern here is the determination of the nature of the chemical groups and their concentration at the interface. Ordinarily, more than one analytical technique is required [21].

Because the functional groups are generally present in small concentrations in relation to the total amount of latex (ca. 60 millequivalents per liter maximum), sensitive analytical methods are needed. Palit and coworkers have developed a dye-partition technique which they and others have successfully used to determine both the nature and concentration of various groups attached to polymer chains [28, 12]. The polymer is separated, dried and dissolved in a suitable solvent. This solution is equilibrated with an aqueous solution of a dye which is substantive only to the functional groups to be determined. Photometry of the polymer solution then gives a measure of the concentration of these groups. By proper selection of dyes, the method has the advantage of specificity for the functional groups desired. Other techniques which have been used include X-ray fluorescence analysis (for sulfur) [8], neutron activation analysis (for oxygen and sodium) [21], and radiotracer methods, but from these analyses of polymer in solution, no information is obtained about the placement of the groups.

To determine the concentration of groups at the interface, conductometric titration of the counter-ions is the method of choice [8], and has been employed for many years for classical colloids as well [29]. All extraneous ions should be removed and the counterions must be titratable. In a negatively charged sol this is easily accomplished by treatment with a mixed-bed ion exchange resin in the H^+ and OH^- forms [25]. The H^+ ions will then be equal in number to the fixed groups at the interface and may be 'counted' by conductometric titration. Potentiometry has also been employed, but apparently is not as sensitive [8, 21]. Titration with a surface active substance of opposite charge until rapid coagulation has been induced [21] or until a change in slope of the equilibrium adsorption isotherm is observed has also proven successful (Figure 11a) [30].

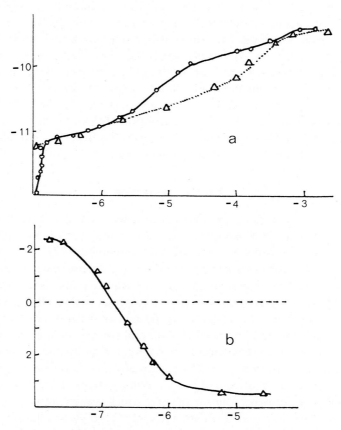

Fig. 11.   (a) Isotherms for adsorption of cetyl trimethyl ammonium (CTA) ions onto: ----△---- poly-styrene, and ——○—— polystyrene latex particles. *Abscissa:* log molar conc'n CTA. *Ordinate:* log moles adsorbed cm^{-2}. (b) Electrophoretic mobility of polystyrene latex particles as a function of CTA concentration. *Abscissa:* same as in (a). *Ordinate:* log mobility in $\mu ms^{-1} volt^{-1} cm^{-1}$. (Reprinted by permission.)

4. Chemistry at the Interface

There is very little published on chemical reactions occurring at the polymer/water interface in polymer colloids beyond the simple acid-base neutralizations of the counterions mentioned above. Some evidence has been developed to indicate that certain secondary reactions must have occurred either during the emulsion polymerization or subsequently on storage.

4.1. CHEMICAL MODIFICATION OF SURFACE GROUPS

All polystyrene latexes prepared by persulfate initiation may be shown to be negatively charged by electrophoresis. That is, all but one, which was discovered after years of storage at the Dow Chemical Co. to be electrically neutral, but very stable. The surface groups on this now famous latex are evidently all hydroxyl [31]. The sulfate groups originally present all must have hydrolyzed slowly upon standing:

$$\begin{matrix} O \\ \sim\!\sim\!\sim OSO^-Na^+ + H_2O \rightarrow \qquad OH + Na^+HSO_4^- . \\ O \end{matrix} \qquad (R\text{-}7)$$

Carboxyl groups are also found at the interfaces of many persulfate-initiated polystyrene colloids [32]. Evidently the hydroxyl groups present may be oxidized at the elevated temperatures of the polymerizations by persulfate, which is known to be a strong oxidizing agent:

$$\sim\!\sim\!\sim CH_2OH + H_2O + 2S_2O_8^= \rightarrow 4HSO_4^- + \qquad COOH . \qquad (R\text{-}8)$$

Reactions R-7 and R-8 are the only ones for which there is good evidence for the chemical modification of interfacial functional groups. The possibilities for other reactions of course are numerous and represent an inviting area for further research.

4.2. ADSORPTION OF SURFACE-ACTIVE MOLECULES

The chemical and electrical nature of the interfacial polyelectrolyte may be modified by adsorption of amphipathic molecules. In this respect they are to be distinguished from ordinary polyelectrolytes. A couple of examples may serve to illustrate:

(1) A polystyrene latex containing interfacial carboxyl groups was equilibrated with varying, but extremely small concentrations of cetyl trimethyl ammonium (CTA) bromide in solution (10^{-8} to 10^{-3} molar). The amount of CTA ion adsorbed was measured by a radioactive tracer technique and the electrophoretic mobility also determined. The results are shown in Figure 11. Two kinds of adsorption are observed, the first involving the reaction between CTA ions and the surface carboxyl groups, and the second in which the long chain alkyl groups of the CTA adsorbs onto hydrophobic styrene. Thus the initial steep slope in the adsorption isotherm (Figure 11a) corresponds to the rapid decline in surface charge manifested in the electrophoretic mobility (Figure 11b). Furthermore the point of zero charge corresponds to the 'knee' in the adsorption curve. After that point the curve for adsorption onto the latex

particles is almost the same as that for the pure polystyrene [30]. The area occupied by the adsorbed cations at saturation varied from 35 to 47 Å^2 per group, indicating a high degree of *orientation* normal to the surface [30], another unique feature of interfacial polyelectrolytes.

(2) In another study of the same type of latex, Ottewill and Vincent have shown that even with a non-ionic adsorbate there may be interactions with the surface ionic groups. They studied ethanol, *n*-propanol and *n*-butanol and concluded that the initial adsorption, at low concentrations of alkanol, probably involved ion-dipole association of hydroxyl with surface carboxylate anions. This would result in the hydrophobic 'tails' of the alkanols being oriented towards the aqueous phase and, more importantly, a desorption of counterions from the double layer, thus affecting the surface electrical potential, Ψ_0 [33].

4.3. IMMUNOCHEMISTRY

Polymer colloids have been used to detect extremely small concentrations of antigens in a modification of the 'passive hemagglutination technique', in which the polymer particles are substituted for red blood cells. The principle involved is essentially the 'amplification' of the results of a chemical reaction by attaching to a relatively small molecule, in this case the antigen, a huge 'tag', the latex particle. The system must be so designed that when the desired chemical reaction occurs, i.e. antibody-antigen bonding, flocculation ('agglutination') of the colloid results. This produces a pronounced change in the light scattering intensity of the system, thus visualizing a very small chemical change. For instance, the flocculation of a colloid containing 10^6 particles, or 10^{-17} moles of particles per liter is easily detectable. Classically, the antigen is bound to the surface of red cells by use of bis-diazotized benzidine, BDB, [34], as illustrated in Figure 12. When a dilute suspension of this material is brought into

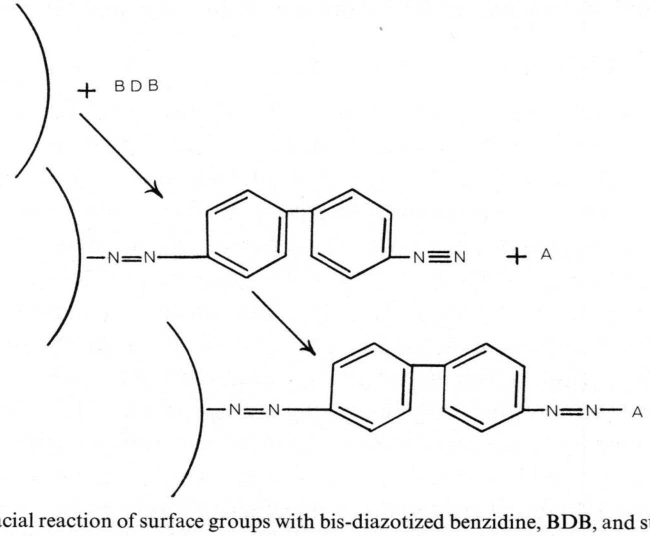

Fig. 12.   Interfacial reaction of surface groups with bis-diazotized benzidine, BDB, and subsequently with antigen protein, *A*, to produce conjugate antibody-specific flocculation.

contact with the conjugate polyvalent antibody, binding of particles together, or 'bridging' via the antibody molecules results and flocculation and sedimentation ensue. It is possible by this means to detect as little as 9 ng of antibody [35, 36].

As an example, human chorionic gonadotropic hormone, HCG, may be attached to the surface of latex particles and thereby serve as a sensitive reagent for the detection of *anti*-HCG factor, which is always present in the urine of a woman eight to ten days after the first missed menstrual period in pregnancy. A simple pregnancy test may thus be performed on a single drop of urine mixed with two drops of the interfacial poly-electrolyte reagent.

Antigen-coated polymer colloid particles may also be useful in the histological localization of antibodies. By placing the big 'tag' onto the antigen and then bathing a sample of tissue or a suspension of cells with this colloidal reagent, latex particles will be affixed only at those sites where antibody molecules are situated in the cellular matter. Subsequent electron microscopy easily reveals the exact location of the antibodies (Figure 13), thus unfolding many secrets of biological processes [37]. This is a more versatile adaptation of the older ferritin-labelling technique [38].

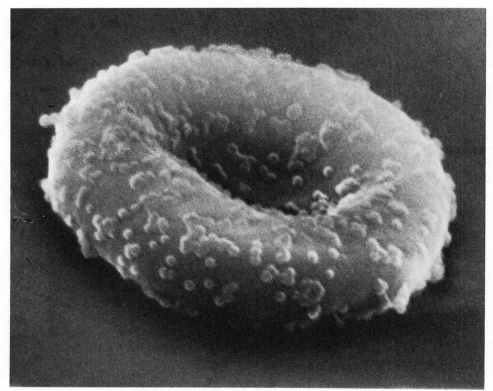

Fig. 13.   Red blood cells labelled with rabbit antibovine erythrocyte antiserum followed by goat anti-rabbit antibodies covalently bonded to latex particles 1300 Å in diameter. Only partial labelling of the cell is observed due to the low concentration of latex conjugate used. (Reproduced by courtesy of R. S. Molday. W. J. Dreyer, A. Rembaum, and S. P. S. Yen.).

4.4. Heterogeneous catalysis

Interfacial polyelectrolytes appear to be natural candidates for use as heterogeneous catalysts both in aqueous and organic media [39]. Simple solid acids and bases often exhibit rate enhancement and occasional stereoselectivity [40]. The enzymes, of course, with their colloidal particle sizes and sophisticated structure possess these characteristics in the highest degree. Polymer colloids will fall somewhere in between. By judicious choice of initiator, rate of initiation, fluid medium and comonomers, by ion exchange, and by chemical modification of surface groups it is now possible to place almost any kind of chemically reactive or catalytic group at the interface in various combinations and at various surface densities. The possibilities for studying the mechanisms of heterogeneous catalysis and for developing new catalysts is apparently boundless.

References

1. Fitch, R. M. and Tsai, C. H.: in R. M. Fitch (ed.), *Polymer Colloids*, Plenum Press, 1971, pp. 73–116.
2. Harkins, W. D.: *J. Am. Chem. Soc.* **69**, 1428 (1947).
3. Smith, W. V. and Ewart, R. H.: *J. Chem. Phys.* **16** (6), 592 (1948).
4. Stockmayer, W. H.: *J. Polym. Sci.* **24** (106), 314 (1957).
5. O'Toole, J. T.: *J. Appl. Polym. Sci.* **9**, 1291 (1965).
6. Ugelstad, J., Mørk, P. C., Dahl, P., and Rangnes, P.: *J. Polym. Sci.* Part C, No. 27, 49 (1969).
7. Fitch, R. M. and Shih, L. B.: *Kolloid Z. Z. Polym.*, in press.
8. Hul, H. J. van den and Vanderhoff, J. W.: in R. M. Fitch (ed.), *Polymer Colloids*, Plenum, New York, 1971, pp. 1–27.
9. Morton, M., Kaizerman, S., and Altier, M. W.: *J. Colloid Interface Sci.* **9**, 300 (1954).
10. Mysels, K. J.: *Introduction to Colloid Chemistry*, Interscience, New York, 1959.
11. Verwey, E. J. W. and Overbeek, J. Th. G.: *The Theory of Stability of Lyophobic Colloids*, Elsevier, Amsterdam, 1948.
12. Dunn, A. S. and Chong, L. C. H.: *Brit. Polymer J.* **2**, 49 (1970).
13. LaMer, V. K. and Dinegar, R. H.: *J. Am. Chem. Soc.* **72**, 4847 (1950).
14. Goodwin, J. W., Hearn, J., Ho, C. C., and Ottewill, R. H.: *Brit. Polymer J.* **5**, 347 (1973).
15. Suzuki, A., Ho, N. F. H., and Higuchi, W. I.: *J. Colloid Interface Sci.* **29** (3), 552 (1969).
16. Fritsche, P. and Ulbricht, J.: *Faserforsch. U. Textiltechn.* **14**, 320 and 517 (1963).
17. Baxendale, J. H., Evans, M. G., and Park, G. S.: *Trans. Faraday Soc.* **42**, 155 (1946).
18. Walling, C.: *Free Radicals in Solution*, John Wiley, New York, 1957.
19. Pramanick, D. and Saha, M. K.: *Indian J. Chem.* **4** (6), 253 (1966).
20. Dainton, F. S., Seaman, P. H., James, D. G. L., and Eaton, R. S.: *J. Polym. Sci.* **34**, 209; and **39**, 279 (1959).
21. Greene, B. W.: *J. Colloid Interface Sci.* **43** (2), 449 and 462 (1973).
22. Greene, B. W., Sheetz, D. P., and Filer, T. D.: *J. Colloid Interface Sci.* **32**, 90 (1970); Greene, B. W. and Saunders, F. L.: *ibid.* **33**, 393 (1970).
23. Wright, H. J., Bremmer, J. F., Bhimani, N., and Fitch, R. M.: U.S. Pat. 3, 501, 432 to Cook Paint and Varnish Co.
24. Vanderhoff, J. W., van den Hul, H. J., Tausk, R. J. M., and Overbeek, J. Th. G.: in G. Goldfinger (ed.), *Clean Surfaces: Their Preparation and Characterization for Interfacial Studies*, Marcel Dekker, New York, 1970, p. 15.
25. McCann, G. D., Bradford, E. B., van den Hul, H. J., and Vanderhoff, J. W.: in R. M. Fitch (ed.), *Polymer Colloids*, Plenum, New York, 1971, p. 29.
26. Schenkel, J. H. and Kitchener, J. A.: *Nature* **182**, 131 (1958).
27. Collins, E. A.: in G. Poehlein (ed.), *Advances in Emulsion Polymerization and Latex Technology*, Lehigh University, Bethlehem, Pa., June 11–15, 1973, pp. 227–280.
28. Roy, G., Mandal, B. M., and Palit, S. R.: in R. M. Fitch (ed.), *Polymer Colloids*, Plenum, New York, 1971, p. 49; Palit, S. R. and Mandal, B. M.: *J. Macromol., Sci.-Revs. Macromol. Chem.* **C2**, 225 (1968).

29. Bruyn, H. de and Overbeek, J. Th. G.: *Kolloid Z.* **84**, 186 (1938).
30. Connor, P. and Ottewill, R. H.: *J. Colloid Interface Sci.* **37** (3), 642 (1971).
31. Vanderhoff, J. W.: private communication.
32. (a) Ottewill, R. H. and Shaw, J. N.: *Kolloid Z. Z. Polym.* **218**, 34;
    (b) Shaw, J. N. and Marshall, M. C.: *J. Polym. Sci.* **A**, 449 (1968).
33. Ottewill, R. H. and Vincent, B.: *J. Chem. Soc., Faraday Trans. I* **68**, 1533 (1972).
34. (a) Pressman, D., Campbell, D. H., and Pauling, L.: *J. Immunol.* **44**, 101 (1942);
    (b) Stavitsky, A. B. and Arguilla, E. R.: *ibid.* **74**, 306 (1955).
35. Roberts, A. N. and Haurowitz, F.: *J. Immunol.* **89**, 348 (1962).
36. Goodman, H. C. and Bozicevich, J.: in J. F. Ackroyd (ed.), *Immunological Methods*, F. A. Davis, Philadelphia, 1964, p. 93.
37. Molday, R. S., Dreyer, W. J., Rembaum, A., and Yen, S. P. S.: *Nature* **249**, 81 (1974).
38. dePetris, S., Karlsbad, G., and Pernis, B.: *J. Expt'l. Med.* **117**, 849 (1963).
39. Fitch, R. M., McCarvill, W., and Gajria, C.: results to be published.
40. Tanabe, K.: *Solid Acids and Bases*, Academic Press, New York, 1970.

POLYELECTROLYTE 'CATALYSIS' IN IONIC REACTIONS

NORIO ISE

Dept. of Polymer Chemistry, Kyoto University, Kyoto, Japan

Abstract. The rate of chemical reactions can be changed by two factors; the rate constant and reactant concentrations. Simple electrolytes can accelerate or decelerate interionic reactions through the rate constant term (primary salt effect). Polyelectrolyte influence (both acceleration and deceleration) on ionic reactions, which is much larger than the simple electrolyte effect, is suggested to be due to changes of the rate constant induced by high electrostatic potentials of macroions. In addition to the electrostatic interactions, the role of hydrophobic interactions and of hydrogen-bonding interactions is dissected for a variety of combinations of 'catalyst' polyelectrolytes and reactants. Polyelectrolyte influence on elementary processes is discussed for equilibrium reactions. The forward and backward steps are affected in a different proportion; the use of the term 'polyelectrolyte catalysis' or 'polymer catalysis' is not justified. The polyelectrolyte acceleration of interionic reactions between like-charged reactants is shown to be associated with decreases in the entropy ($\Delta S^\dagger$) and enthalpy ($\Delta H^\dagger$) of activation. The $\Delta S^\dagger$ decrease is suggested to result from desolvation of the reactant ions by the electrostatic field of macroins. The limitation of the so-called 'concentration' effect and of the Brønsted theory is discussed.

Recently, keen attention has been paid to the 'catalytic' action of high molecular weight compounds [1]. It has been found that, under appropriate conditions, these compounds are much more efficient in accelerating chemical reactions than corresponding low molecular weight compounds. In our group, intensive studies have been carried out mainly on interionic reactions and polyelectrolyte influence thereon. In this article, we would like to review the recent progress of our work; the present review is by no means comprehensive. The relevant contributions from other laboratories will be mentioned sufficiently to place our research in perspective.

1. 'Catalysis' by Low Molecular Weight Electrolytes

Electrostatic interactions give a large deviation from ideality in equilibrium properties of solutions containing low molecular weight electrolytes. This deviation was most successfully disposed of by the Debye-Hückel theory [2]. According to this theory, the ionic species are not distributed in solution in a random manner, but form an ionic atmosphere structure, and the thermodynamic properties such as the activity coefficient of solvent (or the osmotic coefficient), the mean activity coefficient of solute, and the heat of dilution, decrease linearly with the square root of the concentration, in conformity with experimental observations.

The electrostatic interactions are also responsible for unusual, kinetic features of interionic reactions. The change of the rate constants of these reactions with ionic concentration was accounted for by Brønsted in terms of the Debye-Hückel theory and an activated complex theory [3]. For a reaction between two ions, $A+B\rightleftarrows$ $\rightleftarrows X \rightarrow C+D$, ($X$; the activated complex, C and D; the products), the rate constant is

$$k_2 = \frac{kT}{h} K \frac{f_A \cdot f_B}{f_X} = \frac{kT}{h} \frac{f_A \cdot f_B}{f_X} e^{\Delta S^\dagger/R} e^{-\Delta H^\dagger/kT}, \tag{1}$$

Alan Rembaum and Eric Sélégny (eds.), Polyelectrolytes and Their Applications, 71–96. All Rights Reserved.
Copyright © 1975 by D. Reidel Publishing Company, Dordrecht-Holland.

where K is the equilibrium constant between X and the reactants, f the activity coefficient (exactly, the single-ion activity coefficient) of the species indicated by the subscript, $\Delta S^{\ddagger}$ and $\Delta H^{\ddagger}$ are the entropy and enthalpy of activation, h is the Planck constant, k the Boltzmann constant, and T the temperature. If the activity coefficients are estimated from the Debye-Hückel theory in dilute regions, we have for aqueous solutions at 25 °C

$$\log k_2 = \log(kT/h)\, K + 1.018\, Z_A \cdot Z_B I^{1/2}, \tag{2}$$

where Z_A and Z_B are the valencies of the reactants, and I is the ionic strength. Figure 1

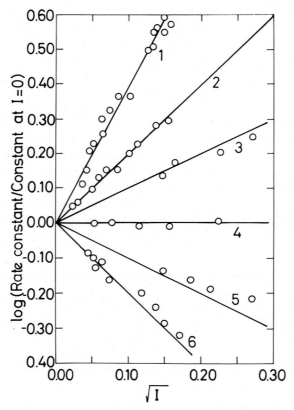

Fig. 1.   The primary salt effect-variations of reaction rates with ionic strength. The circles are observed values; the straight lines are theoretical (Equation (2)).

(1) $2[\mathrm{Co(NH_3)_3Br}]^{2+} + \mathrm{Hg}^{2+} + 2\mathrm{H_2O} \rightarrow 2[\mathrm{Co(NH_3)_5H_2O}]^{3+} + \mathrm{HgBr_2}$
(2) $\mathrm{S_2O_8^{2-}} + 2\mathrm{I}^- \rightarrow \mathrm{I_2} + 2\mathrm{SO_4^{2-}}$
(3) $[\mathrm{NO_2NCOOC_2H_5}] + \mathrm{OH}^- \rightarrow \mathrm{N_2O} + \mathrm{CO_3^{2-}} + \mathrm{C_2H_5OH}$
(4) Cane sugar inversion
(5) $\mathrm{H_2O_2} + 2\mathrm{H}^+ + 2\mathrm{Br}^- \rightarrow 2\mathrm{H_2O} + \mathrm{Br_2}$
(6) $[\mathrm{Co(NH_3)_5Br}]^{2+} + \mathrm{OH}^- \rightarrow [\mathrm{Co(NH_3)_5OH}]^{2+} + \mathrm{Br}^-$

(Taken from W. J. Moore, *Physical Chemistry*, Pretice-Hall, Inc., Englewood Cliffs, N.J., Third Edition, p. 369.)

gives the experimentally found variation of the rate constant in comparison with the theoretical prediction (Equation (2)). For reactions between species of the same sign, the rate enhancement is observed by an increase in the ionic strength (or by addition of electrolytes). On the contrary, reactions between species of the opposite signs are decelerated. For all cases, the agreement between the theory and experiment is excellent.

It should be remembered that the rate enhancement by low molecular weight electrolytes is accounted for in terms of the changes in the rate constant. These changes result from the variation of the activity coefficients of the reactants and critical complex; the free energies of the reaction systems are varied by addition of low molecular weight electrolytes. In this interpretation, the concentrations of reactants are assumed not to be changed by addition of low molecular weight electrolytes. This is ultimately untrue. According to the Debye-Hückel theory, the local ionic concentration is quite different from the bulk concentration. From the physical standpoint, it is reasonable that the concentration of cations is higher around anions than at the infinite distance. Nonetheless, such a heterogeneity is ignored in the Brønsted theory. This is because it is impossible to define and measure the *local* ionic concentration, and is acceptable because the valency of the low molecular weight electrolyte is not high so that the concentration fluctuation is small.

2. 'Catalysis' by High Molecular Weight Electrolytes (Polyelectrolytes)

2.1. 'CATALYSIS' BY ELECTROSTATIC INTERACTIONS

On the basis of this primary salt effect described above, it was thought interesting to inquire the influence of polyelectrolyte addition to reaction systems. The main motivation for this was our conclusion from the measurements of the mean activity coefficient of polyelectrolytes that the ionic distribution in polyelectrolyte-containing solutions is like a more or less regular lattice structure. This is quite different from the ionic atmosphere structure mentioned above for low molecular weight electrolytes. Thus it was expected that polyelectrolytes would exert an influence different from low molecular weight electrolytes on reactions between small ions. The second motivation was experimental findings by some researchers [4] that, under appropriate conditions, high molecular weight compounds were more effective in accelerating chemical reactions than corresponding low molecular weight analogs. In most of these earlier studies, ion-molecule reactions were investigated, in which polymer 'catalysts' interacted with the electrically uncharged substrate through forces other than electrostatic ones (such as hydrophobic forces) and with the ions through electrostatic forces. It is evident that interionic reactions could be influenced more drastically by polyelectrolytes than the ion-molecule reactions. The third motivation was our interest in biologically important ionic reactions. For example, DNA synthesis is believed to be an anion-anion reaction according to the experimental conditions employed by Kornberg, Weiss and associates [5]. Because of the electrostatic repulsive forces between the reactants, this reaction should be very slow. In addition to the repulsive forces between mononucleotides and growing daughter DNA chains, there exist likewise repulsive interactions between the mononucleotides and parent DNA chains. Consequently, the DNA replication is quite unfeasible. In reality, however, it takes place easily *in vivo*. It is interesting to dissect the mechanism of the replication from the physico-chemical point of view.

We studied, first, substitution reactions [6]. These reactions were investigated earlier thoroughly by LaMer [7], Brønsted [8] and Olson [9], It was found out that cationic polyelectrolytes

$$CH_2BrCOO^- + S_2O_3^{2-} \rightarrow CH_2(S_2O_3)COO^{2-} + Br^- \qquad \text{(A)}$$

$$2[Co(NH_3)_5Br]^{2+} + Hg^{2+} + H_2O \rightarrow 2[Co(NH_3)_5H_2O]^{3+} + HgBr_2 \quad \text{(B)}$$

such as polyethylenimine hydrochloride [PEI·HCl] and copolymer of dimethyl-

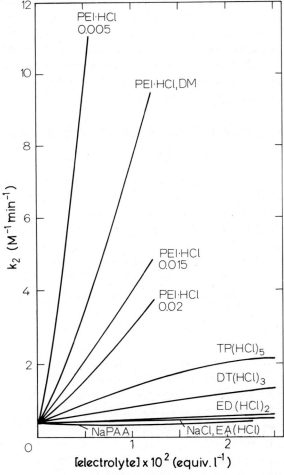

Fig. 2. The rate enhancement of the bromoacetate-thiosulfate ion reaction by various electrolytes (25°C). (Taken from Reference 6.)

diallylammonium chloride and sulfur dioxide [DM] accelerated the reaction (A) whereas the reaction (B) was enhanced by anionic polyelectrolytes such as sodium polystyrenesulfonate [NaPSt], and sodium polyethylenesulfonate [NaPES]. Figure 2 gives the second-order rate constant (k_2) of the reaction (A) in the presence of various electrolytes. It is seen that low molecular weight analogs, tetraethylenpentamine [TP(HCl)$_5$], diethylentriamine [DT(HCl)$_3$], ethylendiamine [ED(HCl)$_2$], and ethylamine [EA(HCl)] give much smaller acceleration than PEI · HCl, and the acceleration becomes smaller with decreasing molecular weight. *Polyelectrolytes are thus shown to be more efficient in accelerating ionic reactions than corresponding low molecular weight substances.* This conclusion was always confirmed for all systems investigated, as far as interionic reactions were concerned.

Polynucleotides are polyelectrolytes. Thus the reaction (B) was enhanced by DNA, and synthetic polynucleotide [10]. One interesting feature is that, at a polynucleotide concentration of 8×10^{-5} eq. l^{-1}, the acceleration factors (k_2/k_{2o}; the ratio of the rate constants with and without polyelectrolyte) were 40, 10 and 10 for poly U, poly G and DNA, whereas the factors were 1 and 0.6 for poly A and poly C, respectively. It was suggested that the deceleration observed for poly C was brought about by the positively charged state under the experimental conditions, and poly A was at the isoelectric point.

In the above study, our purpose was to see whether the polyelectrolyte influence can be accounted for in terms of the primary salt effect. Since the Debey-Hückel theory is not valid for polyelectrolyte-containing solutions the Brønsted equation (Equation (3)) was checked, instead of Equation (2) [11]. The experimental data of the mean activity

$$k_2/k_{2o} = f_A f_B / f_X \tag{3}$$

coefficient of solutes [γ] in a ternary system polyelectrolyte-simple electrolyte-solvent showed that the interaction parameters (β_{ij}) were in the order

$$|\beta_{22}| \gg |\beta_{23}| = |\beta_{32}| > |\beta_{33}|, \tag{4}$$

where the subscripts 2 and 3 denote the polyelectrolyte and low molecular weight electrolyte, respectively, and

$$\beta_{ij} = \partial \ln \gamma_i / \partial \ln m_j, \tag{5}$$

where m_j is the molar concentration of the component j. At low concentrations, furthermore, the interaction parameters β_{ij}'s were all negative. From this observation, it followed that addition of polyelectrolyte decreased the mean activity coefficient of low molecular weight electrolytes more drastically than addition of low molecular weight electrolyte, and the higher the valency of the low molecular weight electrolytes the larger the extent of the decrease. Since the same trend can be expected also for f (the single-ion activity coefficient), the addition of polyelectrolyte lowers f_A and f_B, and even more appreciably f_X, since the critical complex is more highly charged than the reactants for reactions between species of the same sign. Thus we have $k_2/k_{2o} \gg 1$

for polyelectrolytes, as was observed. Thus our conclusion was that the observed acceleration could be basically explained by the primary salt effect [14].

The aquation reactions of $[Co(NH_3)_5Br]^{2+}$ were systematically studied most recently by us [15]. The inducing cations were Tl^{3+}, Hg^{2+} and Ag^+, and much high concentrations of polyelectrolytes were used. Figure 3 gives the acceleration by three polyelectrolytes, namely polyphosphate (PP), PSt and PES. Clearly PP is the most effective and PSt is the least. This reflects the charge density of the polyelectrolytes. The higher the charge density, the larger the acceleration. Figure 4 shows that the

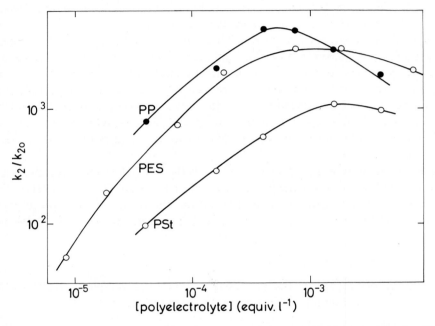

Fig. 3.   Acceleration factor for the Hg^{2+}-induced aquation on polyelectrolyte concentration (25 °C). $[Co(NH_3)_5Br^{2+}]=6\times10^{-5}$ M, $[Hg^{2+}]=5\times10^{-4}$ M, $k_{20}=8.7$ $M^{-1}s^{-1}$. (Taken from Reference 15.)

higher the valency of the inducing cation, the larger the acceleration [16]. This is quite reasonable in the light of the electrostatic interactions of the macroions with the inducing cations becoming stronger with increasing valency of the latter. The electrostatic interactions, which are responsible for the acceleration, can be weakened by addition of low molecular weight electrolytes. As shown in Figure 5, the acceleration factor became smaller with increasing concentration of $HClO_4$. The same trend was also observed by addition of $NaClO_4$.

Electron-transfer reactions between Co-complexes and reductants are being studied in this laboratory. As was noted by Gould [17] earlier, who studied electron-transfer reactions between pentammine cobalt complexes and Cr(II), the main factor in determining the acceleration is the valencies of the oxidants and reductants, as shown in Figure 6, which gives the polyelectrolyte influence on an outer-sphere electron-transfer between $Ru(NH_3)_6^{2+}$ and $Co(III)$ complexes [18]. Because of the solubility and stability of polyelectrolyte solutions containing the reductants, the kinetic study was carried out in the presence of a fairly large amount of foreign salts. Thus, the acceleration factor was 10^2 at the largest for the present reaction system. For other

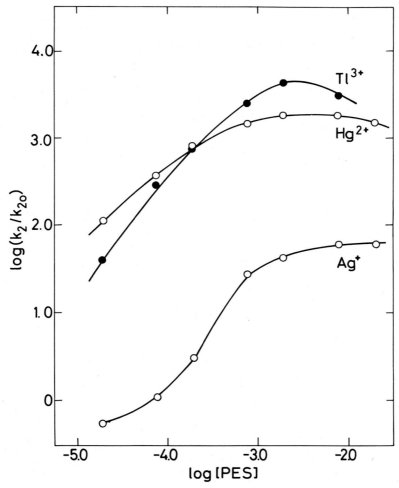

Fig. 4. Acceleration factors by polyethylenesulfonate of the metal-induced aquation of $Co(NH_3)_5Br^{2+}$ (20 °C). $[Co(NH_3)_5Br^{2+}] = 6 \times 10^{-5}$ M, [inducing cation] $= 1 \times 10^{-3}$ eq. 1^{-1}, $[HClO_4] = 0.05$ N. (Taken from Reference 15.)

electron-transfer processes between $[Co(en)_2Cl_2]^+$ and Fe^{2+} [19], $[Co(NH_3)_5Br]^{2+}$ and Fe^{2+} [19], $[Co(NH_3)_5N_3]^{2+}$ and V^{2+} [20], and $[Co(en)_2Cl_2]^+$ and V^{2+} [20], similar values ($10^2 \sim 10^3$) were found as the acceleration factor by PES.

Organic reactions cannot escape from the polyelectrolyte influence, if they are interionic ones. The bimolecular elimination reactions of chloro-fumarate and maleate were enhanced by poly (diethyldiallylammonium chloride) whereas it was hardly affected by tetramethylammonium chloride [21]. Probably because of the separation of the electric charges on the substrate,

$$^-OOC\text{---}C=C\text{---}COO^- + OH^- \rightarrow {}^-OOC\text{---}C\equiv C\text{---}COO^- \tag{C}$$
$$\quad\quad\; |\quad| $$
$$\quad\quad\; H\quad Cl$$

the acceleration factor was only about 10 under the experimental conditions used.

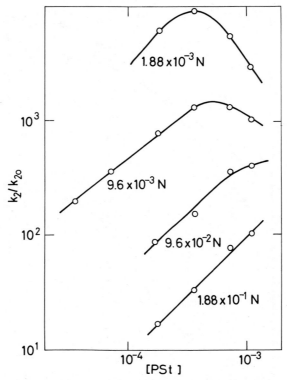

Fig. 5. Acceleration factors by polystyrenesulfonate of the Hg^{2+}-induced aquation of $Co(NH_3)_5Br^{2+}$ (30 °C). $[Co(NH_3)_5Br^{2+}] = 6.7 \times 10^{-5}$ M, $[Hg^{2+}] = 1.32 \times 10^{-4}$ M. $[HClO_4]$ from top to bottom: 1.88×10^{-3} N, 9.6×10^{-3}, 9.6×10^{-2}, 1.88×10^{-1}. (Taken from Reference 15.)

It is now clear that interionic interactions between like-charged ions can be accelerated by polyions of the opposite charge much more strikingly than by low molecular weight electrolytes, under appropriate conditions. The acceleration was suggested to be explainable in terms of the primary salt effect. However, the argument was quite qualitative. This is due to the fact that the critical complexes are electrically charged. For example, a reaction between species of a valence of $Z_A (>0)$ and those of a valence of $Z_B (>0)$ is expected to give $(Z_A + Z_A)$-valent critical complexes. As far as only low molecular weight electrolytes are involved, the physico-chemical properties of the complexes of a high valence can be evaluated by using the Debye-Hückel theory, no matter how they cannot be separated. According to this theory, the activity coefficient is determined by the valence of ionic species and the ionic strength for a given solvent and temperature, if the solution is dilute enough. However, the situation becomes

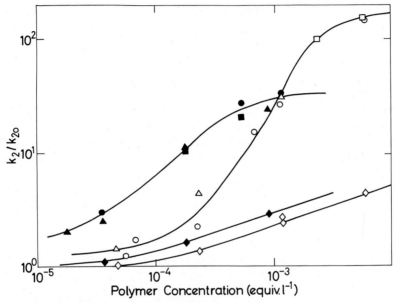

Fig. 6.   Acceleration factors of the electron-transfer reaction between Co(III)-complexes and $Ru(NH_3)_6^{2+}$ by polyethylenesulfonate (25°C). (Taken from Reference 18.)

$[NaCl] = 0.2$ M, $[HCl] = 2 \times 10^{-4}$ M, $[Co(III)] \sim 5 \times 10^{-5}$ M, $[Ru(II)] \sim 5 \times 10^{-4}$ M

Co-Complex		$k_{20}(M^{-1}s^{-1})$
$\bigcirc$, $\bullet$	$Co(en)_2pyCl^{2+}$	9.3×10^2
$\triangle$, $\blacktriangle$	$Co(en)_2H_2OCl^{2+}$	3.5×10^2
$\square$, $\blacksquare$	$Co(en)_2NH_3Cl^{2+}$	3.6×10
$\diamond$, $\blacklozenge$	$Co(en_2)Cl_2^+$	4.4×10^3

The blank symbols are for PES and the filled ones for PSt.

suddenly intricate if polyelectrolytes coexist. For example, the usual definition of the ionic strength does not reflect the physical reality in solutions; because the ionizable groups are bound to each other by covalent bonds, the local concentration of the ionic charges inside the polymer domain is quite different from the average concentration in the bulk phase. Furthermore, the extremely high electrostatic potential of macroions invalidates the well-known linear approximation, $e\psi/kT \ll 1$ (ψ: electrostatic potential, e: electric charge), which is a basic assumption in the Debey-Hückel theory, except for highly dilute solutions. Thus, the theory cannot be applied for polyelectrolyte-containing solutions. In the absence of a complete theory, therefore, theoretical evaluation of the properties of the critical complex in polyelectrolyte solutions is impossible at present. This situation impedes quantitative discussion on the polyelectrolyte acceleration.

In order to overcome this difficulty, to study reactions between oppositely charged ions is advantageous, and even more so when the valencies of cations and anions are equal with opposite signs. In these reactions, the critical complex to be formed has no net charge as a result of compensation. Then, its interaction with polyelectrolyte is weak, so that its activity coefficient can be safely assumed to be unity at low concentrations.

The reaction system we studied [22] is

$$NH_4^+ + OCN^- \rightleftharpoons NH_3 + HCNO \rightleftharpoons (NH_2)_2CO \tag{D}$$

which is exactly as old as the organic chemistry [23]. Warner and Stitt observed that the rate constant decreased with increasing ionic strength and attributed this trend to the primary salt effect [25]. Similarly, the reaction was hindered by addition of cationic and anionic polyelectrolytes. The deceleration factor k_{20}/k_2 was 2.5 by sodium polyacrylate (NaPAA) at a concentration of 0.200 eq. l^{-1}. At the same concentration, NaCl decelerated the reaction by a factor of 1.3. Clearly, *polyelectrolytes are also more efficient in retarding the interionic reaction than low molecular weight electrolytes.*

TABLE I

Observed and calculated deceleration factor
by sodium polyacrylate of the $NH_4^+OCN^-$-
urea conversion (50 °C)

[NaPAA]	$\log(k_2/k_{20})$	
(equiv. l^{-1})	obsd.	calcd.
0	0	0
0.056	−0.140	−0.145
0.070	−0.158	−0.174
0.11	−0.253	−0.251
0.22	−0.421	−0.408

Table I shows the observed deceleration by NaPAA and the theoretical value calculated by the Brønsted equation (Equation (3)). In the calculation, the activity coefficients of NH_4^+ and OCN^- in the ternary system H_2O—NaPAA—NH_4OCN were assumed to be equal to those of Na^+ and Cl^- in the system H_2O—NaPAA——NaCl. The activity coefficient of the critical complex was set to be unity, which is quite reasonable. The agreement between the theory and experiment, which is gratifying, implies that the observed polyelectrolyte influence can be accounted for by Equation (3), or can be ascribed to the primary salt effect.

Comments are necessary on the reason for the agreement mentioned above. It should be recalled that the Brønsted equation was constructed on an assumption that interactions in solutions were relatively weak. In his original paper, Brønsted did not mention limitation in detail, but he expected the theory to be valid to the same extent as the Debye-Hückel theory. This means that the theory can be applied satisfactorily below 10^{-3} M for $1-1$ type electrolytes. Consequently, uses of high concentrations of reactants and polyelectrolytes should be avoided, if the check of the Brønsted equation is aimed at, as in our case. In our eralier study, the concentration of poly-electrolytes was kept at a level as low as possible. This choice brought about naturally a relatively small acceleration or deceleration, but made it possible to discuss whether the influences are the primary salt effect.

The next point to be made is that only the electrostatic interactions are involved in the urea conversion. The Brønsted theory takes into consideration only this type of interactions explicitly. This situation is believed to be one reson for the agreement demonstrated in Table I.

2.2. 'CATALYSIS' BY ELECTROSTATIC AND HYDROPHOBIC INTERACTIONS

The importance of the hydrophobic interactions in 'catalytic' influence of micelles and polyelectrolytes was noted by several investigators [1a, 29–32]. In our study, also,

hydrophobic polyelectrolytes were found to exert more pronounced, and sometimes unique influence on reactions involving hydrophobic substrates than hydrophillic polymers. The spontaneous hydrolysis of dinitrophenylphosphate dianions was accelerated by cationic copolymers of diallylammonium chloride and sulfur dioxide [33]. [For the chemical structures, see the formula above.] The acceleration factor varied with copolymers; a copolymer containing methyl (R_1) and nitrobenzyl (R_2) groups [MNBz] accelerated the reaction 60-fold. If R_1 is methyl and R_2 is benzyl [MBz], the largest acceleration was 16-fold. The factor was 5.5 for $R_1 = R_2 =$ ethyl [DE] and 4 for $R_1 = R_2 =$ methyl [DM]. No acceleration was observed for NaCl, tetramethylammonium chloride, polyvinylpyrolidone and polyethylene glycol. It is seen that the acceleration became smaller with decreasing hydrophobicity of the sub-stituent group of the copolymers.

Acceleration through the hydrophobic interactions is believed to be due to a change of water structure. Since the hydrophobic interactions are short-ranged, this kind of acceleration was expected not to appear when group or groups indifferent to reactions exist at such positions that the accumulation or the critical complex formation is interferred. This was probably the case for the alkaline fading reaction of phenolphthalein, which is an anion-anion reaction [34]. If polyvinylpyridine [PVP] quaternized with benzyl group [BzPVP] is added, the observed rate constant was 0.0064 and ~ 0 $M^{-1} s^{-1}$ at polyelectrolyte concentrations of 2.63×10^{-4} and 1.05×10^{-3} eq. l^{-1}, respectively. Under the same condition without the polyelectrolyte, the rate constant was 0.015 $M^{-1} s^{-1}$. In the presence of less hydrophobic polycations, namely PVP quaternized with ethyl groups (C_2PVP), the rate constant was 0.037 $M^{-1} s^{-1}$. Evidently, *too strong hydrophobic interactions sterically hindered the OH^- attack on the dye ions.*

The cooperative action of the electrostatic and hydrophobic interactions was clearly demonstrated by hydrophobic polycations in the CN^--adduct formation of β-nicotinamide adenine dinucleotide [β-NAD] [35]. This coenzyme carries two negatively charged phophate groups and one positive charge in the alkaline region. Thus, as a whole, it carries one negative charge. In the presence of hydrophobic cationic polyelectrolytes such as PVP quaternized with cetyl (5%) and benzyl (95%) groups [C_{16}Bz PVP], PVP quaternized with butyl group [C_4PVP], and BzPVP, the acceleration factors were 100 at pH = 9.2.

If substrates contain hydrophobic moiety, interionic reactions between oppositely charged ionic species, which should be decelerated by polyelectrolytes according to the electrostatic model, can be even accelerated by hydrophobic polyelectrolytes. Examples are alkaline fading reactions of triphenylmethane dyes [34], which take place between dye cations and hydroxyl ions to form carbinols. In addition to three benzene rings, these dyes have alkyl groups;

	R_1	R_2	R_3
EV	$N(C_2H_5)_2$	$N(C_2H_5)_2$	$(C_2H_5)_2$
CV	$N(CH_3)_2$	$N(CH_3)_2$	$(CH_3)_2$
BG	H	$N(C_2H_5)_2$	$(C_2H_5)_2$
MG	H	$N(CH_3)_2$	$(CH_3)_2$

ethyl violet (EV) has six ethyl groups, crystal violet (CV) six methyl groups, brilliant green (BG) four ethyl groups and malachite green (MG) four methyl groups. The hydrophobicity is expected to decrease in the order EV > CV > BG > MG. Addition of hydrophobic cationic polyelectrolytes and surfactant enhanced the fading reaction whereas anionic ones retarded it. The polyelectrolyte influence on the fading reactions is shown in Figures 7 and 8. For both EV and MG, the reactions were accelerated by hydrophobic cationic polyelectrolytes and surfactant (CTABr, cetyltrimethyl-

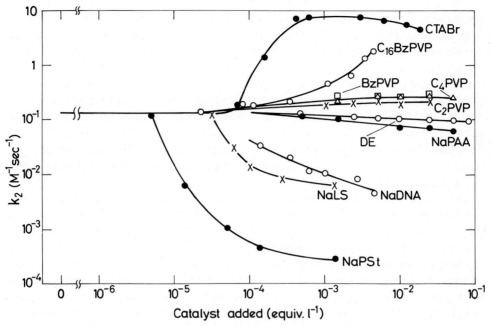

Fig. 7. Polyelectrolyte influence on the fading reaction of ethyl violet (30 °C). $[EV] = 1.05 \times 10^{-5}$ M, $[OH^-] = 1.05 \times 10^{-2}$ M. (Taken from Reference 34.)

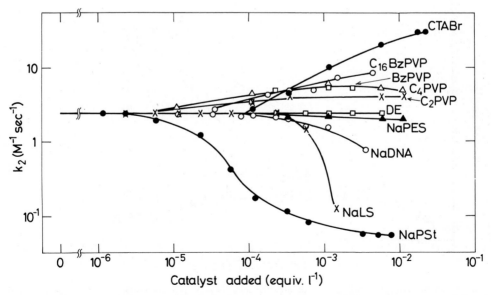

Fig. 8. Polyelectrolyte influence on the fading reaction of malachite green (30 °C). $[MG] = 1.25 \times 10^{-5}$ M, $[OH^-] = 3.49 \times 10^{-2}$ M. (Taken from Reference 34.)

ammonium bromide) whereas it was decelerated by anionic ones [NaLS; sodium laurylsulfate]. The acceleration is in sharp contrast with the polyelectrolyte influence on the urea conversion, which was decelerated by both cationic and anionic polyelectrolytes.

It is reasonable to ascribe the observed acceleration to hydrophobic interactions between the substrate dye cations and the polyelectrolytes and to electrostatic attractive forces between OH^- and the polycations. These two interactions accumulate each substrate in the vicinity of macroions so that the reaction can be accelerated [36]. The most important aspect in the observed acceleration is that the hydrophobic interactions overwhelmed the electrostatic repulsive forces between the dye cations and the polycations in the present reaction system.

The strong retardation by NaPSt is due to the simultaneous contribution of the hydrophobic and electrostatic interactions between the polyanions and dye cations (deceleration factor 10^3). The OH^- is repelled by the electrostatic repulsion by the polyanions. Other polyanions, which lack the hydrophobic groups, decelerate the reactions much more moderately than NaPSt, because only the electrostatic interactions are operating. The strength of the hydrophobic interactions depends on the hydrophobicity of the dye cations. Thus, the ratio of the highest rate in the presence of CTABr to the lowest one in the presence of NaPSt amounts to 10^5. The ratio for MG, which is least hydrophobic among the dyes studied, is about 10^3.

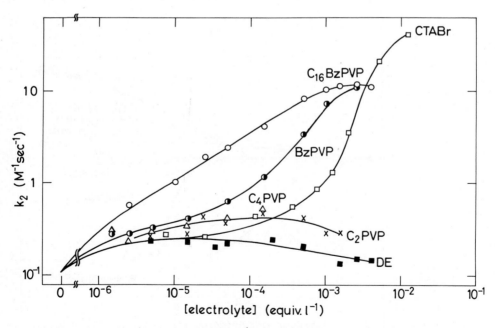

Fig. 9.   Polyelectrolyte influence on the alkaline hydrolysis of p-nitrophenyl palmitate (PNPP) in the binary mixture of 30% ethanol and water. (30 °C). $[PNPP] = 4 \times 10^{-5}$ M, $[NaOH] = 10^{-3}$ M. (Taken from Reference 37.)

Thought not interionic interactions, alkaline hydrolyses of *p*-nitrophenyl esters (acetate, propionate, valerate, caprylate, laurate, and palmitate) were subject to the hydrophobic interactions. Addition of polycations or cationic micelles (CTABr) enhanced the reaction as shown in Figure 9 [37].

This acceleration is clearly due to the hydrophobic interactions between the polyelectrolytes and the esters and the electrostatic interactions between the polyelectrolytes and OH^-. When the hydrophobicity of the polyelectrolytes and the esters becomes weaker, the acceleration becomes smaller.

The cooperative influence of the electrostatic and hydrophobic interactions was demonstrated also in other reactions such as a reaction between chloroacetic acid and thiosulfate [38], an $S_N Ar$ reaction between 3, 5-dinitro-4-chlorobenzoate and OH^- (reaction (F)) [39], and a reaction between 5,5'-dithiobis (2-nitrobenzoate) and CN^- (reaction (G)) [40]. The more hydrophobic the cationic polyelectrolytes, the more strongly the reaction was enhanced. Figure 10 gives the second-order rate constant of the reaction (G). A highly hydrophobic cationic polyelectrolyte, MBz, accelerated the reaction by a factor of 10^3.

2.3. 'CATALYSIS' BY ELECTROSTATIC AND HYDROGEN-BOND INTERACTIONS

Kern [41] reported that PES was only as effective as hydrochloric acid in the hydrolysis of cane sugar. Cation exchanger (crosslinked polystyrenesulfonic acid) was found to be very slightly more effective in the hydrolysis of cane sugar than sulfuric acid [42]. In the hydrolysis of starch, PSt was less efficient than hydrochloric acid [43]. The results are quite reasonable because these polyelectrolytes supply the protons but have no interactions with sugar or starch. Since the functional group of these substrates is hydroxyl group, it is easily expected that polyelectrolytes capable of forming hydrogen bonds with the substrate can be effective accelerators of the reaction. Thus the acid hydrolysis of dextrin was followed in the presence of copolymers of vinyl alcohol and ethylene sulfonic acid [44]. The acceleration observed was not large but

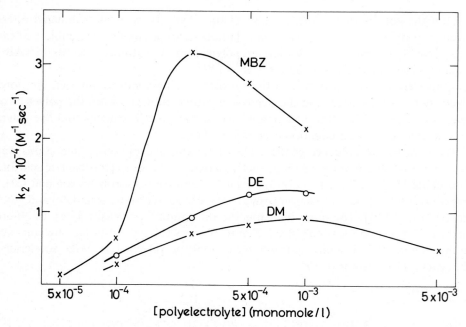

Fig. 10. Polyelectrolyte influence on the cleavage reaction of 5,5′-dithiobis (2-nitrobenzoate) [DTNB] by
CN⁻ (30 °C). [DTNB] = 6×10^{-5} M, [KCN] = 5×10^{-4} M, pH = 10.6. (Taken from Reference 40.)

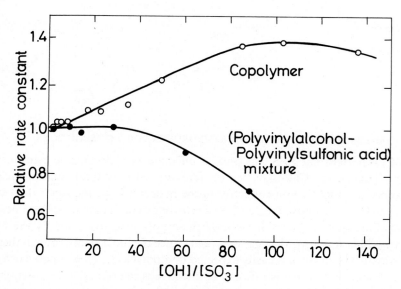

Fig. 11. Rate constant of the hydrolysis of dextrin in the presence of copolymers of varying contents of
hydroxyl groups (80 °C). (Taken from Reference 44.)

real. Figure 11 shows the specific rate constant (k_{cat}) observed with the copolymers relative to that found with sulfuric acid. The relative rate constant increased with increasing vinylalcohol content in the copolymer at first and passed through a maximum. However, the mixture of polyvinylalcohol and polyethylenesulfonic acid retarded the reaction. It is tempting to suggest that the copolymers interact with the substrate by hydrogen-bond formation and simultaneously with the protons so that acceleration results. As was the case with bifunctional catalysis, an optimum vinylalcohol content (about 100 moles per sulfonic group) exists. The presence of too many hydroxyl groups lowers the charge density of the copolymer, causing smaller rates.

An interesting feature of this reaction system is that the rate-concentration profile showed the Michaelis-Menten behavior, as shown in Figure 12. In Table II are tabulated the kinetic quantities defined in Equation (6), which reads

$$S + C \underset{\overleftarrow{k}}{\overset{\overrightarrow{k}}{\rightleftarrows}} SC \overset{k_s}{\rightarrow} C + P, \tag{6}$$

where S and C denote the substrate and catalyst, SC is the complex, P is the product,

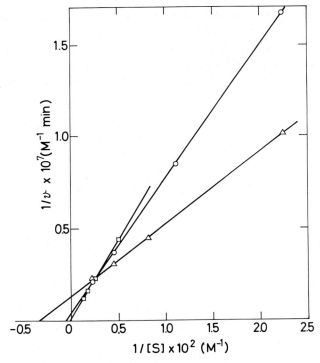

Fig. 12.   Lineweaver-Burk plot of the kinetic data of the dextrin hydrolysis in the presence of a copolymer of vinyl alcohol and ethylene sulfonic acid (80 °C). Copolymer contained 1.0 mole % sulfonic groups and 99 mole % vinyl alcohol groups. [Copolymer] $= 5.0 \times 10^{-3}$ mole 1^{-1}. Copolymer: △ No. 3, ○ No. 5, □ No. 8. (Taken from Reference 44.)

NORIO ISE

TABLE II

Kinetic parameters of the hydrolysis of dextrin in the presence of
copolymers of vinyl alcohol and ethylene sulfonic acid

Copolymer[a]	$V_{max} \times 10^6$ (M min^{-1})	$k_c \times 10^{-4}$ (1 min^{-1})	K_s (M)
No. 3 (1 mole%)	0.77	1.54	0.030
No. 5 (2.8 mole%)	2.50	5.00	0.22
No. 8 (10.5 mole%)	9.10	18.2	0.78

Copolymer concn.; 2.13×10^{-3} M for No. 3.
$\qquad\qquad\quad 4.55 \times 10^{-4}$ M for No. 5.
$\qquad\qquad\quad 1.40 \times 10^{-4}$ M for No. 8.

[a] The figure in the brackets denotes the content of vinylsulfonic acid
group in the copolymers.

and K_s is the Michaelis constant $(= [\bar{k} + k_c]/\bar{k})$. The trend for K_s to increase with
increasing content of vinylsulfonic acid group is quite reasonable, since the vinyl
alcohol units are responsible for 'binding' of the substrates. The k_c shows the same
tendency as K_s; the smaller K_s the smaller k_c. The complex formed is expected to have
a 'tight' structure so that the substrates are not at a favorable position which is
accessible to the protons attracted by the sulfonic acid group. The similar situation
is expected when short-ranged interactions are operating as a driving force of the substrate binding. As a matter of fact, Shinkai and Kunitake reported [45] the same
trend for the nitroacetoxybenzoate in the presence of copolymers of methylvinylimid-
azole and vinylpyrolidone or acrylamide [46].

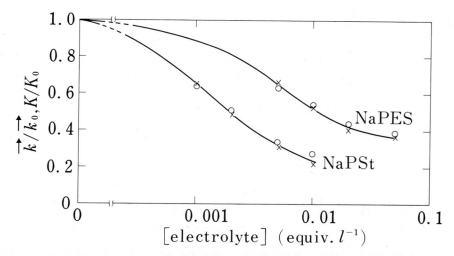

Fig. 13. Polyelectrolyte influences on the forward reaction rate constant $(\vec{k})$ and on the equilibrium constant (K) of the CN$^-$-adduct formation of NAA (25 °C). [NAA] $= 1.98 \times 10^{-4}$ M, [KOH] $= 1.0 \times 10^{-3}$ M.
$\bigcirc$: $\vec{k}/\vec{k}_0$; $\times$: K/K_0. (Taken from Reference 48.)

2.4. ARE POLYELECTROLYTES REALLY CATALYSTS?

Whatever the driving forces are, it is now clear that ionic reactions can be accelerated or decelerated by polyelectrolytes. However it was on the *overall* reaction rate that polyelectrolytes were demonstrated to have an influence. By the principle of microscopic reversibility, at least two elementary processes, forward and backward, are involved in any reaction. Thus the next question is: which of the forward or backward process was affected? To answer this question, we studied the following equilibrium reaction [47, 48]. Since the reaction (H) is the one between oppositely charged ionic species, addition of polyelectrolytes retarded the forward process. Figure 13

$$\text{(H)}$$

shows the polyelectrolyte influence on the second-order rate constant of the forward process. Simultaneously, the equilibrium constant $K(=\vec{k}/\overleftarrow{k})$ became smaller with polyelectrolyte addition; the equilibrium shifted toward the reactant side. A noteworthy trend is that the $\vec{k}$ and K were affected in the same proportion. In other words, the backward step was not influenced at all. This is quite reasonable because this process is a decomposition of a neutral compound, which cannot appreciably be influenced by polyelectrolytes.

The forward process of an equilibrium reaction between β-nicotinamide adenine dinucleotide (Coenzyme I) and CN^- was accelerated by polycations; $\vec{k}/\vec{k}_o$ was 20 by an addition of $C_{16}BzPVP$ at a concentration of 7×10^{-3} eq. 1^{-1} [35]. This is quite a reasonable result because this is an interionic reaction between anions. At the same concentration, $\overleftarrow{k}/\overleftarrow{k}_o$ *was* 3.

The different influence on the two elementary processes leads us to a conclusion that *the polyelectrolytes should not be regarded as catalysts*, if we follow earlier definition by Ostwald [49] that a catalyst is a substance which influences the backward and forward processes in the same proportion. It seems to be the recent usual practice to use the terminology 'polymer catalyst', when polymer shows a rate-enhancing ability. To our knowledge, however, no work was reported on the influence of polymers on equilibrium reactions, or in other words, on polymer effect on elementary processes. Thus, the use of the term 'polymer catalyst' is not yet justified.

2.5. THERMODYNAMIC CONSIDERATION ON CAUSES OF 'CATALYTIC' ACTION OF
     POLYELECTROLYTES

In the above sections, the contributions of electrostatic, hydrophobic, and hydrogen-bonding interactions in the 'catalytic' influence of polyelectrolytes in various types of reactions have been discussed.*

In this section, we discuss the polyelectrolyte influence from a thermodynamic point of view. Table III gives the thermodynamic parameters of the NAA–CN⁻ reaction in the presence and absence of polyelectrolytes. Figure 14 is the free energy diagram. The free energy of the reaction (ΔG) was increased. In spite of these changes, the difference $\Delta G^{\ddagger} - \Delta G$ ($= \Delta G_{-}^{\ddagger}$, the free energy of activation of the backward reaction) remained unchanged, indicating that the backward reaction was not influenced at all. Obviously,

TABLE III

Thermodynamic parameters for the NAA–CN⁻ reaction (25 °C)

	None	Polyelectrolytes		
		DE	NaPES	NaPSt
$\Delta G^{\ddagger}$	17.5	18.0	18.1	18.0
$\Delta H^{\ddagger}$	9.1	8.3	9.4	9.2
$\Delta S^{\ddagger}$	−28	−33	−29	−29
ΔG	−3.1	−2.6	−2.5	−2.6
ΔH	−9.8	−7.5	−7.7	−6.6
ΔS	−23	−16	−17	−14
$\Delta G^{\ddagger} - \Delta G$	20.6	20.6	20.6	20.6

[NAA] $= 1.98 \times 10^{-3}$ M, [KCN] $= 4.0 \times 10^{-3}$ M, [KOH] $= 1.0 \times 10^{-3}$ M, [DE] $=$ [NaPES] $= 0.02$ eq. l^{-1}, [NaPSt] $= 0.002$ eq. l^{-1}.

The units of free energies and enthalpies are in kcal mole^{-1} and the entroies in e.u.

* Most recently we have studied the charge-transfer interactions between polyelectrolytes and several substrates [50]. The complex formation of tryptamine hydrochloride with polyvinylpyridine quaternized with butyl chloride [C$_4$PVP] is due to charge-transfer interactions, which overwhelm the electrostatic repulsive forces between the two cationic substrates. Addition of potassium chloride and calcium chloride (0.5 eq. l^{-1}) increased the complex formation constant (K) (from 13) to 37 and 57 (1 eq.$^{-1}$) at [C$_4$PVP] $= 4 \times 10^{-3}$ eq. l^{-1} and [tryptamine] $= 4.8 \times 10^{-2}$ M. If the $\overleftarrow{k}$ was not influenced, the shift of the equilibrium implies an increase in $\overrightarrow{k}$. This is consistent with the primary salt effect. On the other hand, a cationic polyelectrolyte, DE, was rather insensitive to K, as was the case with an anionic polyelectrolyte, polyacrylate, in the bromoacetate-thiosulfate reaction (see Figure 2). The complex formation of indole acetate with propylchloride-quaternized polyvinylpyridine [C$_3$PVP], C$_4$PVP and benzylchloride-quaternized polyvinylpyridine [BzPVP], which is a reaction between oppositely charged species, was hindered by potassium chloride, calcium chloride, and DE, as was observed for the urea conversion. The formation constant K was decreased from 98 M^{-1} for C$_4$PVP to about 20 M^{-1} by addition of 3×10^{-2} eq. l^{-1} of DE, KCl and CaCl$_2$. If $\overleftarrow{k}$ was not changed, $\overrightarrow{k}$ was decreased by these electrolytes. In the complex formation of (cationic) C$_4$PVP and (neutral) L-tryptophan, the formation constant K was practically indifferent to addition of the electrolytes. Electrolyte influence on these three reactions indicates the importance of the electrostatic interactions.

the polyelectrolyte addition decreased the free energy of the reactant from G_2 to G_1, causing a decrease from ΔG to ΔG_p. The deceleration observed was brought about by an increase in $\Delta G^{\ddagger}$. In other words, the polyelectrolyte influence is due to a thermodynamic disturbance on the reactants. This is consistent with our above-mentioned interpretation of the polyelectrolyte influence in terms of the primary salt effect; actually the kinetic effects could be accounted for by the activity coefficients of the reactants* (see Equation (3)).

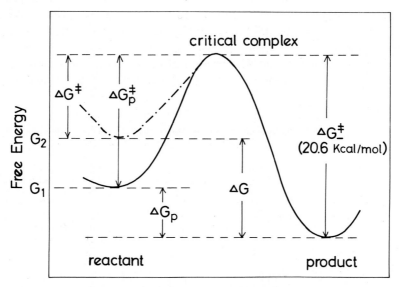

Fig. 14.   Free energy diagram of the equilibrium reaction of NAA with CN^-. The subscript 'p' indicates the presence of polyelectrolytes.

It has to be pointed out that, with a few exceptions, the acceleration by polyelectrolytes was associated with decreases in $\Delta H^{\ddagger}$ and $\Delta S^{\ddagger}$. Table IV gives the thermodynamic parameters for the aquation reactions of $Co(NH_3)_5Br^{2+}$ induced by Ag^+. Similar decreases in $\Delta H^{\ddagger}$ and $\Delta S^{\ddagger}$ were found for various reactions: the Hg^{2+}-induced aquation of $Co(NH_3)_5Br^{2+}$, the S_NAr reaction of dinitrochlorobenzoic acid with OH^- [51], the hydrolysis of 2,4-dinitrophenyl phosphates [reaction (E)] [33], the outer-sphere electron-transfers between Co-complexes [$Co(NH_3)_5N_3^{2+}$, $Co(NH_3)_5Br^{2+}$, $Co(en)_2Cl_2^+$] and $Ru(NH_3)_6^{2+}$ or V^{2+} [8, 20] [en: ethylenediamine], the polyvinyl-imidazole-accelerated solvolysis of p-nitrophenylacetate [52], the coupling reactions of dinitrofluorobenzene with aminoacids [53], dipeptides [53] and aniline [54], the lignin sulfonic acid-accelerated hydrolysis of methyl acetate [55], and the hydrolysis of nitrophenyl esters [37]. The opposite tendency (acceleration caused by increases

---

* In the case of reactions between similarly charged ionic species, the activity coefficients of not only the reactants but also the critical complex were varied.

in $\Delta H^{\dagger}$ and $\Delta S^{\dagger}$) was observed for the Tl^{3+}-induced aquation of $Co(NH_3)_5Br^{2+}$ [15], for the electron transfer by the bridge mechanism between Co-complexes [$Co(NH_3)_5Cl^{2+}$, $Co(NH_3)_5Br^{2+}$, $transCo(en)_2Cl^+$, $cisCo(NH_3)_4(N_3)_2^+$, $trans$-$Co(NH_3)_4(N_3)_2^+$] and Fe^{2+} [19, 56] for the electron transfer between $Co(phen)_3^{3+}$ and $Co(phen)_3^{2+}$ [phen: o-phenanthroline] [57], and for the hydrolysis of dipeptides by Dowex-50 [58]. These reactions are characterized by the very strong hydration of the reactant, by the bridge mechanism, by highly hydrophobic groups in the substrate, or by the three-dimensional structure of the 'catalyst'. We are inclined to regard the increases in $\Delta H^{\dagger}$ and $\Delta S^{\dagger}$ found for these reaction systems as exceptional; We believe that the decreases in $\Delta H^{\dagger}$ and $\Delta S^{\dagger}$ are essentially valid in representing the acceleration by polyelectrolytes.

It is tempting to suggest that dehydration of the reactant ions is responsible for these changes. The dehydration might be possible in the extremely high electrostatic

TABLE IV

Thermodynamic parameters for the Ag^+-induced aquation
reaction of $Co(NH_3)_5Br^{2+}$ (25 °C)

Electrolyte	Concn. × 10⁴ (eq. 1^{-1})	$\Delta H^{\dagger}$ (kcal mole^{-1})	$\Delta S^{\dagger}$ (e.u.)	$\Delta G^{\dagger}$ (kcal mole^{-1})
–	–	14.4	−15	18.9
NaPSt	0.3	12.3	−19	18.1
	3.0	3.3	−41	15.5
	30	12.6	−13	16.6
NaNO₃	90	20.8	+8	18.4

$[Ag^+] = 7.5 \times 10^{-4}$ M, $[Co(NH_3)_5Br^{2+}] = 6 \times 10^{-5}$ M.

field in the vicinity of macroions. If the dehydration takes place, the entropy of the reactant should be increased. If the critical complex is not influenced, as was the case for the NAA–CN⁻ reaction, this increase should result a decrease in $\Delta S^{\dagger}$. Furthermore, the dehydration should lower the enthalpy barrier, causing a decrease in $\Delta H^{\dagger}$.

The interpretation in terms of the dehydration turns out to be useful also for the deceleration effect by polyelectrolytes observed for the ammonium cyanate-urea conversion. In this case, $\Delta H^{\dagger}$ was hardly affected by polyelectrolyte addition whereas $\Delta S^{\dagger}$ was decreased. For example $\Delta H^{\dagger}$ and $\Delta S^{\dagger}$ were 22.9 kcal mole^{-1} and -2 e.u. in the absence of polyelectrolytes. If sodium polyacrylate (NaPAA) was added at a concentration of 0.1 equiv 1^{-1}, these parameters were 20.3 kcal mole^{-1} and -12 e.u. [22].*

The acceleration observed for the hydrolysis of dinitrophosphate dianions [33] (reaction (E)) was found to result from decreases in $\Delta H^{\dagger}$ and $\Delta S^{\dagger}$. Bunton et al. have claimed [59] that this is a spontaneous process and the acceleration by addition of cationic micelles can be caused by a change of phosphorous-oxygen bond angles

---

* The dehydration should decrease $\Delta H^{\dagger}$. This decrease is noticeable in the urea conversion, though rather small ($22.9 - 20.3 = 2.6$ kcal mole^{-1}).

toward of the methaphosphate ion. It is not unreasonable to suppose that this change can be facilitated by dehydration of the reactant ions.

Quite unfortunately, the dehydration of ions in polyelectrolyte solutions has not been studied, though it is physically plausible especially for ions of relatively low surface potentials. Future study is certainly necessary.

The trend for $\Delta H^{\dagger}$ and $\Delta S^{\dagger}$ of increasing with polyelectrolyte addition invites further comments. When hydrophobic substrates stabilize very strongly the iceberg structure of water, it might be possible that attractive forces between the substrates are so strong that coulombic influence of macroions is no more important.* This is supposedly the case for the electron-transfer reaction between $Co(phen)_3^{3+}$ and $Co(phen)_3^{2+}$. In this system the dehydration of reactant ions does not play an important role. The critical complex, which is penta-valent, is subject to much stronger influence of macroions than tri- or divalent reactants. It is reasonable that the critical complex can exist in a region much closer to the macroions than the reactants. Thus, liberation of water molecules can take place with greater ease from the iceberg structure around the critical complexes, causing an increase in $\Delta S^{\dagger}$. The $\Delta S^{\dagger}$ increase in the presence of three-dimensional polyelectrolytes is also reasonable. In the network structure, the electrostatic potential is quite high, but the influence on reactants is more or less symmetrical; the desolvation does not take place. The $\Delta S^{\dagger}$ increase observed for the Tl^{3+}-induced aquation of $Co(NH_3)_5Br^{2+}$ was again associated with the property of this cations being dehydrated only with great difficulty.

3. Concluding Remarks

The rate of chemical reaction can be changed by varying (1) the rate constant and/or (2) the concentration of reactant. The primary salt effect, or 'catalytic' influence on ionic reactions by simple electrolytes, is brought about by variation of the rate constant term, as was mentioned above according to Brønsted [3]. On the other hand, catalytic influence by polymeric electrolytes was accounted for in terms of change in the substrate concentration by most of the previous authors [1a–d]. To our knowledge, it has not been pointed out that the concentration factor is responsible for the catalytic action by simple electrolytes. On the other hand, the polyelectrolyte influence has not been discussed in terms of changes of the rate constant except by us [1e].

It should be noted that the concentration effect, though intuitively clear-cut and understandable, has a limit in its validity. It is due to the plain fact that molecules or ions occupy definite exclusion volumes. Because of this exclusion volume effect, the reactants can be 'concentrated' only up to a certain upper limit. According to Koshland [61], the intramolecular reaction rate is only 4.6 times larger than the intermolecular one when the reactants have the same exclusion volumes as the solvent

---

* Remember that addition of highly hydrophobic cationic polyelectrolytes retarded the fading reaction of phenolphthalein.

water and when the coordination number in the close-packed structure is 12. The particular figure (4.6) is evidently too small to account for the observed polyelectrolyte acceleration. Though Koshland's calculation does not hold in our cases, it shows the basic deficiency of the concentration factor. The inadequacy of this factor was recently noted by Morawetz [56].

Thus it is necessary to pay due attention to the rate constant term, in addition to the concentration factor, if the adequate explanation of the polyelectrolyte influence on chemical reactions is sought. As has been mentioned in the present paper, the polyelectrolyte influence was accounted for in terms of the rate constant change by using the Brønsted equation. Because of the limitation of the Brønsted equation, quantitative discussion was restricted to small influence of polyelectrolyte (as seen from the deceleration effect on the ammonium cyanate-urea conversion). One of the most important future problems in this field is to extend the Brønsted equation to account for such huge polyelectrolyte influences as described above.

Finally comments are necessary on a general feature of the polyelectrolyte catalysis. As stated in the introductory part, high molecular weight compounds are more efficient in accelerating chemical reactions under appropriate conditions. It should be noted that this statement is not always justified. For example, polyvinylpyridine is less efficient than 4-picoline in enhancing the solvolysis of dinitrophenyl acetate [62], whereas this polymer is extremely efficient compared to 4-picoline for the solvolysis of 5-nitro-4-acetoxysalicylic acid. Imidazole was 50 times more efficient than polyvinylimidazole in the formation of N-acyl compounds of p-nitrophenyl acetate in dimethylformamide [63]. In the hydrolyses of butylacetate and propylacetate, dodecyl benzenesulfonic acid was 6 and 3 times more efficient than polystyrene sulfonate, respectively [64].

On the other hand, we note that polyelectrolytes are *always* more efficient than the corresponding low molecular weight electrolytes in interionic reactions.* This is evidently due to the long range nature of coulombic interactions acting between polyelectrolyte and ionic substrates. The interactions are strong enough to influence ionic substrates in the vicinity of polyelectrolyte and simultaneously to affect the free energy of the reactants. Other interactions such as hydrophobic ones are short-ranged so that steric hindrance effects might be expected only for polymeric 'catalysts'. Thus, interionic reactions are, generally speaking, much more drastically influenced by polyelectrolytes than ion-molecule reactions.* The situation is suggestive of the fact that many biologically important reactions are interionic ones and take place in the presence of intricate polyelectrolytes, namely enzymes.

* An exceptional case for interionic reactions was reported for oligomers of vinylimidazole, which were less efficient than imidazole in solvolysis of 4-acetoxy-3-nitrobenzenesulfonate [65].
** Recent study by Klotz, Kiefer and others [32] shows that the hydrolysis of 2-hydroxy-5-nitrophenyl sulfonate can be accelerated by a polyethylenimine derivative by a factor of 10^{12}. As far as we know, this is the largest acceleration by synthetic polyelectrolytes, and is presumably related to the branched structure of the polymer.

References and Notes

1. For previous reviews on this subject, see the followings:
   (a) Sakurada, I.: *J. Pure Appl. Chem.* **16**, 236 (1968);
   (b) Overberger, C.: *Accounts Chem. Res.* **2**, 217 (1969);
   (c) Morawetz, H.: *Adv. Catalysis* **20**, 341 (1969);
   (d) Morawetz, H.: *Accounts Chem. Res.* **3**, 354 (1970);
   (e) Ise, N.: *Adv. Polymer Sci.* **7**, 536 (1971).
2. Debye, P. J. and Hückel, E.: *Physik. Z.* **24**, 185 (1923).
3. Brønsted, J. N.: *Z. physik. Chem.* **102**, 169 (1922); **115**, 337 (1925).
4. See, for example, Morawetz, H.: *Macromolecules in Solutions*, Interscience Publishers, New York, N.Y., 1965, Chapter 9.
5. See, for example, Steiner, R. F.: *The Chemical Foundations of Molecular Biology*, D. Van Nostrand Comp., New York, 1965, Chapter 9.
6. Ise, N. and Matsui, F.: *J. Am. Chem. Soc.* **90**, 4242 (1968)
7. LaMer, V. K. and Kamner, M. E.: *J. Am. Chem. Soc.* **57**, 2662, 2669 (1935).
8. Brønsted, J. N. and Livingston, R.: *J. Am. Chem. Soc.* **49**, 435 (1927).
9. Olson, A. R. and Simonson, T. R.: *J. Chem. Phys.* **17**, 1167 (1949).
10. Ise, N.: *Nature* **225**, 66 (1970).
11. For this purpose, addition of a large amount of polyelectrolytes was avoided. The concentrations of polyelectrolytes were kept at levels as low as possible, and at about the same levels as those of the substrates. Thus, the acceleration factor was rather small, definitely smaller than that reported by Morawetz and Vogel [12], who claimed a 2.47×10^5-fold acceleration for the $[Co(NH_3)_5Cl]^{2+}$—Hg^{2+} reaction by polymethacryloxyethylsulfonic acid.
12. Morawetz, H. and Vogel, B.: *J. Am. Chem. Soc.* **91**, 563 (1969).
13. Okubo, T., Ise, N., and Matsui, F.: *J. Am. Chem. Soc.* **89**, 3697 (1967).
14. A more quantitative discussion will be given below for the ammonium cyanate-urea conversion.
15. Ise, N. and Matsuda, Y.: *J. Chem. Soc., Faraday Trans. I.* **69**, 99 (1973).
16. Because of the hydrolysis of Tl^{3+}, at least half of the Tl ions exist in the form of $Tl(H_2O)_5OH^{2+}$. Thus, for the Tl^{3+} system the polyelectrolyte was as effective as for the Hg^{2+}-induced case.
17. Gould, E. S.: *J. Am. Chem. Soc.* **92**, 6797 (1970).
18. Kunugi, S. and Ise, N.: *Z. physik. Chem. NF.* **91**, 174 (1974).
19. Ise, N. and Shikata, M.: unpublished results.
20. Ise, N. and Kim, C.: unpublished results.
21. Ueda, T., Harada, S., and Ise, N.: *Chem. Comm.* **1971**, 99.
22. Okubo, T. and Ise, N.: *Proc. Roy. Soc. (London)* **A327**, 413 (1972).
23. It should be noted that two alternative mechanisms have been proposed, which are equally supported by the kinetic evidence. The ionic mechanism was advocated for example by Walker and Hambly [24], and Warner and Stitt [25], whereas the molecular mechanism was claimed from a stereochemical analysis [26], from a study of the reactions of amines with carbon dioxide [27], and from other sources of information [28].
24. Walker, J. and Hambly, F. J.: *J. Chem. Soc.* **67**, 746 (1895).
25. Warner, J. C. and Stitt, F. B.: *J. Am. Chem. Soc.* **55**, 4807 (1938).
26. Lowry, T. M.: *Trans. Faraday Soc.* **30**, 375 (1934).
27. Jensen, M. B.: *Acta. Chem. Scand.* **13**, 289 (1959).
28. Frost, A. R. and Pearson, R. G.: *Kinetics and Mechanism*, John Wiley & Sons, Inc., New York, 2nd edition, Chapter 12.
29. Duystee, E. F. J. and Grunwald, E.: *J. Am. Chem. Soc.* **81**, 4540 (1959).
30. Winters, L. J. and Grunwald, E.: *J. Am. Chem. Soc.* **87**, 4608 (1965).
31. Albrizzio, J., Archila, J., Rodulfo, T., and Cordes, E. H.: *J. Org. Chem.* **37**, 871 (1972).
32. Kiefer, H. C., Congdon, W. I., Scarpa, I. S., and Klotz, I. M.: *Proc. Nat. Acad. Sci., U.S.* **69**, 2155 (1972).
33. Ueda, T., Harada, S., and Ise, N.: *Polymer J.* **3**, 476 (1972).
34. Okubo, T. and Ise, N.: *J. Am. Chem. Soc.* **95**, 2293 (1973).
35. Okubo, T. and Ise, N.: *J. Biol. Chem.* **249**, 3563 (1974).
36. The accumulation of reactants does not necessarily cause an increase in the collision frequency. This will be discussed in the later section of the present paper.

37. Okubo, T. and Ise, N.: *J. Org. Chem.* **38**, 3120 (1973).
38. Ueda, T., Harada, S., and Ise, N.: *Polymer J.* **6**, 308 (1974).
39. Ueda, T., Harada, S., and Ise, N.: *Polymer J.* **6**, 313 (1974).
40. Ueda, T., Harada, S., and Ise, N.: *Polymer J.* **6**, 326 (1974).
41. Kern, W., Herold, W., and Sherhag, B.: *Makromol. Chem.* **17**, 231 (1955).
42. Hartler, N. and Hyllengren, K.: *J. Polymer Sci.* **55**, 779 (1961).
43. Painter, T. J. and Morgan, W. T.: *Chem. Ind. (London)*, 437 (1961).
44. Arai, K. and Ise, N.: *Makromol. Chem.*, in press.
45. Sinkai, S. and Kunitake, T.: presented at the *20th Annual Meeting of the Society of High Polymers*, Japan (1971).
46. The anti-parallel relation between the two constants was observed in the presence of partially quar-ternized polyvinylimidazole [45]. The substrate binding is caused by the long-ranged electrostatic interactions, so that the complex is 'loose'. The substrates are at positions easily accessible to the neutral imidazole group, a catalytic site. Thus small K_s favors the reactions.
47. Ise, N. and Okubo, T.: *Nature* **242**, 605 (1973).
48. Okubo, T. and Ise, N.: *J. Am. Chem. Soc.* **95**, 4031 (1973).
49. Ostwald, W.: *Phys. Z.* **3**, 313 (1902).
50. Okubo, T. and Ise, N.: unpublished results.
51. Ueda, T., Harada, S., and Ise, N.: publication in preparation.
52. Overberger, C. G., Pierre, T. St., Yaroslavsky, C., and Yaroslavsky, S.: *J. Am. Chem. Soc.* **88**, 1184 (1966).
53. Ueda, T., Harada, S., and Ise, N.: *Polymer J.* **6**, 318 (1974).
54. Ueda, T., Harada, S., and Ise, N.: publication in preparation.
55. Suzuki, K. and Taniguchi, K.: presented at the *21st Annual Meeting of the Society of Polymer Science*, Japan (1972).
56. Morawetz, H. and Gordimer, G.: *J. Am. Chem. Soc.* **92**, 7532 (1970).
57. Brückner, S., Crescenzi, V., and Quadrifoglio, F.: *J. Chem. Soc. A*, 1168 (1970).
58. Whittaker, J. R. and Deatherage, F. E.: *J. Am. Chem. Soc.* **77**, 3360 (1955).
59. Bunton, C. A., Fendler, E. J., and Fendler, J. H.: *J. Am. Chem. Soc.* **89**, 1221 (1967).
60. Bunton, C. A., Fendler, E. J., Sepulveda, L., and Yang, K.-U.: *J. Am. Chem. Soc.* **90**, 5512 (1968).
61. Koshland, D. E., Jr.: *J. Theoret. Biol.* **2**, 75 (1962).
62. Letsinger, R. L. and Savereide, T. J.: *J. Am. Chem. Soc.* **84**, 3122 (1962).
63. Machida, K., Kurosaki, K., and Okawara, M.: presented at the *21st Annual Meeting of the Society of Polymer Science*, Japan (1972).
64. Nakamura, Y., Miyake, K., and Hirai, H.: presented at the *21st Annual Meeting of the Society of Polymer Science*, Japan (1972).
65. Overberger, C. G. and Okamoto, Y.: *Macromolecules* **5**, 363 (1972).

MECHANISTIC CONSIDERATIONS OF
CYCLOCOPOLYMERIZATION AND SOME PROPERTIES
OF 'PYRAN' COPOLYMER

GEORGE B. BUTLER

Director, Center for Macromolecular Science, University of Florida, Gainesville, Fla. 32611, U.S.A.

Abstract. The regularly alternating $1:2$ cyclocopolymer of divinyl ether and maleic anhydride, often referred to as 'Pyran' copolymer, has been found to possess interesting anti-tumor properties and to be an interferon inducer. A variety of copolymers of this and similar types have been evaluated by the Drug Research and Development Branch of the National Institutes of Health. The most significant results of this evaluation study will be reported. It is the purpose of this paper to review some of the mechanistic aspects of cyclocopolymerization in general and to relate these theories to the structure of pyran copolymer, which was the first cyclocopolymer discovered. Those 1,4-dienes which readily undergo cyclocopolymerization with suitable alkenes also form charge-transfer complexes with the same alkenes. These results are consistent with participation of the $1:1$ charge-transfer complex of divinyl ether and maleic anhydride in this system in a copolymerization process with a second molecule of maleic anhydride to form the cyclocopolymer. Previous studies have indicated the copolymer to have a rather broad molecular weight distribution. In view of the extensive interest in the physiological properties of this material, further studies on the molecular weight and molecular weight distribution, including use of gel permeation chromatographic methods of fractionation have been undertaken and reported. Attempts have been made to fractionate pyran cyclocopolymer as the polyanion by GPC on Deactivated Porasil columns without success. Sucessful fractionation was attained on the dimethyl ester of the cyclocopolymer on Styragel columns, and comparisons were made between the copolymer prepared in this manner and the copolymer prepared by cyclocopolymerization of divinyl ether and dimethylfumarate. This paper reviews some of the results of these studies.

A number of papers have been published from this laboratory [1] dealing with the subject of cyclocopolymerization; the observations which led to the original proposal and some of the evidence obtained so support this proposal were published in two papers which were presented at earlier meetings [2, 3]. A review paper [4] was published which included only the structure of the divinyl ether (DVE) – maleic anhydride (MA) copolymer and a limited number of intrinsic viscosity values of some of the other copolymers. Even though thr DVE–MA copolymerization was observed to yield soluble, noncross-linked copolymer [2] before the postulation of the alternating intra-intermolecular chain propagation (now known commonly as 'cyclopolymerization') to explain the failure of 1,6-dienes to yield cross-linked polymers [5], it was only after unquestionable evidence had been obtained for the cyclopooymerization mechanism [6] that a plausible explanation for this most unusual behavior in the copolymerization of 1,4-dienes with alkenes was forthcoming. A satisfactory mechanism for this unusual behavior must account for (1) the failure of the system to cross-link in accordance with the widely accepted (at the time) theory of Staudinger [7]; (2) the essential absence of carbon-carbon double bond content in the copolymer; (3) polymer composition equivalent to a diene-olefin molar ratio of $1:2$, and (4) essentially quantitative conversion of monomers to copolymer.

Although copolymers of certain 1,6-dienes with a number of well known olefinic monomers have now been reported [8], a fundamental assumption in copolymerizations of this type is that the intramolecular cyclization step involves only the 1,6-diene and all members of the cyclic structure are contributed by this comonomer. The DVE–MA copolymerization represents the first example of a cyclopolymerization in which more than one comonomer is involved in the cyclization step which in turn results in formation of cyclic structures containing members contributed by more than one comonomer. The mechanism proposed here (Scheme I) assumes that one comonomer, the 1,4-diene, contributes four members to the developing cyclic structure while the other comonomer, the alkene, contributes the ramaining two members.

Scheme I

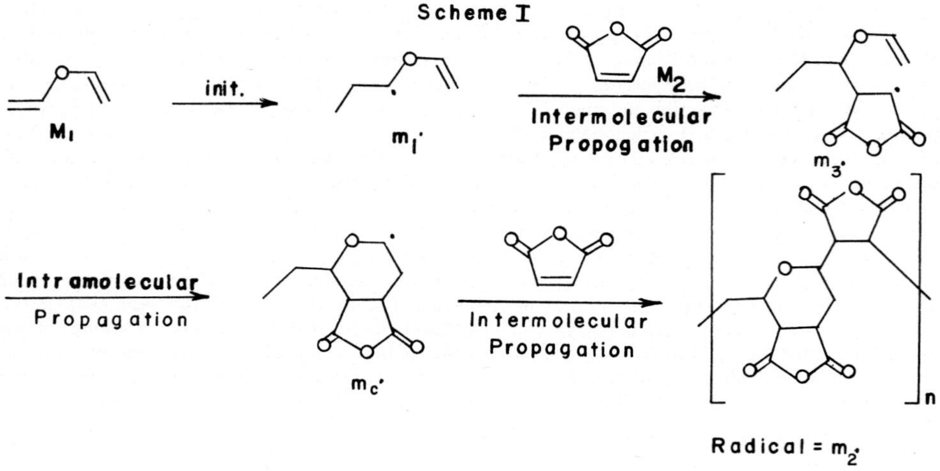

The proposed structure for the DVE–MA copolymer is supported by (1) the elemental analysis for the copolymer obtained at high conversion is consistent; (2) the IR spectrum is essentially devoid of residual double bond absorption and contains characteristic absorption bands for cyclic anhydride and six-membered cyclic ether structures; (3) the copolymer composition of 1:2 molar ratio of DVE to MA is consistent with the known reactivity ratios of these types of monomers; and (4) the presence of the cyclic ether group has been confirmed by chemical evidence which involved cleavage by hydriodic acid and incorporation of iodine into the polymer. Similar but not such extensive evidence has been obtained for the other copolymers reported [9].

An interesting property found to be associated with certain of these copolymers, particularly the DVE–MA copolymer, is their physiological action. The DVE–MA copolymer of $\bar{M}_n = 15000$–20000 has been found by the Cancer Chemotherapy National Service Center, National Institutes of Health, to be a promising anti-tumor agent [10]. For example, in certain tests, the weight of the tumor developed by the test animal was only 11% of that of the control animal. Also, more recent results [11] have shown in both laboratory and clinical studies that this copolymer possesses the ability to induce interferon generation.

Interferon, first discovered in 1957 [12], is a substance of protein-like structure which is produced in the cells of vertebrates in response to viral infection, and possesses antiviral action. Today, it is generally accepted that interferons play an essential role in the formation of a host's nonspecific resistance to superinfection with a second virus [13]. Thus, the role of interferon in combating viral infections may be similar to that of antibodies toward bacterial infections. It is immediately apparent that a successful and readily available interferon inducer, such as the divinyl ether-maleic anhydride cyclocopolymer (DVE–MA) may be, could play an extremely significant role not only in aiding in recovery from a viral infection, but in its use in prior generation of interferon to prevent a viral attack.

The results of more recent investigations on 'Pyran' copolymer will be presented in a paper by Dr Regelson at this symposium.

Anti-tumor activity of varying degree appears to be a general property of copolymers related in structure to 'Pyran' copolymer. Some similar copolymers whose preparation and structures are shown in Schemes II and III have been evaluated by

SCHEME II

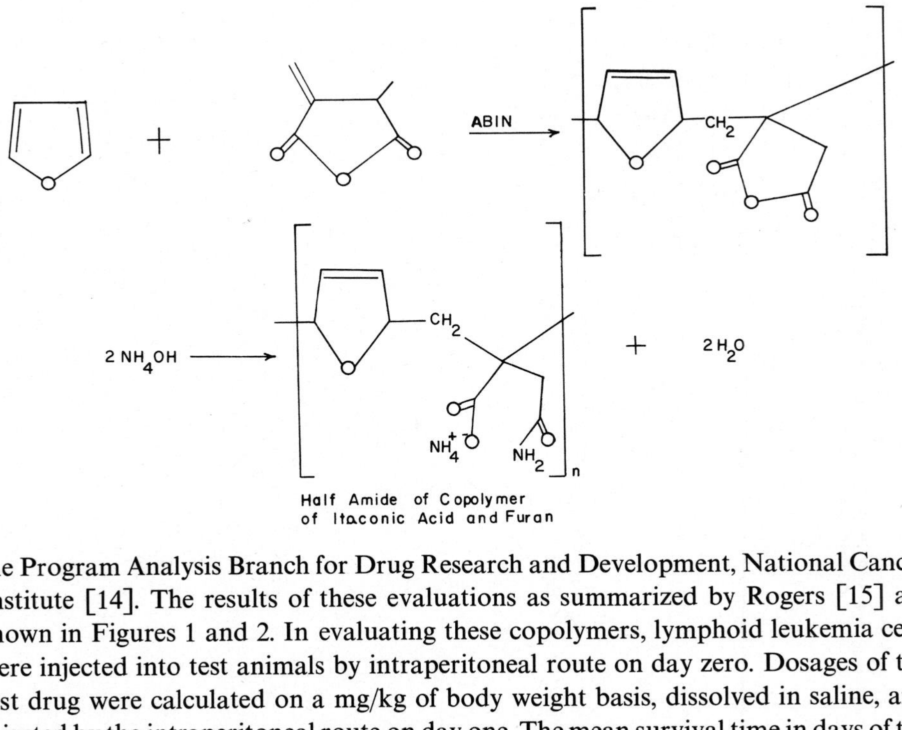

Half Amide of Copolymer
of Itaconic Acid and Furan

the Program Analysis Branch for Drug Research and Development, National Cancer Institute [14]. The results of these evaluations as summarized by Rogers [15] are shown in Figures 1 and 2. In evaluating these copolymers, lymphoid leukemia cells were injected into test animals by intraperitoneal route on day zero. Dosages of the test drug were calculated on a mg/kg of body weight basis, dissolved in saline, and injected by the intraperitoneal route on day one. The mean survival time in days of the test group and the control group was calculated and a ratio of test animals to control animals was calculated. In all tests the animals were evaluated at five days, for survival, as a measure of drug toxicity. All data presented in this paper represent 6/6 survivors in each test group.

SCHEME III

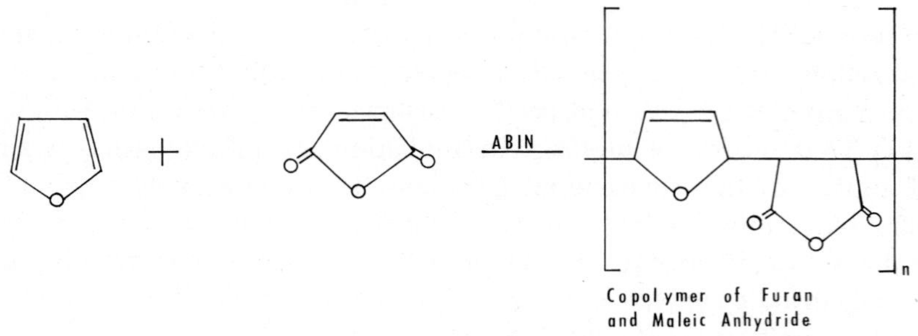

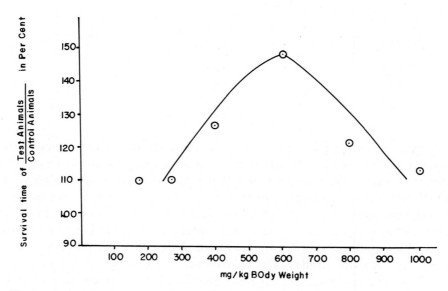

Copolymer of Furan
and Maleic Anhydride

Barton *et al.* [17] have derived a general copolymer equation for the cyclocopolymerization of 1,4-dienes (M_1) and monoolefins (M_2). The kinetic scheme considered is shown in Equations (1)–(9).

$$m_1^\cdot + M_1 \xrightarrow{k_{11}} m_1^\cdot \tag{1}$$
$$m_1^\cdot + M_2 \xrightarrow{k_{12}} m_3^\cdot \tag{2}$$
$$m_3^\cdot \xrightarrow{k_c} m_c^\cdot \tag{3}$$
$$m_3^\cdot + M_1 \xrightarrow{k_{31}} m_1^\cdot \tag{4}$$
$$m_3^\cdot + M_2 \xrightarrow{k_{32}} m_2^\cdot \tag{5}$$
$$m_c^\cdot + M_1 \xrightarrow{k_{c1}} m_1^\cdot \tag{6}$$
$$m_c^\cdot + M_2 \xrightarrow{k_{c2}} m_2^\cdot \tag{7}$$
$$m_2^\cdot + M_1 \xrightarrow{k_{21}} m_1^\cdot \tag{8}$$
$$m_2^\cdot + M_2 \xrightarrow{k_{22}} m_2^\cdot \tag{9}$$

Fig. 1.   Anti-tumor evaluation results on the half amide of the itaconic acid-furan copolymer.

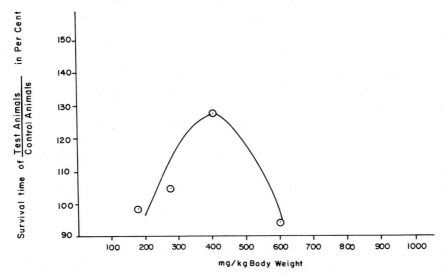

Fig. 2. Anti-tumor evaluation results on copolymer of furan and maleic anhydride [14].

which led to Equation (10):

$$n = \frac{(1+r_1 x)\,\{1/[M_2]+(1/a)\,(1+x/r_3)\}}{(1/a)\,\{(x/r_3)+(r_2/x)+2\}+(1/[M_2])\,\{1+(1+r_2/x)\,(1+r_c x)^{-1}\}}, \qquad (10)$$

where $x = [M_1]/[M_2]$, $r_1 = k_{11}/k_{12}$, $r_2 = k_{22}/k_{21}$, $r_3 = k_{32}/k_{31}$, $r_c = k_{c1}/k_{c2}$, $a = k_c/k_{32}$.

Equation (10) is a differential copolymer composition equation which is applicable to the proposed scheme of cyclocopolymerization. The equation may be applied by putting $n \simeq [m_1]/[m_2]$, the fractional ratio of monomers combined in the copolymer at low conversions.

A similar Equation (11) can be derived relating the relative rate of addition of diene and the rate of cyclization, as follows.

$$\begin{aligned} d[M_1]/d[M_c] &= [(K_{11}+K_{12})/K_{12}K_c]\,(K_{31}+K_{32}+k_c) \\ &= (r_1 x+1)\,\{([M_1]/r_3 a)+([M_2]/a)+1\}. \end{aligned} \qquad (11)$$

Equation (11) applies at low conversions, where $d[m_1]/d[m_c] \simeq [m_1]/[m_c]$, the ratio of the total fraction of diene (unsaturated and cyclic) to the fraction of diene in cyclized units, in the copolymer.

If precise analytical methods are available for determining both the total fraction of diene in the copolymer and the fraction of either cyclic units or pendant vinyl groups, then by making a series of such measurements for different initial monomer feed compositions, values for r_1, r_3, and a could be obtained from Equation (11). Then the remaining two parameters, r_2 and r_c could be obtained from Equation (10).

In certain special cases Equation (10) may be approximated to simpler forms, as in the following examples.

(a) If $k_c \gg k_{32}$ so that a is very large and cyclization is the predominant reaction of the radicals $m_3^{\cdot}$, then Equation (10) gives:

$$n = (1 + r_1 x)(1 + r_c x)/[r_c x + (r_2/x) + 2]. \tag{12}$$

This is equivalent to considering the addition of monoolefin to diene radicals to be a concerted bimolecular step proceeding through a cyclic transition state and producing the cyclic repeating unit.

(b) If in addition there is a strong alternating tendency so that $(r_1, r_2, r_c) \rightarrow 0$ than Equation (12) reduces in the limit to $n = \frac{1}{2}$. This predicts an alternating copolymer composition of 2:1 molar in contrast to 1:1 for the similar limiting case of the classical binary copolymer composition equation.

(c) If the diene has a negligible tendency to add to its own radicals and $r_1 \simeq r_c \simeq 0$, and there is also predominant cyclization, then Equation (12) gives:

$$n = /[(r_2/x) + 2]. \tag{13}$$

A plot of $1/n$ against $1/x$ should be linear with a slope r_2 and an intercept 2.0.

The application of each of these special cases of the theory was shown for experimental systems by these authors. An example of special case (a) is the divinyl ether-acrylonitrile system shown in Figure 3.

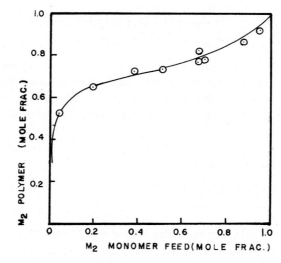

Fig. 3.   Divinyl ether (M_1)-acrylonitrile (M_2) copolymers. Points are experimental. Line calculated for:
$r_1 = 0.024$; $r_2 = 0.938$; $r_c = 0.017$.

The divinyl ether-maleic anhydride system represents an example of special case (b) in which there is a strong alternating tendency and the resultant copolymer composition is 1:2 in DVE:MA.

The 1,4-pentadiene-acrylonitrile system (Figure 4) represents an example of special case (c) in which a plot of $1/n$ against $1/x$ is linear with slope r_2 and an intercept of approximately 2.0.

Two later papers [18] described some additional cyclocopolymers derived from 1,4-dienes and alkenes.

Butler and Joyce [19] presented evidence that 1,4-dienes and alkenes which most readily undergo cyclocopolymerization also form charge-transfer complexes with each other. In the case of the divinyl ether-maleic anhydride system, the complex

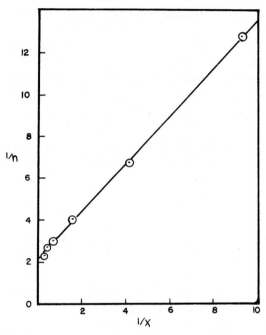

Fig. 4.   1,4-Pentadiene (M_1)-acrylonitrile (M_2) copolymers. $r_2 = 1.129$.

formed is a 1:1 complex with the diene playing the role of the electron donor and maleic anhydride the electron acceptor. These authors found the equilibrium constant for formation of the complex with this pair to be 3.6×10^{-2}. Some additional equilibrium constants were also reported for diene-alkene pairs which undergo cyclocopolymerization. It was also proposed by these authors that the 1:1 molecular complex may undergo copolymerization with an additional molecule of maleic anhydride to account for the molar ratio of 1:2 in VE:MA in the copolymer.

Previous authors [20] had proposed that the molecular complex formed between donor-acceptor olefin pairs might participate in the copolymerization mechanism to account for the regularly alternating free-radical initiated copolymerization between styrene and maleic anhydride. This explanation was supported later by Barb [21], while other authors [22, 23] explained the alternating free-radical copolymerization as resulting from an electron-transfer interaction between the growing radical and the monomer in the transition state of propagation. Styrene is not the only monomer leading to 1:1 copolymer structure; examples of compounds which form such polymers with maleic anhydride, fumaronitrile, etc. include also ethylene [24, 25], vinyl ethers [26–29], and ω-phenylalkenes [30, 31], as well as many others. The copolymerization of styrene and maleic anhydride in different solvents was recently reinvestigated [32] and the formation of charge-transfer complexes between the comonomers and their participation in the alternating free-radical copolymerization was extensively studied by Iwatsuki and Yamashita [26–28]. These authors conclude that the alternating copolymerization could be reduced to a homopolymerization of the charge-transfer complex formed between the comonomers and that the terpolymerization of these alternating copolymerizable monomers with a third monomer which has little or no interaction with either monomer of the pair could be reduced to a copolymerization of the charge-transfer complex formed between the alternating copolymerizable monomers and the third monomer.

The theoretical aspects of charge-transfer complexes have been discussed by Briegleb [33], Andrews and Keefer [34], by Mulliken and Person [35], and by Foster [36]. Formation of the charge-transfer complex is generally regarded as involving an electron transfer from the highest occupied molecular orbital of the donor (D) to the lowest unoccupied molecular orbital of the acceptor (A) and can be represented as shown in Scheme IV.

Scheme IV

$$D+A \rightleftharpoons [D, A] \rightleftharpoons [D \cdot^+, A \cdot^-] \begin{array}{c} \longrightarrow \ ^+D-A^- \\ \rightleftharpoons D \cdot^+ + A \cdot^-. \\ \longrightarrow \ \cdot D-A \cdot \end{array}$$

A few spontaneous copolymerizations between exceptionally reactive donor: acceptor olefinic pairs have been observed. Miller and Gilbert [37] observed that vinylidene cyanide spontaneously copolymerized with vinyl ethers when the two monomers were mixed at room temperature. Yang and Gaoni [38] observed that 2,4,6-trinitrostyrene as the acceptor monomer spontaneously copolymerized with 4-vinylpyridine as the donor monomer when the two were mixed at room temperature. Butler and Sharpe [39] reported that divinyl ether and divinyl sulfone spontaneously copolymerized upon monomer mixing. Thus, the participation of the charge-transfer complex in the copolymerization mechanism of such strong electron donor: electron acceptor monomer pairs appears to have considerable support.

Butler and Campus [40] undertook a study to provide further evidence of the formation of the charge-transfer complex between the comonomers and of its participation in the cyclocopolymerization. The 1,4-diene used was divinyl ether (DVE) and the monoolefins were maleic anhydride (MA) and fumaronitrile (FN). The results of the determination of the composition of the charge-transfer complex formed between DVE–MA are shown in Figure 5. Acrylonitrile (AN) was used as the third monomer in the terpolymerization experiments. For comparison, the results of a complex study of styrene-maleic anhydride and ethyl vinyl ether (EVE)-maleic anhydride were also reported. The results of determination of the equilibrium constants for the DVE–MA and styrene-MA complexes by the NMR method are shown in Figure 6. The results

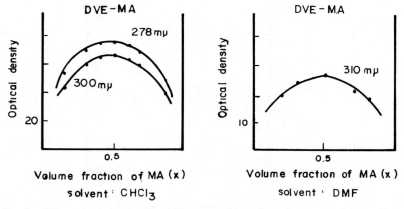

Fig. 5.   Ultraviolet determination of the composition of the charge-transfer complexes.

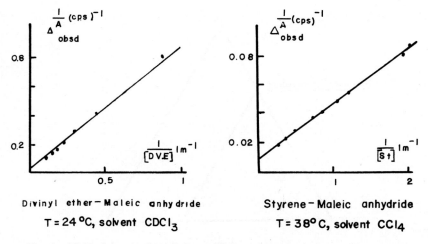

Fig. 6.   NMR determination of the equilibrium constant of complexation of the
charge-transfer complexes.

observed and the proposed cyclocopolymerization mechanism are consistent with participation of the charge-transfer complex as a distinct species in the copolymerization. It was the purpose of this investigation to determine whether there was a dilution effect on the relative reactivities of the monomers in support of the charge-transfer participation concept, and whether the results of a suitable terpolymerization study would also support this postulate. In the divinyl ether-fumaronitrile system, the maximum rate of copolymerization occurred at a monomer feed ratio of 1:2 and the composition of the copolymer was also 1:2 at a total monomer concentration of 3 mole/l as shown in Figure 7. However, when the concentration was progressively lowered to 0.5 mole/l at the same monomer feed ratio, the fumaronitrile content of the copolymer decreased in a linear manner. In a series of terpolymerization experi-

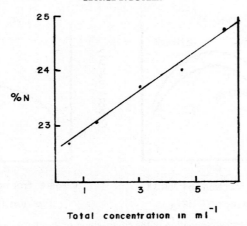

Fig. 7.   Effect of dilution on the copolymerization of divinyl ether (M_1) with fumaronitrile (M_2).

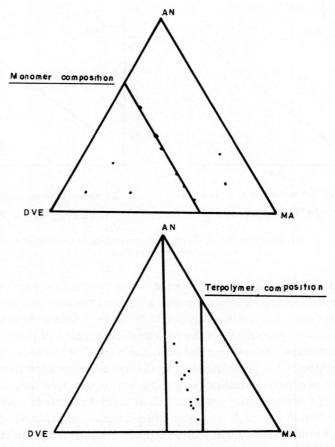

Fig. 8.   Radical terpolymerization of divinyl ether (DVE)-maleic anhydride (MA)-acrylonitrile (AN).

ments with the divinyl ether-maleic anhydride-acrylonitrile system, it was shown that the divinyl ether-maleic anhydride ratio in the terpolymer was always less than 1:1 and had an upper limit of 1:2, regardless of the feed ratio of the termonomers as shown in Figures 8 and 9. These results are consistent with the participation of the charge-

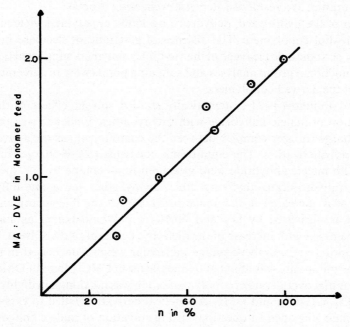

Fig. 9.   Variation of the 1:2 DVE–MA content on the terpolymer (*n*) vs the MA:DVE
ratio in the monomer feed.

transfer complex of divinyl ether and maleic anhydride in a copolymerization process with either maleic anhydride or acrylonitrile as the comonomer.

The geometry of these complexes is unknown as well as whether one or both double bonds of the diene are involved in the complex: however, the structures I and II may be considered [40]:

In a series of more recently published papers, solvent effects, kinetics, steric effects, and ultraviolet initiation in the region of absorption of the complex were studied. Guilbault and Butler [41] found chloromaleic anhydride to copolymerize with divinyl ether to form soluble copolymers of 1:1 composition, having no residual unsaturation. A bicyclic structure was proposed in which the polymer backbone consists only of divinyl ether units. The ease with which the copolymers underwent dehydrohalogenation indicates that the hydrogen and chlorine atoms on the anhydride unit are in a trans configuration as a result of a stepwise cyclization process.

Oxidation of the hydrolyzed, dehydrohalogenated copolymers yielded the corresponding vic-diol copolymers. The absence of a significant decrease in molecular weight upon periodic acid cleavage of the vic-diol copolymers supported the proposed structure. Functional group analyses and softening points were in agreement with the structures of the derived copolymers.

Butler and Fujimori [42] systematically studied solvent effects in the cyclocopolymerization of maleic anhydride with divinyl ether. Evidence was presented to support a charge-transfer complex between the comonomers as the active species in the cyclocopolymerization. The equilibrium constants (K) of charge transfer complexation with maleic anhydride were measured in n-heptane by UV spectrophotometry for tetrahydrofuran, ethyl vinyl ether, divinyl ether, furan, and dihydropyran. The values of K increased in the above order. K_{eq} of the maleic anhydride-divinyl ether pair was measured by UV and NMR in polyhaloalkanes as solvents. The K-values decreased with increase of the dielectric constant of the solvent. The rate of copolymerization and number-average molecular weights decreased in more polar solvents. The initial rate was about 100 times faster in $CHCl_3$ than in DMF. In dilute solution the initial overall rate of copolymerization was maximum in divinyl ether-rich feed in $CHCl_3$, CH_2Cl_2, and DMF. A kinetical derivation failed to explain the rate profile. Though still open to question, strong solvation of maleic anhydride by the solvent was proposed as an explanation. The overall rate of copolymerization was proportional to one-half order of the AIBN concentration in DMF. The overall energy of activation was 27 kcal/mole in $CHCl_3$ and in DMF. Thermal autopolymerization, photopolymerization, and γ-ray polymerization of the maleic anhydride-divinyl ether pair in bulk gave the same 2:1 copolymer. Retardation of the rate by hydroquinone was observed in thermal autopolymerization. The dative state of the charge-transfer complex was considered to be the initiating species.

Zeegers and Butler [43] studied the kinetics of the AIBN-initiated copolymerization of divinyl ether and ethyl vinyl ether with maleic anhydride in seven different solvents. The yield at 100% conversion as a function of the feed composition when the total monomer concentration was kept constant gave confirmation of the composition of these copolymers: Divinyl ether: maleic anhydride = 1:2 and ethyl vinyl ether: maleic anhydride = 1:1. The study of the initial rate as a function of the feed composition made it possible to determine the relative values of the different propagation rate constants consistent with a mechanism by successive and selective additions. In the ethyl vinyl ether-maleic anhydride system, addition of ethyl vinyl ether is slower than addition of maleic anhydride; in the divinyl ether-maleic anhydride system, the addition of divinyl ether is slower than addition of the first molecule, while addition of the second maleic anhydride molecule is slower than the first one. The study of dependence of the monomer concentration, of the AIBN concentration, and of the efficiency of the initiator, on the rate of polymerization, showed that the true order of monomer concentration is close to 1, while its apparent order varies from 1 to 2. From all the kinetic data it was concluded that the mechanism of these copolymerizations can be explained without relying upon the concept of participation of the charge-transfer complex formed between the monomers. However, participation of the complex in a competing mechanism with the above cannot be completely excluded.

Fujimori and Butler [44] found that tetrahydronaphthoquinone and dimethyl tetrahydronaphthoquinone formed donor-acceptor complexes with divinyl ether, the latter being the electron donor. Since participation of such complexes has been considered in the cyclocopolymerization of 1,4-dienes with monoolefins such as divinyl ether-maleic anhydride and divinyl ether-fumaronitrile systems, radical copolymerization of tetrahydronaphthoquinone and dimethyl tetrahydronaphtho-quinone with divinyl ether was studied. It was found that these copolymers have constant 1:1 composition regardless of the feed composition. The terpolymerization of divinyl ether-tetrahydronaphthoquinone-dimethyl tetrahydronaphthaquinone con-firmed the 1:1 donor-acceptor composition in the polymer. Integration of the NMR spectrum was used in determining the copolymer composition. Spectroscopic data suggested a cyclized repeating unit in which the copolymer main chain consists of only divinyl ether units. There is a marked difference between these copolymers and the typical cyclocopolymers, such as those of divinyl ether-maleic anhydride and divinyl ether-fumaronitrile, in which the copolymer main chains consist of divinyl ether and the comonomer alternately, with the overall composition being 1:2. These results were interpreted in terms of the steric effect of the bulky acceptor monomers and the electronic interaction between the comonomers. Competition between an acceptor monomer and the charge-transfer complex toward the cyclized divinyl ether radical in the propagation step appears to favor the charge-transfer complex.

Zeegers and Butler [45] obtained a soluble cyclocopolymer of 1:2 composition by UV irradiation of divinyl ether and fumaronitrile in methanolic solution. By appro-priate calculations based on the results of the rate observed under different concen-tration conditions and with different UV filters, it was shown that both the complex formed between divinyl ether and fumaronitrile, and the noncomplexed species (divinyl ether, fumaronitrile, and methanol) are able to initiate polymerization by light excitation. A study was completed on the same polymerization in the same solvent initiated by a radical initiator. The characteristics of the polymers are the same as those of the photoinitiated polymers. The kinetic results were similar when the polymers were obtained by both kinds of initiation, indicating a similar polymeriza-tion mechanism.

Acrylonitrile is known to form a cyclocopolymer with 1,4-dienes such as divinyl ether and 3,3-dimethyl-1,4-pentadiene with radical initiators. Since acrylonitrile has a high tendency toward homopolymerization, the copolymers are not of regular structure. Lewis acids such as $ZnCl_2$ and $Al(Et)_3$ were used by Fujimori and Butler [46] to increase the e-values of acrylonitrile and methacrylonitrile through complexation. Acrylonitrile, methacrylonitrile, and 2- and 4-vinylpyridine were copolymerized with divinyl ether and 1,4-pentadiene with Lewis acids. In all cases the rate of copolymerization was much enhanced and the alternating tendency of the cyclopolymer increased with the amount of added Lewis acids. A 1:2 divinyl ether:acrylonitrile alternating cyclopolymer was obtained spontaneously or with AIBN and $Al(Et)_3$ in hexane. Also 1:2 alternating cyclocopolymer was successfully obtained in acetone by using a large amount of $ZnCl_2$. The identification of charge-transfer complexation between the divinyl ether and the 2:1 acrylonitrile-zinc chloride complex, and between the 1-hexene and the 2:1 acrylonitrile-zinc chloride complex may support the participation of a charge-transfer complex between all 1,4-dienes studied and the monoolefin-Lewis acid complexes in the cyclocopolymerization mechanism to increase the rate and the alternating tendency.

Fujimori and Butler [47] measured the equilibrium constants for charge-transfer complex formation between divinyl ether and several substituted maleic anhydrides and 2-cyclopentene-1,4-dione in $CHCl_3$ by use of UV spectroscopy. The copolymerization of divinyl ether with these acceptor monoolefins produced regular cyclocopolymers of constant 1:1 or 1:2 (divinyl ether:monoolefin) composition regardless of the feed composition. Comparison of the charge-transfer complexation and the cyclocopolymerization lead to the following conclusions: (1) A strong charge-transfer complex gives regular cyclocopolymer of constant 1:1 composition having a copolymer backbone made up of only 1,4-diene untis; (2) when a monoolefin is unreactive (often sterically), the 1:1 cyclocopolymer is produced; and (3) if charge-transfer complexation is weak and the monoolefin is reactive toward radicals (but not so reactive as to homopolymerize easily), a 1:2 alternating cyclocopolymer is produced. A facile and quantitative elimination of hydrogen halides with dilute aqueous NaOH solution was found. Comparison of the elimination reactions for divinyl ether-chloromaleic anhydride, divinyl ether-bromomaleic anhydride, and divinyl ether-dichloromaleic anhydride 1:1 regular alternating cyclocopolymers lead to a conclusion which supports the six-membered ring structure of the repeating cyclic unit formed through the intramolecular cyclization in the cyclocopolymerization.

Allen and Turner [48] have reported the results of a study of some physical properties of copolymers of DVE–MA. One sample having an intrinsic viscosity of 0.280 dl/g. (acetone, 30°) was found to have a number-average molecular weight, $\bar{M}_n$, of 47 500 and a weight-average molecular weight, $\bar{M}_w$, of $250\,000 \pm 25\,000$ for an $\bar{M}_w/\bar{M}_n$ ratio of about five. The first sample was fractionated by use of a stepwise continuous solvent-gradient column method from acetone solution with hexane as precipitant to obtain nine fractions varying in intrinsic viscosities (acetone, 30°) from 0.019 dl/g. for the lowest molecular weight fraction to 0.493 dl/g. for the highest molecular weight fraction as shown in Figure 10.

These authors observed that intrinsic viscosities and osmotic pressures of the DVE–MA copolymer when measured in DMF showed unusual slopes, suggestive of polyelectrolyte behavior. Their intrinsic viscosity data are shown in Figure 11 and their osmotic pressure data are shown in Figure 12.

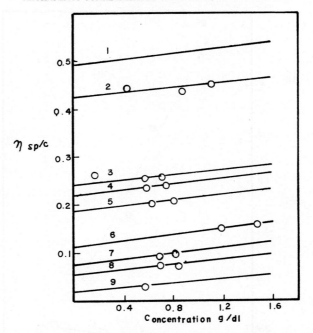

Fig. 10. Intrinsic viscosities of divinyl ether maleic anhydride cyclocopolymer Sample No. 24 fractions in acetone at $30 \pm 0.1\,°C$. Curve 1 was calculated.

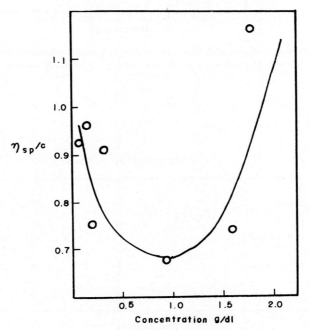

Fig. 11. Intrinsic viscosity of divinyl ether maleic anhydride cyclocopolymer Sample No. 24 in dimethyl formamide at $30° \pm 0.1\,°C$.

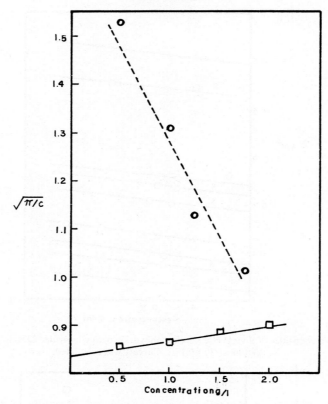

$\sqrt{\pi/c}$

Concentration g/l

Fig. 12.   Osmotic pressure results of divinyl ether maleic anhydride cyclocopolymer Sample No. 25 in dimethyl formamide (dotted curve) and in acetone (solid curve).

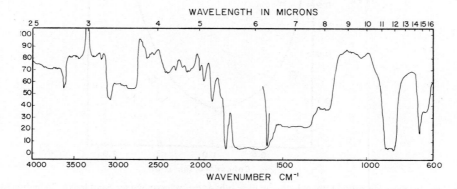

Fig. 13.   Infrared spectrum (No. 2317) of Maleic anhydride-DMF reaction product in DMF.

Butler and Wu [49] in an investigation directed toward development of a gel permeation chromatography method for fractionation of the DVE–MA copolymer, observed that infrared spectra of both maleic anhydride and DVE–MA copolymer in DMF showed the presence of strong carboxylate anion at 1400 cm^{-1} and around 1600 cm^{-1}. These results confirm the proposal by Allen and Turner [48] that DVE–MA copolymer indeed performs as a polyelectrolyte in DMF. The IR spectra of maleic anhydride in DMF and DVE–MA copolymer in DMF are shown as Figures 13 and 14, respectively.

The structure of the reaction product of DVE–MA with DMF is suggested by Butler and Wu [49] to be either A or B as shown in Scheme V.

Scheme V

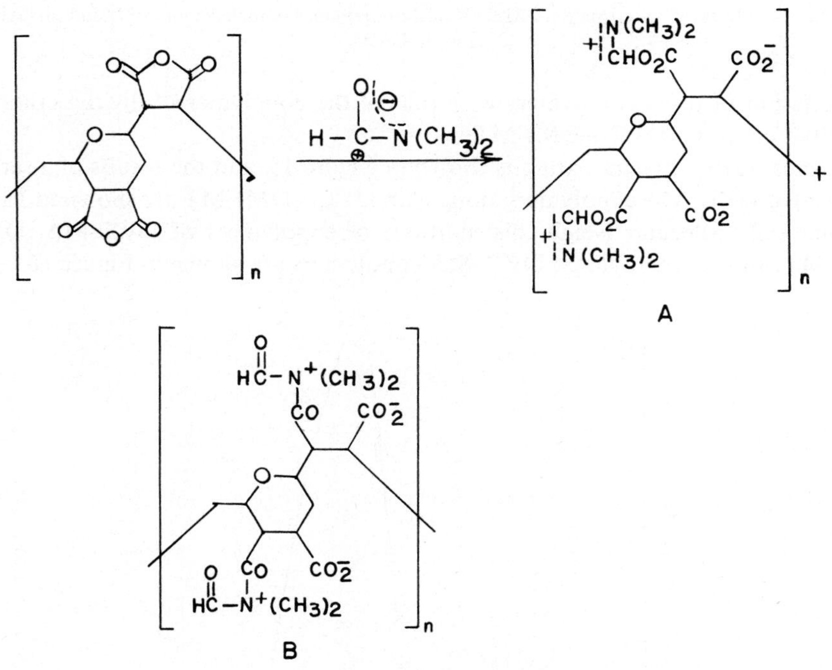

The present evidence indicates that molecular weight is an important factor in controlling the interferon-inducing property or the toxicity factor of the DVE–MA copolymer. The development of a method which would permit fractionation of the copolymer into fractions of narrow molecular weight distribution for further evaluation as an interferon inducer was the purpose of the investigation undertaken by Butler and Wu [49]. Two copolymer samples having $\eta_{inh} = 1.56$ dl/g and 0.15 dl/g were subjected to fractionation as the polyelectrolyte on one 3 ft column each, in series, of Type AX, BX, CX, DX and EX Deactivated Porasil, without much success. However, by conversion of the copolymers to their dimethyl ester, it was possible to fractionate these derivatives on standard Styragel columns, and to compare the

GEORGE B. BUTLER

WAVELENGTH IN MICRONS

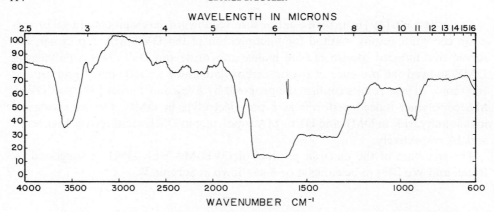

Fig. 14.   Infrared spectrum (No. 2391–3) of the copolymer of maleic anhydride and divinyl ether in DMF.

fractionation of these copolymers with that of the copolymer of divinyl ether with dimethyl fumarate (DVE + DMFMT).

A typical GPC chromatogram is shown in Figure 15, and the results of a series of methylated DVE–MA copolymers along with DVE + DMFMT are shown in Table 1. The integral molecular weight distributions of copolymers of DVE–MA, DVE + DMFMT, and the methylated DVE–MA copolymers are shown in Figure 16.

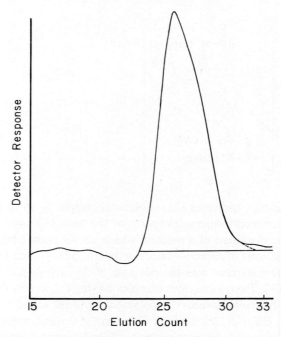

Fig. 15.   Typical GPC Chromatogram (GPC Set 65) of dimethyl ester of copolymer of divinyl ether and maleic anhydride (No. 16 DME 2).

TABLE I

Analyses of copolymer DME series (DME = dimethyl ester of copolymer of divinyl ether and maleic anhydride)

	A–DME	B–DME	C–DME	D–DME	E–DME	F–DME	M–DME	N–DME	X–DME	DVE+DMFMT
$\bar{M}_w \times 10^{-5}$	2.22	2.26	2.07	2.15	2.76	2.91	1.38	1.47	3.22	0.993
$\bar{M}_n \times 10^{-5}$	1.47	1.54	1.48	1.59	2.21	2.26	1.08	1.15	1.99	0.759
$\bar{M}_w/\bar{M}_n$	1.51	1.47	1.40	1.35	1.25	1.29	1.28	1.27	1.61	1.31
Peak maximum (elution count)	25.2	25.0	25.0	25.0	24.1	23.9	27.2	25.8	24.8	27.2
Peak base width (elution count)	20.0–32.0	20.5–32.0	20.5–31.0	21.0–31.0	20.5–30.0	20.5–30.0	21.0–32.0	23.0–32.5	19.5–31.0	24.0–34.0

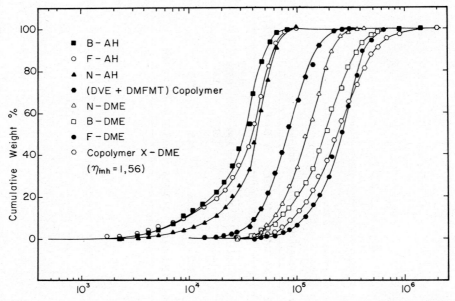

Fig. 16.   Integral molecular weight distribution of copolymers of maleic anhydride and divinyl ether and their methyl esters.

Acknowledgements

Permission to use Figures 3–9 by the publisher of *Journal of Polymer Science*, Figures 10–12 by the publisher of *J. Macromol. Sci.-Chem.*, and Figures 13–16 by Plenum Press is gratefully acknowledged.

References

1. The most recent paper is Fujimori, K. and Butler, G. B.: *J. Macromol. Sci.-Chem.*, **A7**(2), 415 (1973).
2. Butler, G. B.: *Abstracts, 133rd American Chemical Society Meeting*, San Francisco, Calif., April 13–18, 1958, p. 6R. This paper dealt with the original copolymerization of divinyl ether and maleic anhydride (Nov. 22, 1951).
3. Butler, G. B.: *Abstracts, 134th American Chemical Society Meeting*, Chicago, Ill., Sept. 7–12, 1958, p. 32T. This paper dealt with some additional examples of the cyclocopolymerization.
4. (a) Butler, G. B.: *J. Polym. Sci.* **48**, 279 (1960). (b) Butler, G. B.: U.S. Patent, 3, 320, 216 (May 16, 1967); U.S. Patent Reissue 26, 407 (June 11, 1968).
5. (a) Butler, G. B. and Bunch, R. L.: *J. Am. Chem. Soc.* **71**, 3120 (1949); Butler, G. B. and Ingley, F. L.: *J. Am. Chem. Soc.* **73**, 895 (1951). (b) Butler, G. B., Gropp, A. H., Angelo, R. J., Husa, W. H., and Jorolan, E. P.: Fifth Quarterly Report, Atomic Energy Commission Contract AT-(40-1)-1353, Sept. 15, 1953. (c) Butler, G. B.: Gordon Research Conference, Ion Exchange, June 15, 1955 (Science **121**, 574 (1955)). (d) Butler, G. B. and Angelo, R. J.: *J. Am. Chem. Soc.* **79**, 3128 (1957).
6. Butler, G. B., Crawshaw, A., and Miller, W. L.: *J. Am. Chem. Soc.* **80**, 3615 (1958).
7. Staudinger, H. and Heuer, W.: *Ber.* **67**, 1159 (1934).
8. Schuller, W. H., Price, J. A., Moore, S. T., and Thomas, W. M.: *J. Chem. Eng. Data* **4**, 273 (1959). (This paper includes representative examples.)
9. Butler, G. B.: *J. Macromol. Sci.-Chem.* **A5**(1), 219 (1971).
10. Based upon structure submitted to NIH by letter of October 12, 1959, sample submitted on June 14, 1960, and test results of July, 1961.

11. (a) Merigan, T. C.: *Nature* **214**, 416 (1967). (b) Merigan, T. C. and Regelson, W.: *New Engl. J. Med.* **277**, 1283 (1967).
12. Isaacs, A. and Lindenmann, J.: *Proc. Roy. Soc. (London), Ser. B* **147**, 258 (1957).
13. Murthy, Y. K. S. and Anders, H. P.: *Angew. Chem. Internat. Edit.* **9**, 480 (1970).
14. Program Analysis Branch, Drug Research and Development, National Institutes of Health, Building 37, Room 6C09, Bethesda, Maryland 20014.
15. Rogers, C. L.: A report prepared as part of the requirement for a course in pharmacology, College of Medicine, University of Florida, March 1973.
16. Sharabash, M., Butler, G. B., and Badgett, J. T.: *J. Macromol. Sci.-Chem.* **A4** 51 (1970).
17. Barton, J. M., Butler, G. B., and Chapin, E. C.: *J. Polymer Sci., Part A* **3**, 501 (1965).
18. (a) Butler, G. B. and Kasat, R. B.: *J. Polymer Sci., Part A-1* **3**, 4205 (1965). (b) Butler, G. B., Vanhaeren, G., and Ramadier, M. F.: *J. Polymer Sci., Part A-1* **5**, 1265 (1967).
19. Butler, G. B. and Joyce, K. C.: *J. Polym. Sci.* **C22**, 45 (1968).
20. Bartlett, P. D. and Nozaki, K.: *J. Am. Chem. Soc.* **68**, 1495 (1946).
21. Barb, W.: *Trans. Faraday Soc.* **49**, 143 (1953).
22. Lewis, F. M., Walling, C., Cummings, W., Briggs, E. R., and Mayo, F. R.: *J. Am. Chem. Soc.* **70**, 1519 (1948).
23. Walling, C., Briggs, E., Wolfstein, K., and Mayo, F. R.: *J. Am. Chem. Soc.* **70**, 1537 (1948).
24. Flett, L. H. and Gardner, W. H.: *Maleic Anhydride Derivatives, Reactions of the Double Bond*, Wiley, New York, 1952, p. 26.
25. Hanford, W. E.: U.S. Pat. 2, 378, 629 (1945); *Chem. Abstr.* **39**, 4265 (1945).
26. Iwatsuki, S., Yamashita, Y., *et al.*: *Kogyo Kagaku Zasshi* **67**, 1464, 1467, 1470 (1964); *ibid.* **68**, 1138, 1963, 1967, 2463 (1965); *ibid.* **69**, 145 (1966).
27. Iwatsuki, S., Yamashita, Y., *et al.*: *Makromol. Chem.* **89**, 205 (1965); *ibid.* **102**, 232 (1967).
28. Iwatsuki, S. and Yamashita, Y.: *J. Polym. Sci. A-1* **5**, 1753 (1967).
29. Kimbrough, R. D.: *J. Polym. Sci. B* **2**, 85 (1964).
30. Martin, M. M. and Jensen, N. P.: *J. Org. Chem.* **27**, 1201 (1962).
31. Aso, C. and Kunitake, T.: *Bull. Chem. Soc. Japan* **38**, 1564 (1965).
32. Tsuchida, E., Ohtani, Y., Nakadai, H., and Shitohara, I.: *Kogyo Kagaku Zasshi* **70**, 573 (1967).
33. Briegleb, G.: *Elektronen-Donator-Acceptor Komplexe*, Springer Verlag, Berlin-Göttingen Heidelberg, 1961.
34. Andrews, L. J. and Keefer, R. M.: *Molecular Complexes in Organic Chemistry*, Holden-Day, Inc., San Francisco, 1964.
35. Mulliken, R. S. and Person, W. B.: *Molecular Complexes*, Wiley-Interscience, New York, 1969.
36. Foster, R.: *Organic Charge-Transfer Complexes*, Academic Press, New York, 1969.
37. Miller, F. F. and Gilbert, H.: Canadaian Pat. 569, 262 (To B. F. Goodrich, Jan. 20, 1959).
38. Yang, N. C. and Gaoni, Y.: *J. Am. Chem. Soc.* **86**, 5023 (1964).
39. Butler, G. B. and Sharpe, A. J., Jr.: *Polymer Letters* **9**, 125 (1971).
40. Butler, G. B. and Campus, A. F.: *J. Polymer Sci., Part. A-1* **8**, 545 (1970).
4.1. Guilbault, L. J. and Butler, G. B.: *J. Macromol. Sci.-Chem.* **A5(7)**, 1219 (1971).
42. Butler, G. B. and Fujimori, K.: *J. Macromol. Sci.-Chem.* **A6(8)**, 1553 (1972).
43. Zeegers, B. and Butler, G. B.: *J. Macromol. Sci.-Chem.* **A6(8)**, 1569 (1972).
44. Fujimori, K. and Butler, G. B.: *J. Macromol. Sci.-Chem.* **A6(8)**, 1609 (1972).
45. Zeegers, B. and Butler, G. B.: *J. Macromol. Sci.-Chem.* **A7(2)**, 349 (1973).
46. Fujimori, K. and Butler, G. B.: *J. Macromol. Sci.-Chem.* **A7(2)**, 387 (1973).
47. Fujimori, K. and Butler, G. B.: *J. Macromol. Sci.-Chem.* **A7(2)**, 415 (1973).
48. Allen, V. R. and Turner, S. R.: *J. Macromol. Sci.-Chem.* **A5(1)**, 229 (1971).
49. Butler, G. B. and Wu, C.: in N. M. Bikales (ed.), *Water-Soluble Polymers*, Vol. 2, Polymer Science and Technology, Plenum Press, New York, 1973, p.

ZWITTERION SILANE-MODIFIED POLYMER LATEXES

EDWIN P. PLUEDDEMANN

Dow Corning Corporation, Midland, Mich., U.S.A.

Abstract. Addition of aqueous solutions of alkoxysilanes to polymer latexes gives a complex physical-chemical system that allows most efficient utilization of silanes as polymer modifiers. Silanes may be used to improve solvent resistance and weathering of polymer films, but they are most effective in imparting adhesion. Under optimum conditions the silanols do not migrate into the polymer phase, but remain in the water phase or on the particle surface. Methyltrimethoxysilane and Dow Corning® Z-6020 (a diamine-functional silane) have most favorable solubility characteristics for modifying polymer latexes. Simple addition of a trace of Z-6020 to resinous polymer latexes often imparts good water-resistant adhesion to most mineral surfaces. Rubbery polymer latexes require a silane-modified resinous or tacky phase at a mineral surface. Silane-modified tackifiers or plasticizers are often effective as additives to thermoplastic rubber latexes for adhesion of air-dried films. Silane-modified resin precursors are most effective for adhesion of films cured at elevated temperatures. One of the most promising resin precursors is a silane-modified Dow Zwitterion resin that is completely soluble in the aqueous phase but cures "in situ" to form a very water-resistant primer when cured on mineral surfaces in contact with the polymer latex.

1. Introduction

Silicone modification of alkyds, epoxies, acrylics and other coating resins are well-known for imparting improved gloss retention and weather resistance [1]. Copolymers of structure $(\text{Siloxane}) \equiv \text{SiOCH}_2\text{-(Resin)}$ are prepared in organic solvents with relatively large proportions of silicone, since performance is approximately proportional to the silicone content.

Although aqueous suspension copolymerization of vinyl monomers and reactive trialkoxysilanes generally leads to cross-linked polymers [2], it has been observed that silanol-modified emulsified copolymers are quite stable and do not cross-link until the polymer films dry. Silanol groups apparently are stabilized by orientation at the polymer-water interface and act as hydrophilic emulsion stabilizers comparable to carboxyl groups [3]. After the latex is deposited as a film, the silanol groups condense to form siloxanemodified water-resistant coatings.

Silane-modified emulsion copolymers may be prepared by direct emulsion copolymerization of suitable silanes with other monomers, but it would be much more convenient to modify existing emulsion polymers by simple additions of silanes.

Additions of alkoxysilanes to aqueous polymer emulsions give a complex physical-chemical system that may undergo marked changes in the latex and in deposited films with aging, but allows separate deposition of silane and polymer from a single solution.

The nature of aqueous solutions of silane coupling agents has been studied rather extensively [4]. Neutral alkoxysilanes hydrolyze rapidly at pH 4 to alkylsilane triols, and then condense slowly to oligomeric organosiloxanols.

Alan Rembaum and Eric Sélégny (eds.), Polyelectrolytes and Their Applications, 119–128. All Rights Reserved.
Copyright © 1975 by D. Reidel Publishing Company, Dordrecht-Holland.

$$RSi(OMe) \xrightarrow[\text{fast}]{H_2O} RSi(OH)_3 + 3\ MeOH$$

$$RSi(OH)_3 \xrightarrow{\text{slow}} \underset{\underset{OH}{|}}{HOSi} \overset{R}{\underset{\underset{OH}{|}}{|}} -OSi\overset{R}{\underset{\underset{OH}{|}}{|}} -O-Si\overset{R}{\underset{\underset{OH}{|}}{|}} -OH + H_2O.$$

The initial silanetriols are generally soluble in water, and cannot be extracted from aqueous solutions by toluene. After condensation has proceeded sufficiently, siloxanols go through an oil-soluble stage before they cross-link to insoluble resins.

Gamma aminopropyl trialkoxysilanes hydrolyze rapidly in water but do not form insoluble oligomeric siloxanols. Instead, they form stable low-viscosity aqueous solutions that do not go through an oil soluble stage, and even in aged solutions the siloxanes cannot be extracted from water by toluene. Various cyclic Zwitterion structures have been postulated for the aqueous silane solutions. The gamma nitrogen could conceivably form a six-membered cyclic ammonium silanolate or a five-membered cyclic penta-coordinate silicon compound. When filmed onto a surface from water, and dried, the complex soluble structures cross-link to infusible siloxanes with strong adhesion to most minerals.

Requirements for adhesion of organic polymers to mineral surfaces in the presence of silane coupling agents have been well defined [5]. Water-resistant bonds to most mineral surfaces will be obtained if two requirements are met:

(1) The resin is modified with a silane to present a silanol surface to the mineral adherend. Chemical reaction of silane with resin is preferred, although solution compatibility may be sufficient with certain thermoplastics.

(2) The morphology of the interface allows equilibrium bonding of silanols with the surface. This requires a rigid or tacky (but not rubbery or flexible) polymer at the interface.

Silane modification of a thermoplastic resin latex, therefore requires only a small amount of silane coupling agent with such solubility characteristics that it will not migrate into the latex particle, but when filmed on a surface will have sufficient compatibility with the polymer film to bond to it.

Simple addition of silanes to a rubbery polymer latex cannot impart water-resistant bonds to a mineral unless the silane deposits a resinous or tacky layer at the interface. A silane-modified resin or plasticizer will be effective if it can be retained as a separate phase in the emulsion and deposited as a primer layer on the mineral.

The adhesion of a flexible polymer to a mineral surface through a silane-modified resinous or tacky phase differs from the use of resinous cross-linkers for reactive thermoplastics in that only the polymer at the interface is modified to obtain adhesion without changing the properties of the film itself.

Studies with silane-modified latexes suggest that low molecular weight, water-soluble polyelectrolytes deposit preferentially on a mineral surface to give a well-ordered layer of modifier between the latex polymer and the mineral surface. Physical properties of this intermediate layer could be controlled to give the desired rigid or tacky morphology if a silane-modified low molecular weight resin precursor could be dispersed in the water phase of the latex without migrating into the latex polymer particle. Dow's Zwitterion resin XD-8-157L described in U.S. Patent 3,660,431 seems to fit this concept perfectly. This material is a 30% solution in methanol of the structure:

The Zwitterion resin precursor is very soluble in water and retains its ionic form until water is removed. Dried films are very hard and tough with good resistance to solvents. Simple modification with a water-soluble silane provides a film with strong water-resistant adhesion to minerals.

Flexible acrylic polymers in the form of latexes are used as pigment binders on fiberglass, polyester, and other fabrics that are not readily dyed. They are also used to prepare caulks and adhesives. Very soft, tacky acrylic polymers have direct adhesion to most surfaces and need no adhesion promoters. Hard acrylic polymers are readily modified with simple compatible silanes to obtain good adhesion to mineral surfaces: Soft, flexible, non-tacky acrylics have many desirable properties, but do not have good direct adhesion to most solid surfaces and the bonds to mineral surfaces are not water-resistant.

For application in pigment binders and latex caulks it is important that any modification to obtain adhesion does not reduce the flexibility of the bulk of polymers.

2. Experimental

2.1. SILANE-MODIFIED RESIN LATEXES

The complex interplay of water-soluble silanes with polymer particles in a resin latex are conveniently followed by observing the adhesion of plaster on primed glass microscope slides (Table I). Wet plaster, dried against glass, has virtually no adhesion to glass or silane-treated glass, good dry adhesion to latex-treated glass, and both dry- and wet-adhesion to proper combinations of silane and latex. Observations summarized in Table I are in agreement with the concept of a two-phase system, with sequential deposition of silane, followed by latex polymer on the glass surface.

Mixtures of water-soluble silanes and typical latex polymers were checked as primers on glass for plaster adhesion. Conditions that allow the silane to remain in aqueous solution (Figure 1), or to deposit only on the latex particle surface are demonstrated by primer films that impart instant water-resistant adhesion of plaster

TABLE I

Adhesion to glass microscope slides

Primer on glass Dow latex 209 + 10% silane	Adhesion of plaster	
	Dry	Wet
No primer	Very poor	Very poor
Z-6020 – alone	Poor	Poor
Latex – alone	Good	Poor
Z-6020, followed by latex	Good	Good
Latex, followed by Z-6020	Poor	Poor
Z-6020 and polymer in organic solvent	Poor	Poor
Fresh Z-6020 solution in latex	Good	Good

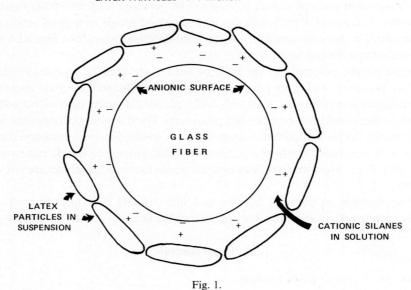

Fig. 1.

to glass. Slow migration of condensing siloxanols into the latex particles results in loss of primer effectiveness as the latex ages. (Table II) Silanols with large neutral organic groups are most compatible with the organic polymer and are readily lost in the latex particle. Methylsilanols are unique, probably due to the very poor solubility of methylsiloxanes in most organic polymers. As methylsilanols condense, they apparently deposit on the latex particle surface where they are available for bonding. Gamma aminopropyl silanes also are generally effective as latex modifiers, since they hydrolyze to complex cyclic structures of low molecular weight that retain their water solubility and do not migrate into the latex particle.

For some applications it was desired that surfaces could be pre-primed and allowed

TABLE II

Silane-modified (10%) D.L. 209 as primers for plaster
(Primer dried 15 min at room temp.)

Silane added	Primer aged in vial wet-adhesion of plaster		
	Initially	4-Weeks	5-Months
None	P	P	P
$MeSi(OMe)_3$	F–G	F–G	F–G
$ViSi(OMe)_3$	F–G	F	F
$PhSi(OMe)_3$	P	P	P
$ClPrSi(OMe)_3$	F	v. poor	v. poor
$HSPrSi(OMe)_3$	F–G	F–G	F
$CH_2=C(CH_3)COO(CH_2)_3Si(OMe)_3$	F–G	P	F
$\overset{O}{\overset{/\backslash}{CH_2CHCH_2O(CH_2)_3Si(OMe)_3}}$	F–G	P	P–F
$Me_2N(CH_2)_3Si(OMe)_3$	P	P	P
$MeHN(CH_2)_3Si(OMe)_3$	G	P	P
$H_2N(CH_2)_3Si(OEt)_3$	F–G	P	–
$H_2NCH_2CH_2NH(CH_2)_3Si(OMe)_3$	G	G	G

P = poor, F = fair, G = good.

to age indefinitely before applying plaster or mortar. Primer films of Elvacet 81-900 and of Dow Latex 209 with added 2% $MeSi(OMe)_3$ or Z-6020 were allowed to stand on glass at room temperature for up to 6 months before applying wet plaster. The Elvacet primers (PVAc) improved with age, while the styrene-butadiene resin (latex 209) film lost its ability to impart adhesion of plaster to glass.

The aging effect was accelerated by drying the primed glass in a 100° oven (Table III).

Various types of latex polymers were compared as primers with 10% added Z-6020. The latex polymers are described in Table IV. Polyvinyl acetate and a Dow Styrene-butadiene latex looked most promising when modified with Z-6020. Several of the latexes were good primers on glass for initial adhesion of plaster, but failed within an hour in water. Among the silane-modified latexes, those that gave very soft films or very hard films were not effective as primers for plaster. Plaster developed very good dry adhesion to soft primer films, but the primer lost adhesion to glass when wet. Very hard primers developed good dry and wet adhesion to glass, but plaster didn't develop any bond to the primer.

It appears that carboxyl-modification is necessary for bonding polymer latexes through amine-functional silanes. A series of experimental styrene-butadiene-acid terpolymers were compared as film formers and primers with 5% added Z-6020 (Table V). Difunctional acids gave the most stable mixes, but poorest water resistance. Polymers with acrylic acid modification were best primers for plaster. Alkylmaleate modified polymers had poor stability with silane, and deposited rubbery films with poor wet adhesion to glass.

TABLE III

Adhesion of plaster to aged primer films
(2% silane in resin latex)

Primer	Wet-adhesion[a] of plaster to aged primer age of primer film on glass (room temp.)				
	1 hr	1 day	1 week	1 month	6 months
Elvacet – MeSi(OMe)$_3$	P	F	F–G	G	G
Elvacet – Z-6020	P	P	–	F	G
D.L. 209 – MeSi(OMe)$_3$	G	G	F	F	P
D.L. 209 – Z-6020	G	G	–	F	P

Primer	Wet adhesion of plaster to primed glass (dried at 100°C)				
	2 min	10 min	30 min	1 hr	2 hrs
Elvacet – MeSi(OMe)$_3$	P	G	G	–	G
Elvacet – Z-6020	G	G	F–G	F	P
D.L. 209 – Z-6020	G	G	G	F–G	F
D.L. 209 – MeSi(OMe)$_3$	G	G	F–G	P	F

[a] Plaster dried 3 days on primed glass and tested after soaking in water 1 day.

TABLE IV

Latex polymers for plaster adhesion

Latex	Composition	Film	Wet adhesion[a]	
			Primer[a]	Plaster
Airco Flexbond 150	ViOAc copolymer	Sticky	Poor	Fair
Aircoflex 510A	Ethylene-ViOAc	Rubberty	V. Poor	Poor
Borden Polyco 804 PL	ViOAc copolymer	Tough	Poor	Fair
DuPont Elvacet 81-900	ViOAc	Hard	Fair	*Good*
Dow D.L. 234	S–B Resin	V. Hard	Good	V. Poor
D.L. 209	S–B Resin	Tough	Fair	*V. Good*
Naugatex 3902	S–B Rubber	Flexible	Poor	Fair
R&H Rhoplex AC-35	Acrylic	Flexible	Fair	Poor
Rhoplex AC-201	Acrylic	V. Hard	Good	V. Poor
D.C. XR6-0832	Silicone-acrylic	Tough	Fair	Poor
Goodrich Hycar 2600 × 146	Acrylic	Sticky	V. Poor	V. Poor
Mill Onyx Latex 5A	Polyurethane	Tough	Fair	Fair

[a] Wet adhesion after 24 hrs in water.
[b] Primer contains 10% Z-6020 based on latex solids.

An optimum silane-modified resin latex has been given a Dow Corning experimental product designation, XZ-8-5080 Primer XZ-8-5080 applied to glass was effective instantly when wet plaster was applied to wet primer. Addition of XZ-8-5080 to plaster mixes, however, did not impart any adhesion to glass and caused loss in strength of the

TABLE V
Silane-modified (5%) SBR Latexes as Primers[b]

Polymer composition				Latex stability with silane	Wet adhesion of primer on glass		Wet adhesion of plaster
Styrene	Butadiene	Acid	Acid[a]		Dry 100°,	Dry 150°	
70	20	5,5	A, MA	4	4	4	1
64	36	–	None	4	4	4	1
58	40	2	Acrylic	4	4	4	4
58	38	4	Acrylic	3	4	4	4
57	39	4	Fumaric	4	1	3	1
58	39	3	Itatonic	4	2	3	2
58	39	3	Fumaric	4	2	3	2
55	41	3, 1	A, MA	3	4	4	4
47	50	3	HM	2	1	1	2
47	50	3 ...	Itatonic	4	2	3	2
46	49	5	HM	1	1	1	2
24[c]	?	?	?	3	1	3	2

[a]   A = Acrylic.
  MA = Methacrylic.
  HM = Hexylmaleate.
[b] Rating from 1 (poor) to 4 (excellent).
[c] Goodrite 2570 × 1 B.F. Goodrich Chemical Co.

plaster. Primer XZ8-5080 was not only effective on glass, but also gave good adhesion of plaster to steel, aluminum, ceramic tile, cedar wood, styrafoam, and urethane foam.

Thickness of XZ8-5080 film required to impart good wet adhesion of plaster to glass was estimated by spreading 2 drops of primer at various concentrations on microscope slides of 10 cm² area.

Estimated primer thickness on glass	Adhesion of plaster	
	Dry	Wet
30 microns	G	G
5 microns	G	G
0.5 microns	G	G
0.05 microns	P	P

It is suggested that XZ8-5080 will be an effective general primer for plaster to materials of construction. It may be diluted (from 50% solids) with water to 2–5% solids and sprayed on surfaces at the construction site before plastering.

Our tests with 'Thinset' mortar on ceramic tile indicated that priming the tile with XZ8-5080 did improve adhesion, but unprimed adhesion was so good that our qualitative tests were inconclusive. The effect of adhesion of mortar to styrafoam was more dramatic. Half of the smooth surface of a styrafoam plank was primed with XZ8-5080 while the other half remained unprimed. Within 15 minutes a fresh mix of

'Thinset' mortar was troweled on the surface. After 3 days, adhesion was compared qualitatively.

	Adhesion of mortar to styrafoam	
	Primed	Unprimed
Dry	Very good	Fair
Wet 1 hour	Good	Poor
Wet 1 day	Good	Very poor

In addition, XZ8-5080 is an effective primer on glass for various thermoplastic rubbers and resins. The major contribution of observations on plaster adhesion, however, may be in pointing the way for general modification of polymer latexes with silanes. Silane-modified latexes generally are good film formers, and if the films are hard – they have good wet adhesion to glass. Precautions should be taken to keep the silane in the water phase until the material is filmed out as a coating.

2.2. SILANE-MODIFIED FLEXIBLE POLYMER LATEXES

Rohm and Haas, Rhoplex AC-35 was selected as a typical flexible acrylic latex. Films of AC-35 on glass or other mineral surfaces are clear, tough and flexible, with a Tukon hardness of about 1, but generally have poor adhesion, especially after soaking in water. Addition of 5% silane (as 50% dispersions in water) to AC-35 generally did not improve wet adhesion to glass.

TABLE VI

Silane-modified Zwitterion resin (SDZ) in acrylic latex AC-35
Adhesion of (6%) modified AC-35 to glass
(Dry 1 hr, 100°, soak on water 1 day)

Function on Si	SDZ	Silane alone
None	Poor	Poor
CH_3^-	Poor	P-Fair
$CH_2=CH^-$	Poor	Poor
$H_2N(CH_2)_3^-$	Fair	Poor
$H_2NCH_2CH_2NH(CH_2)_3^-$	Exc.	Poor
$Me_2N(CH_2)^-$	Exc.	–
$MeHN(CH_2)_3$	Exc.	–
Epoxy	Exc.	Poor
$HSPr^-$	Exc.	Poor

% Z-6020 in XD-5157L	R.T. cure	100° cure
None	Poor	Poor
0.2	Poor	Poor
0.5	Poor	Fair
1.0	Fair	Good
2.0	Fair	Exc.
4.0	F-Good	Exc.

Among various silane-modified resin precursors tested as additives for adhesion, the polyfunctional Zwitterion resins were, by far, the best. The Dow Zwitterion resin XD-8157L filmed on glass cures at 100° to a very hard, clear amber film with strong dry adhesion, but poor wet adhesion to glass.

Addition of 1–2% Z-6020 to the XD-8157L solution gives a product (B-2014-110) that is stable in alkaline solutions, but cures rapidly as films on glass with excellent resistance to water. The film on glass is so hard that is is a poor primer for most polymers. The polymers cannot interdiffuse with the primer molecules and develop no adhesion to the primer film.

Addition of silane modified XD-8157L to AC-35 gave a stable emulsion that discolored slightly, but gave clear films on glass. Various silanes (5%) added to XD-8157L were mixed with AC-35 to give a 6%-modified latex. Films on glass were dried 1 hour at 100 °C. All films were clear, flexible, and had excellent dry adhesion to glass. Adhesion after soaking 1 day in water is indicated in Table VI. Z-6020 looked as good as any of the silane additives, and was compared at various levels of addition to

TABLE VII

Adhesion of acrylic latex (AC-35) film to surfaces
(film dried 30 min at 100 °C)

Surface	Latex alone	5% SDZ in latex
Glass	0	3
Ceramic tile	0	3
Tin-coated steel	1	3
Cold-rolled steel	0	3
Stainless steel	0	3
Chromium plated steel	0	3
Aluminum	0	3
Titanium	0	3
Galvanized steel	1	3
Magnesium	1	3
Polyester laminate	1	3
Epoxy laminate	1	3
Phenolic laminate	1	3
Mylar film	0	3
Nylon film	1	2

0 = Poor dry adhesion.
1 = Good dry adhesion, but poor wet adhesion.
2 = Good dry adhesion, fair wet adhesion.
[3] = Good dry adhesion, good wet adhesion.

XD-5157L. From 1% to 2%, Z-6020 seemed to be adequate to impart wet adhesion to AC-35 films dried at 100 °C.

Acrylic latex AC-35 with 5% modifier (1% Z-6020 in XD-5157L) was filmed on various surfaces as indicated in Table VII. Films dried at 100° had outstanding adhesion on all surfaces tested.

A comparable Zwitterion resin precursor derived from a low molecular weight novalak was also very effective as an additive when modified with a trace of silane.

A thermoplastic Zwitterion resin XD-8156L related to XD-8157L did not impart wet adhesion to AC-35 even with 2% Z-6020 modification.

3. Discussion

Adhesion of plaster to glass, primed with silane-modified latexes, is a rapid test to discern the complex interplay of water-soluble silanes with polymer particles in a latex. Conditions that allow the silane to reamin in aqueous solution, or to deposit only on the latex particle surface are demonstrated by air dried films that impart instant water-resistant adhesion of plaster to glass.

Slow migration of condensing siloxanols into the latex particles result in loss of primer effectiveness as the latex ages. Methyltrimethoxysilane and Z-6020 have the optimum solubility characteristics for most latex systems.

After the primer film dries, there is a slow reorientation of the silane-modified film as shown by plaster adhesion. Methylsiloxanes migrate to the glass surface to cause improved adhesion. This migration may take several months at room temperature, but is accomplished in a few minutes at 100°C. Migration of hydrolyzed Z-6020 in dried films tends toward silane modification of the polymer air interface in such a manner as to be detrimental to plaster adhesion.

A product, XZ8-5080, comprising 2% Z-6020 in Dow Latex 209 is recommended as a general instant primer for plaster. Wet plaster applied to virtually any surface primed with 2–5% dispersion of XZ8-5080 solids will develop a strong water-resistant bond. The primer may also be useful in adhesion of thermoplastic rubbers to mineral surfaces.

Our theory of adhesion of flexible polymers to mineral surfaces was useful in predicting methods of obtaining water-resistant adhesion of flexible polymer latexes. Mobile, low molecular weight resin precursosrs appear to mix with a typical acrylic latex to give stable emulsions that deposit the resin precursor preferentially on a mineral surface to give a resinous primer action without impairing the flexibility of the acrylic film. With proper silane modification the added resin precursor imparted good wet-adhesion as well as initial dry adhesion.

References

1. Brown, L. H.: in R. S. Myers and J. S. Long (eds.), *Treatise on Coatings*, Vol. 1, Part III, Chapter 13, Marcel Dekker Inc., N. Y. 1972.
2. Buening, R. and Koetzsh, H. J.: *Angew. Makromol. Chem.* **13**, 89 (1970).
3. Thames, S. F. and Evans, J. M.: *J. Paint Tech.* **43** (No. 558), 48 (1971).
4. Plueddemann, E.: *SPI 24th Conf. on Reinforced Plastics*, Paper 19-A, 1969.
5. Plueddemann, E.: *Stevens Institute of Tech. Symposium on Adhesive-Bonded Structures* (1972), *J. Paint Tech.* **42**, No. 550, 600 (1970).

PART II

BIOMEDICINE

THE ROLE OF MOLECULAR WEIGHT IN
PHARMACOLOGIC AND BIOLOGIC ACTIVITY
OF SYNTHETIC POLYANIONS

W. REGELSON, P. MORAHAN, and A. KAPLAN

Division of Medical Oncology, Depts. of Medicine, Microbiology and Surgery
Medical College of Virginia, Virginia Commonwealth University, Richmond, Va., U.S.A.

Abstract. Pyran Copolymer (Divema, NSC46015*), a synthetic polyanion, divinyl ether maleic anhydride copolymer, has achieved clinical trial because of antitumor activity and its role as an anti-viral
agent. Most of the previous biologic and clinical work with this agent has been done with pyran copolymer
of an average molecular weight of 18 000. (NSC46015).

Different molecular weight fractions of pyran copolymer have been made available by Hercules.[†] Biologic experience with these fractions in mice indicates that one can separate toxicologic and physiologic
effects by close control of the molecular weight in the samples studied. In general, the higher the molecular
weight, the more toxic pyran is in mice as measured by increased lethality, decreasing body weight and a
related increase in liver and splenic mass as well as in increase in lung weight and decrease in thymic
weight. Toxicologically, the higher molecular weight fractions are associated with increasing sensitivity of
mice to the lethal action of bacterial endotoxin, as well as interference with the metabolism of barbiturate
and aminopyrine. Increases are seen in serum glutamic pyruvate transaminase activity related to administration of higher molecular weight fractions, but whether this is true enzyme induction or hepatic toxicity
has not yet been established.

Higher molecular weight fractions inhibit phagocytosis; in contrast, lower molecular weight fractions
stimulate phagocytic response. These observations, we feel, explain the biphasic effect on phagocytic response reported for the 18 000 molecular weight material (NSC46015) that has been clinically tested. The
concept of a rebound response for this agent and other macromolecular complexes is, therefore, a result
of biologic summation of molecular weight distribution of low and high molecular weight fractions in a
given polymer sample.

Immunologic stimulation, measured by increases in circulating antibody to sheep red blood cells is seen
for the high molecular weight pyran samples, but more work has to be done in this area with lower molecular weight fractions. Dose timing in regard to antigen administration is critical to the response generated
and this has to be examined for each fraction studied.

While anti-tumor activity is seen for all molecular weight fractions studied, of clinical importance to
anti-tumor activity, the therapeutic index improves as a consequence of the utilization of lower molecular
weight fractions which have potent anti-tumor activity with minimum toxicity.

Polyanions are derived from natural and synthetic sources. This paper is concerned
with those polyanions that enter into biologic function through distribution or entry
throughout the host. These polyanions are similar to certain proteins, glycoproteins,
or polynucleotides which modulate a variety of biologic responses related to host
defense reactions, including resistance to virus, bacterial, protozoal, and fungal
infection. Immunologic and hormonal responses, inflammation, wound repair, blood
clotting, and tissue damage are also subject to the action of these macromolecules. In
this regard, in some ways polymers mimic the action of infecting organisms and thus
modify their action on the host.

Polymers can also interact with the host through blocking or activating enzymes,

[†] Provided by David Breslow, Hercules, Wilimgton, Delaware, and the Drug Development Branch of the
National Cancer Institute, Bethesda, Maryland.

and hormones, or combining with toxic materials; in this manner their role in biology is of increasing importance. However, whatever benefits are to be gained in the biologic use of polymers must be balanced by their acute or delayed toxicity which we are now finding can be a function of molecular weight.

The action of polyanions as mitotic inhibitors and their functional role in growth processes has recently been reviewed by us (Regelson, 1974a, b; 1973a; 1970a; 1968a, b) as has the role of polynucleotides in immunology (Plescia and Braun, 1968; Braun and Firshein, 1967) and virus resistance (Beers and Braun, 1971). Recently, Norby (1971) has confirmed earlier work that heparin possesses growth regulating activity in tissue culture through effects on mitosis and death rate. There has also been related but independent research of the inhibitory effect of heparin and other anticoagulants on the growth and spread of tumor metastasis and the penetration of chemotherapeutic agents into the tumor environment (Regelson, 1974a; 1968a, b; Elias et al., 1973).

Another group of polyanions with biologic activity are polyphosphates, which are present in bacteria. In *Corynebacterium diphtheriae*, polyphosphate accumulation correlates with the cycle of cell division. The action of uranium and divalent cations in inhibiting the glucose metabolism of yeast and bacteria is beleived to be mediated via complexing with polyphosphates at the cell surface (Rothstein and Meier, 1951). Polymers of phosphoric acid have detergent activity, as they are able to form stable soluble complexes with most cations (Gasselin et al., 1952). Pharmacologically, systemic administration of polyphosphates in mammals shows toxicity similar to that of polysulfonates on systemic administration (Ebel, 1959), but their systemic degradation (Ebel, 1959; Mattenheimer, 1956; Gassalin et al., 1952) may make them less toxic on prolonged administration. Calcium pyrophosphate is phagocytosed by macrophages, and can produce lysosomal breakdown with subsequent synovial or mesothelial inflammation resembling that seen in gout ('pseudogout') (Riddle et al., 1966). This has been found to be pertinent to joint destruction and inflammation, and similar phagocytic uptake of polyanions and phagosomal leakage of enzymes could be responsible for tumor desruction which has also been seen for polyphosphates.

Of recent interest to both detoxification and therapeutic specificity, adriamycin, a cationic antibiotic that intercalates into DNA, has been complexed to deoxyribonucleic acid with enhancement of clinical antitumor activity. The theory behind this mechanism relates to increased endocytic uptake with subsequent lysozomal breakdown of DNA in tumor cells resulting in release of the active adriamycin (Trouet and DeDuve, 1973). This concept of complexing a biodegradable polyanion to any biologically active cationic compound for changes in pharmaco-kinetics has great clinical promise.

Apart from the above effects on living cells, polyanions have been shown to effect viruses. Heparin was used in the crystallization and isolation of tobacco mosaic virus, and a number of polyanions have been shown to inhibit viruses *in vitro* and *in vivo*. However, systematic searches for effective antiviral polyanions utilizing selective enzyme inhibition or stimulation have had little success. Recently, the polyanion, pyran copolymer, has been shown to inhibit RNA dependent DNA polymerase

(Chirigos, 1973), while pyran and polyethylene and polystyrene sulfonate are potent stimuli to ribonucleic acid polymerases (Miller, Berlowitz, and Regelson, 1972). In regard to antiviral action, what seems to be important is an effect on cell surface (Vaheri, 1964). There is also the little understood area regarding the inhibition or enhancement of the virulence of viruses by polyanions *vis-à-vis* polycations (Regelson, 1970; 1968a; Vaheri, 1964). Of tremendous clinical potential is the prolonged protective action of synthetic polyanions when given prior to virus inoculation (Merigan and Finkelstein, 1968; Morahan, Regelson, and Munson, 1972). This has established impetus for assaying the fundamental role for polyanions in controlling host resistance to a variety of pathophysiology.

1. Antitumor Activity of Polyanions

Significant to the development of rationale chemotherapy was the systematic study of anionic dyes by Paul Ehrlich and his co-workers. Trypanocidal dyes, such as isamine blue, showed clinical antitumor activity which was initially related to stimulation of phagocytic 'pyrrhol' cells.

Studies with heparin, a native acid mucopolysaccharide, primarily developed for its anticoagulant and lipolytic activity (important to the prevention of atherosclerosis), showed the drug could also inhibit inflammatory and immunologic processes. Heparin's ability to alter the isoelectric point of protein provoked speculation (Fisher, 1936) that it interfered with the nutritional role of protein. Fisher found that heparin inhibited the proliferation of yeast, as well as regeneration of cell syncytial injury *in vitro*. The anticoagulant effect of heparin and its possible effects on calcium binding was also related to inhibition of tumor growth (Regelson, 1968a, b).

Our interest in this field began with attempts to develop synthetic or naturally derived substitutes for native heparin which possess antimitotic activity (heparinoids). These also include inorganic polyphosphates, heteropolymolybdates, (Carruthers and Regelson, 1963) and phosphotungstates (Jasmin *et al.*, 1974). Clinically, these compounds (polysulfonates, pyran copolymer and related polycarboxylic maleic anhydride (MA) copolymers), which were to serve as lipolytic agents or anticoagulants, produced alopecia, ulceration of the G.I. tract, and osteoporosis leading to pathologic fractures. Testing of these heparinoids for antitumor activity has led to more recent systematic clinical trial of two agents: polyethylene sulfonate and the polycarboxylic polymer, pyran copolymer or Divema, derived from divinyl ether maleic anhydride (NSC 46015). Pyran is still under clinical investigation.

Pyran copolymer or Divema, derived from divinyl ether maleic anhydride (NSC 46015) induces low levels of the antiviral substance interferon and can either suppress or stimulate the functional activity of the phagocytic cells of the body (the reticuloendothelial system) (Regelson, 1967, 1968a, b; Merigan and Regelson, 1967). Pyran also alters the immunologic response, increases resistance to a wide variety of microorganisms (*S. aureus*, *D. pneumonia*, *C. neoformans*, and *T. duttoni*), (Regelson, 1967, 1970b; Regelson and Munson, 1970a, b; Regelson *et al.*, 1970a, b; Munson *et al.*,

1970a, b), and most recently, has been shown to enhance vaccine effectiveness (Campbell and Richmond, 1973). In addition, pyran exhibits anticoagulant properties by inhibiting the conversion of fibrinogen to fibrin (Shamash and Alexander, 1969; Roberts, Regelson, and Kingsbury, 1974). In a similar fashion to polystyrene sulfonate or polyethylene sulfonate, pyran can inhibit deoxyribonuclease II, induce hydrophilic gels with isolated nuclei (Tunis and Regelson, 1965), displace histone from nuclear DNA, activate RNA polymerases (Miller and Berlowitz, 1973; Miller, Berlowitz, and Regelson, 1971, 1972), and inhibit RNA dependent DNA polymerase (Chirigos, 1973).

The search for effective antitumor synthetic heparinoids led us to evaluate the relationship of activity to molecular weight of synthetic polyanions derived from ethylene maleic anhydride (E/MA) and polyacrylic acid (Regelson *et al.*, 1960a, b). The higher the molecular weight of the polymer, the more toxic the E/MA copolymer was in both the mouse and dog. Optimum activity was obtained in the series where carboxamide and ionizable carboxyl groups were interdispersed along the polymer backbone. Monomers were inactive and when all carboxyl groups were converted to carboxamides, significant tumor-inhibitory activity was lost. These results are similar to those that have been seen for the antiviral action of these compounds. The most effective regimen was prophylactic treatment. Adequate treatment of mice with E/MA polymers prior to tumor inoculation resulted in inhibition of tumor growth, even when tumor was inoculated as long as one week following drug injection (Regelson, 1967, 1968b).

The inhibitory effect seen on treatment prior to tumor implantation is similar to that observed for agents which non-specifically stimulate host resistance, such as yeast, plant polysaccharides or Bacillus Calmette-Guerin (BCG). Pharmocological evaluation of E/MA fractions of MW 1200–30000 in dogs showed them to possess toxicity that precluded clinical trial.

In addition to divinyl ether maleic anhydride copolymers, such as NSC 46015 (pyran copolymer, Divema) and E/MA synthetic polyanions, a number of polycarboxylic antitumor polyanions have been synthesized by others (Guile, 1962; Butler, 1964; Hodnett, 1969; Ottenbrite, 1974).

2. Pyran Copolymer

The Drug Development Branch of the National Cancer Institute and our own laboratories demonstrated antitumor activity for divinyl ether maleic anhydride copolymers. Samples of polycarboxylic divinyl ether maleic anhydride copolymer hydrolysate (NSC 46015; pyran 3,4-dicarboxylic anhydride, tetrahydro-2-methyl-6-(tetrahydro-2,5-dioxo-3-furyl) of average MW 17000 to 36000 show a broad spectrum of antitumor activity against transplanted tumors and a good therapeutic index.

We have found that pyran copolymer induces production of interferon in both mice and man (Regelson, 1967; Merigan and Regelson, 1967). Pyran has minimal *in vitro* toxicity to tumor cells in contrast to alkylating or other chemotherapeutic agents (Morahan *et al.*, 1974). However, *in vivo* treatment given prior to inoculation of mice with Friend leukemia virus, Lewis lung and the allogeneic sarcoma 180 or Ehrlich ascites tumor, significantly inhibits subsequent tumor growth (Regelson, 1967; Morahan et al., 1974). Similar inhibition of Friend leukemia virus splenomegaly is obtained with methyl vinyl ether and styrene/MA copolymers. Pyran also inhibits tumors induced by Rauscher leukemia virus, polyoma virus, and Gross leukemia virus in normal or thymectomized, immunosuppressed mice (Chirigos et al., 1969; Chirigos, 1970; Hirsch et al., 1973). Pyran treatment of mice beating the LSTRA lymphoma produces apparent cures if the tumor burden is reduced by prior treatment with an alkylting agent (Chirigos, 1973).

Pyran administration after tumor implantation also inhibits the growth of the first generation transplant of the mammary carcinoma (Sandberg and Goldin, 1971), a chemically induced sarcoma (Kapila, Smith, and Rubin, 1971), the Lewis lung and B16 melanoma syngeneic tumors (Morahan et al., 1974) and delays or reduces the onset of dimethylbenzanthracene (DMBA) tumors in the hamster cheek pouch (Regelson and Elzay, 1973). This antitumor action of pyran is not correlated with phagocytic stimulation or interferon inducing capacity (Regelson, Munson and Wooles, 1970; Regelson et al., 1970), but in vitro work indicates that the antitumor action may correlate with an increased macrophage-killing capacity (Kaplan, Morahan, and Munson, 1973).

The route of pyran administration is important, because intravenous pyran can stimulate Friend leukemia virus activity in contrast to intraperitoneal pyran, which is inhibitory (Schuller, Morahan, and Snodgrass, 1973). Paradoxically, certain murine sarcoma virus induced tumors are stimulated by pyran as well as other interferon inducers (Gazdar et al., 1972). This stimulation may be related to the ability of pyran to induce hepato-splenomegaly or splenomegaly in normal mice.

Hepatic and renal damage represent the major toxicity seen at high dosages in the dog and monkey. In mice, ^{14}C labelled pyran of average MW 17000 is lost from blood within 24 hours; at 5 weeks there is 30% residual activity shared between liver and spleen. At 3 months < 10% of its activity is present (Regelson and Munson, 1970a, b; Regelson, Munson, and Wooles, 1970). One sample of average molecular weight (17000–22500) has been in clinical trial in man and its use is limited by occasional serious acute vascular effects. Thrombocytopenia, which is reminiscent of poly-ethylene sulfonate, is also seen (Regelson and Holland, 1962) and hemolysis has been reported in one case (Leavitt, Merigan, and Freeman, 1971). Pyrogenicity at a dose of 12 mg/kg/day has been seen in most patients; similar observations have been shown for polynucleotides in animals (Merigan and Regelson, 1967). Cytoplasmic inclusions following pyran administration have been found in all the nucleated elements of the blood as well as in the phagocytic reticuloendothelial cells of liver, spleen, and bone marrow (Regelson, 1967). The anticoagulation effects are due to interference with fibrin formation (Shamash and Alexander, 1969; Roberts, Regelson, and Kingsbury, 1974), but bleeding has not been a problem.

In an attempt to improve the therapeutic index of pyran copolymer, we have completed studies which indicate that molecular weight can alter the biologic response to pyran with the maintenance of antitumor and phagocytic stimulating activity independent of major toxicity. We feel that these observations will be applicable to experience with other polyanions.

3. The Role of Molecular Weight in Biologic Activity

The pyran copolymer preparation approved for clinical investigation (NSC 46015) was prepared by a peroxide-catalyzed polymerization in benzene, using carbon tetrachloride as a chain-transfer agent to control the molecular weight. This has led to the production of a polymer with a broad molecular weight distribution, which in samples available in the laboratory and clinic has averaged between 17 000 and 23 000 daltons. The current clinical material, available from the National Cancer Institute, has an average molecular weight of 22 500.

Under the direction of Dr David Breslow of Hercules, lower molecular weight samples were provided in an effort to maintain antitumor and antiviral biologic activity with improved therapeutic index (Regelson, Munson, and Morahan, 1973). A variety of fractionation, degradation and synthetic techniques were utilized to produce active antitumor samples of different molecular weights. Intrinsic viscosities and weight-average molecular weights were determined and compared with known polystyrene standards. The techniques of synthesis of these active pyran samples of different molecular weight with different biologic activity have recently been reported (Breslow, Edwards, and Newbury, 1973).

4. Experimental

In the study of the biologic activity of pyran copolymer molecular weight samples, different toxic effects were examined. These included body weight, organomegaly, anemia (Hgb), white blood cell count, sensitization to endotoxin, changes in serum glutamic pyruvate transaminase (SGPT) activity, changes in reticuloendothelial system (RES) function as determined by the rate of carbon clearance, and effects on drug metabolism. The effects of differing molecular weights on toxicity were related to effects of these polymers in modulating host responses to tumor, virus infection and other antigens. With decreasing specific viscosity, the acute intravenous LD_{50}'s in mice increased from 75 to 150 mg/kg.

Molecular weights ranged from 2 500 to values above 50 000. All fractions of pyran (Divema) were compared to a standard parent compound, XA124-177, which has a molecular weight of 32 000 and was prepared in the same way as the clinical NSC 46015 (MW 22 500).

The data presented here relate to the pyran series X176265-67-(1-13). Fractions 1–4 range from 2 500–14 900 daltons and the molecular weights of fractions 5–13 range from 16 000 to > 750 000 daltons. As is to be expected, different series of pyran samples have shown some variation in biologic response, but results of this, and other work reported elsewhere (Breslow, Edwards, and Newbury, 1973), support the contention that molecular weight is the major factor determining the biologic activity of polyanions.

5. Materials and Methods

Pyran solutions were prepared by solubilizing the compound in 0.15 M NaCl in a 60° water bath and subsequently raising the pH to 7.2–7.4 with 1 N NaOH.

Mice used were uniform as to age, sex and weight. Mice used in the toxicologic studies were NYLAR mice provided by the New York State Virus Research Laboratories. C57B1/6 and BDF$_1$ mice were used in the macrophage and immunoadjuvant studies, respectively.

6. Antiviral Activity

Mice were inoculated i.v. with pyran and challenged i.v. 24 hours later with 50 LD$_{50}$'s of encephalomyocarditis (EMC) virus. The EMC virus (ATCC strain VR129) was prepared as a 10% brain homogenate from infected weanling mice. Deaths were recorded daily and percent mortality calculated (Morahan, Regelson, and Munson, 1972; Morahan *et al.*, 1972).

7. Immunoadjuvant Activity

Antibody producing cells to sheep erythrocytes (SRBC) were measured by the hemolytic plaque assay as modified by Cunningham and Szenberg (1968). Mice were inoculated i.p. with 5×10^8 SRBC on day 0, and with 25 mg/kg pyran (NSC 46051) i.v. on days -1, 0, 1 or 2 with respect to SRBC and the splenic plaque forming cells determined on day 3, 4, 5 and 6 post SRBC inoculation.

8. Phagocytic Activity

Mice were inoculated i.v. with a single injection of the pyran sample, 25 mg/kg and at intervals thereafter were injected i.v. with C11/1431 Pelikan carbon. The method used was that of Biozzi *et al.* (1953), wherein colloidal carbon was injected in the tail vein at a dose of 160 mg/kg and serial samples taken for optical density readings. The phagocytic index, *K*, of the carbon in the vascular system was calculated as previously described (Regelson and Munson, 1970; Regelson *et al.*, 1970; Munson *et al.*, 1970).

9. Sensitization of Mice to Endotoxin

Mice were inoculated i.v. with the pyran sample and were given an i.v. challenge 24 hours later with various doses of *Salmonella typhosa* 0901 lipopolysaccharide (Difco). Deaths occurring within 72 hours were recorded. The LD$_{50}$'s of endotoxin in treated and control mice was determined by the Cornfield-Mantel modification of Karber's method (Munson and Regelson, 1971).

10. Inhibition of Microsomal Mixed Functional Oxidase Enzymes *In Vitro*

Mice were inoculated with the polyanion and livers removed 24 hours later. The $9000 \times g$ liver supernatant fraction from five individual livers was incubated for 30 minutes with the substrates together with all the necessary cofactors. The metabolism of aminopyrene to formaldehyde or of aniline to *p*-aminophenol was measured as previously described (Munson, Regelson, and Munson, 1972).

11. Statistical Evaluation

Means with standard deviations and standard errors (SE) were calculated for all data, and the Student 't' test was used to determine significant differences from control values.

12. Results

Table I summarizes changes in whole body weight 1 week after a single i.v. injection of 25 mg/kg of pyran fractions. Untreated mice showed an average weight gain of 2.1 g. Mice receiving the lower molecular weight fractions, 2, 3 and 4 at 25 mg/kg, showed normal weight gains. All other fractions produced marked weight losses.

All fractions induced at least a slight leucocytosis. However, fraction 10 and 12 increased the WBC about 300%. Hemoglobin levels one week post pyran administration were within the normal values for mice treated with fractions 2, 3 and 4 but there was a significant decrease in hemoglobin levels in mice treated with fractions 5–12.

Fractions 2–6 caused no significant increase in liver weight while fractions 8–12 produced hepatomegaly. In regard to spleen weights, control spleens represented 0.34% of the body weight. Mice treated with fractions 2 and 3 showed no significant splenomegaly while fractions 4, 5 and 6 showed a slight increase in spleen weight. Increasing molecular weight fractions 8, 10 and 12 induced marked increases in spleen size. Treatment with fractions 2, 3 and 4 caused no significant alterations in lung weights, while fractions 5, 6, 8, 10 and 12 caused increases in lung weight between 55% and 86%. Thymic atrophy was most pronounced and consistent in the higher molecular weight fractions.

No changes were seen for fraction 2, 3 and 4 in serum glutamic pyruvic transaminase (SGPT) values at 48 hours after drug administration. The higher molecular weight fractions 5–11 and the parent XA124–177 showed significant SGPT elevations. At this time, we cannot say if SGPT elevations are a function of a liver toxicity related to cellular injury and enzyme leak, or a manifestation of true enzyme induction which could be concomitant with stimulating effects on liver mass. In addition, since all SGPT elevations were done at 48 hours after drug administration, it is possible that the lower molecular weight samples showed SGPT effects but of earlier onset.

In relation to effects on drug metabolism, fractions 1–4 showed no effects on the *in vitro* metabolsim of aminopyrene. Fraction 5 showed a 7.1% inhibition, fraction 6 a 30% inhibition and the remaining fractions inhibited microsomal enzymes by a striking 40 and 60 percent.

TABLE I

Pyran

(Divema) fractions	XA124-177	Saline	1	2	3	4	5	6	7	8	9	10	11	12	13
Body weight changes (grams)	ND	+2.1	ND	+1.6	ND	+1.8	+0.2	−2.6	ND	−0.5	ND	−1.7	ND	−3.0	ND
Liver weight (% body weight)	7.1	6.1	ND	6.5	6.7	6.5	6.5	6.6	ND	8.0	ND	8.0	ND	7.5	ND
Spleen weight (% body weight)	1.35	0.34	ND	0.40	0.42	0.63	0.71	0.59	ND	1.46	ND	1.15	ND	0.99	ND
Lung weight (% body weight)	1.29	0.76	ND	0.96	1.15	0.89	1.15	1.31	ND	1.22	ND	1.21	ND	1.43	ND
Thymus weight (% body weight)	0.15	0.24	ND	0.18	0.20	0.22	0.20	0.16	ND	0.18	ND	0.16	ND	0.12	ND
Hemoglobin levels	12.8	16.8	ND	16.4	16.5	16.3	15.3	14.8	ND	12.2	ND	11.7	ND	13.5	ND
WBC/mm^3 × 10^3	9.1	5.5	ND	9.1	9.9	7.3	10.8	9.1	ND	10.6	ND	14.4	ND	13.3	ND
SGPT sigma frankel units	^{a}200.0	32.0	ND	46.0	36.0	52.0	87.0	93.0	ND	96.0	ND	205.0	ND	84.0	ND
Antipyrine metabolism % inhibition	ND	0	0	0	0	0	7.1	29.5	51.8	56.5	46.0	53.0	40.0	45.0	49.5
Carbon clearance rate (K)	ND	0.045	0.073	0.055	0.055	0.075	0.028	ND	ND	ND	ND	ND	0.007	ND	ND
Endotoxin sensitivity LD$_{50}$ (S. typhosa, mg/kg)	0.038	15.0	15.0	15.0	15.0	15.0	3.0	0.38	0.48	0.48	0.48	0.48	0.48	0.82	1.4
Antitumor activity % inhibition, Ehrlich carcinoma	50.0	0	41.0	43.0	49.0	61.0	45.0	45.0	67.0	ND	32.2	ND	75.0	ND	73.0
Antiviral activity % mortality (EMC virus)	ND	90.0	90.0	90.0	80.0	20.0	30.0	10.0	10.0	20.0	20.0	20.0	ND	10.0	20.0

Note: ND indicates not done.

Certain immunoadjuvants such as Bacillus Calmette-Guerin (BCG), the live attenuated tuberculosis vaccine, pertussis vaccine, and antitumor agents as diverse as 5-fluorouracil and vincristine have been shown to sensitize mice to the lethal action of bacterial endotoxins. For this reason, pyran was tested for endotoxin sensitization. Fractions 1, 2 and 3 were not able to sensitize to endotoxin, as seen in LD_{50}'s for endotoxin of greater than 15 mg/kg. Fraction 5, however, did sensitize mice but not to the same extent as fractions 6–13. Fractions 6–11 sensitized mice to endotoxin to the same extent as the parent compound, XA124-177 (Table I). Thus, there appeared to be a correlation throughout these various parameters of increasing toxicity with increasing molecular weight.

In regard to phagocytic stimulation, lower molecular weight fractions 1, 2, 3 and 4, all of which were below 15000 in MW, showed primary stimulation of phagocytosis as measured by carbon clearance at 24 hours. The higher molecular weight fractions resembled the parent compounds NSC 46015 or XA124-177 and showed a biphasic phagocytic response with the initial inhibition followed by an increase back to the normal level or a stimulation in 4–5 days. Fraction 5 represents the first fraction which produced phagocytic depression, and all fractions above this molecular weight showed a similar depression of phagocytosis when evaluation was made 24 hours post pyran administration (Table I). The data suggests that toxicity might be related to the ability of certain molecular weight species of pyran to inhibit phagocytosis.

The first fraction showing antiviral activity was fraction 4, and all subsequent higher molecular weight fractions protected mice from the lethal action of EMC virus (Table I). In contrast, all fractions studied showed effective antitumor activity against the allogeneic Ehrlich solid tumor. The data indicates that there might be different mechanisms of antitumor and antiviral activity for pyran.

In other work, the antitumor activity of pyran may involve macrophage activation. Peritoneal macrophages harvested from C57B1/6 mice 7 days after inoculation with pyran (25 mg/kg i.p.) are cytotoxic in vitro to Lewis lung or B16 melanoma cells while normal macrophages have no effect on cell growth. These pyran activated macrophages are also capable of destroying normal Balb/c 3T3 cells but to a lesser extent than the tumor cells. However, practically no activity is detected against the WI-38 human lung fibroblast cell strain or C57B1/6 secondary mouse embryo cells. Macrophages and not lymphocytes are responsible for tumor kill in this system; we have shown that peritoneal exudate cells depleted of macrophages by treatment with carbonyl iron are not cytotoxic to tumor cells in vitro.

Other studies regarding the mechanism of immunoadjuvant activity of pyran have shown that the drug delays and depresses cell mediated cytotoxicity to the syngeneic DBA/2 P815 mastocytoma in C57B1/6 mice (Baird et al., 1973), and does not increase the level of cytotoxic antibody to this allogeneic tumor antigen. This data fits previous experience which shows that pyran can block immunoadjuvant induced arthritis (Kapusta and Mendelson, 1969). In contrast, we have now confirmed that pyran greatly enhances both the primary and secondary specific antibody response to sheep erythrocytes, as measured by hemolytic plaque formation when one inoculation of 25 mg/kg pyran is given intravenously in a single injection from 1 day prior to 2 days after antigen.

13. Discussion

Pyran fractions of lower molecular weights appear to maintain antitumor activity with better therapeutic index. Toxicity as manifested by weight loss, anemia, leukocytosis, hepatosplenomegaly, prolongation of type I (aminopyrene) and barbiturate metabolism, and endotoxin sensitization could be reduced or eliminated. Reducing molecular weihgt also prevented the 24–48 hour rise in SGPT. Most importantly, the biphasic phagocytic effect (phagocytic inhibition followed by stimulation) produced by the clinical pyran appeared to be a summation effect of the distribution of both high and low molecular weight fractions included in the parent sample, NSC 46015 (Regelson, Munson, and Morahan, 1973; Drummond and Munson, 1973; Breslow et al., 1973). These results were obtained with different molecular weight species of pyran obtained by one fractionation procedure. Variations seen utilizing other fractionation procedures suggest that molecular weight is only one of the variables which effects the activity of pyran, but is by far the predominating biologic effect.

The clinical pyran copolymer has an average molecular weight of 22 500 but it is a mixture of a variety of both high and low molecular weight fractions. The data reported here emphasize the fact that we must now look at the biologic action of pure molecular weight polyanionic species in order to definitively show a structure-activity relationship. This is particularly important to phagocytic stimulation versus inhibition for these polymers, as well as specific toxic or therapeutic effects. From these efforts, we should be able to develop low molecular weight species of defined chemical nature that should possess antitumor activity and macrophage activating capacity with minimal toxicity.

In relation to these observations as discussed previously, pyran can both inhibit and stimulate phagocytosis depending on molecular weight. Oligonucleotides and poly I:C can have similar activity (Regelson et al., 1970a; Freedman and Braun, 1965; Regelson et al., 1970d). Polymethacrylic (PMA) fractions (Van Bekkum and Ross, 1972), sulfated polysaccharides (Sasaki, 1967), and E/MA copolymers can also mobilize leukocytes; PMA fractions ranging from MW of 25000 to 1 000 000 have been termed lymphocyte mobilizing compounds and are under study for antitumor and radiation effects.

It has also been found that RES phagocytic stimulation combined with the use of chelating agents acts to accelerate the loss of plutonium from liver parenchymal and phagocytic cells. In this regard, glucan interfered with the increasing aggregation of plutonium that develops with time in intoxicated mice. Other phagocytic stimulating agents have similar activity (Rosenthal et al., 1967, 1972). Heparin and other polyanions have been shown to distinguish radiated tissue from non-radiated through displacement of nucleoprotein from binding to DNA (Matyasoua et al., 1971).

In regard to the mechanism of antitumor action for pyran and related polyanions, much work has been done in relation to the interferon inducing capacity of pyran.

However, interferon levels do not relate to antitumor or antiviral effects (Regelson, 1968b, 1970; Regelson and Munson, 1970; Regelson, Munson, and Wooles, 1970; Morahan, Regelson, and Munson, 1972).

Pyran has been shown to specifically stimulate 19 S antibody formation as measured by Jerne plaquing of splenic cells (Braun *et al.*, 1970; Regelson *et al.*, 1970b; Baird *et al.*, 1973), and paradoxically aborts adjuvant induced arthritis in rats (Kapusta and Mendelson, 1969), and cytotoxic antitumor action of lymphocytes against the allogeneic tumor cells. The effect of pyran on maintaining the resistance of thymectomized mice to tumor inducing viruses may be meaningful to these observations (Hirsch *et al.*, 1972), but these relationships are unclear.

Pyran resembles the synthetic polynucleotides in regard to its action on the host immune response. The action of poly rI:rC or the more potent polyadenosine: uridine (poly rA:rU) copolymers on immunologic response has been extensively reviewed elsewhere (Beers and Braun, 1971), and we have found similarities and differences between pyran and the polynucleotides. The action of poly rI:rC in relation to immunologic and antiviral effects is only partly related to molecular weight (Morahan *et al.*, 1972a, b), and these differences may relate to systemic biodegradability.

Of great clinical interest as discussed before is the action of pyran as an enhancer of vaccine activity at concentrations which are not toxic to the mice under study, and at levels which have nothing to do with a direct or indirect antiviral effect (Richmond and Campbell, 1973). Similar results have been seen for tungsten-antimoniate (Jasmin *et al.*, 1974). One possible mechanism for this activity of polyanions may relate to coupling of the polyanion to tumor antigen. Soluble antigens incorporated into highly cross-linked synthetic polymers form insoluble antigens, which complex with antibodies (Kent and Slade, 1959; Bernfeld and Wan, 1963) and may heighten the localizing activity of antibody; this could effect tumor growth.

Our preliminary *in vitro* work has shown that macrophage activation can in large measure distinguish tumor (Lewis lung, B16 melanoma) from normal (3T3, secondary mouse embryo and WI-38) cells. This selective enhancement of macrophage-killing capacity may be important to pyran's mechanism of antitumor action. However, pyran also enhances the level of circulating antibody to some antigens and inhibit cellular immunity to allogeneic tumor cells. The interrelationships of the effect of pyran on macrophages, T cells, and B cells need to be elucidated before an immunologic mode of antitumor action can be established. Investigations also should be performed on the role of these polymers as activators or inhibitors of the serum complement system which could also be important for modulating antitumor effects.

In conclusion, all of us should be encouraged to look for a useful place for polyanions in regard to host resistance to tumor growth. This report demonstrates that certain polyanions can be clinically useful when toxicity is removed. Polyanions are controlling factors in cell physiology, and their clinical potential in the control of pathophysiology is only in its infancy.

Acknowledgements

This work was accomplished with the help of grants: N.I.H. CA10537-04 and Hercules, Inc.

References

Baird, L. G., Kaplan, A. M., and Regelson, W.: *Proceedings of the 10th Annual Meeting of the Reticuloendothelial Society*, Williamsburg, Va., 1973.

Beers, R. F. and Braun, W.: *Biological Effects of Polynucleotides*, Springer-Verlag, N.Y., 1970.

Bernfield, P. and Wan, J.: *Science* **142**, 678 (1963).

Biozzi, G.: *British Journal of Experimental Pathology* **34**, 441 (1953).

Braun, W. and Fershein, W.: *Bacteriol. Rev.* **31**, 83 (1967).

Braun, W., Regelson, W., Yasima, Y., and Ishizuka, M.: *Proc. Soc. Exp. Biol. Med.* **133**, 171–175 (1970).

Breslow, D. S., Edwards, E. I., and Newbury, N. R.: *Nature* **246**, 160 (1973).

Butler, G. B.: personal communication, 1964.

Campbell, C. H. and Richmond, J. Y.: *Infec. and Immun.* **7**, 199–204 (1973).

Carruthers, C. and Regelson, W.: *Oncologia* **16**, 101–108 (1963).

Chirigos, M. A.: *Bibl. Haemnt.* **36**, 278 (1970).

Chirigos, M. A.: personal communication, 1973.

Chirigos, M. A., Turner, W., Pearson, J., and Griffin, W.: *International Journal of Cancer* **4**, 267 (1969).

Cunningham, A. J. and Szenberg, A.: *Immunology* **14**, 599 (1968).

Drummond, D. and Munson, A.: *Proceedings of the 10th Annual Meeting of the Reticuloendothelial Society*, Williamsburg, Va., 1973.

Ebel, J. P.: *Bull. Soc. Pharm. (Strasbourg)* **2**, 21 (1959).

Elias, E. G., Sepulveda, F., and Mink, I. B.: *Cancer Chemotherapy Reports*, 1972.

Fisher, A.: *Protoplasma* **26**, 344–350 (1936).

Freedman, H. H. and Braun, W.: *Proc. Soc. Exptl. Biol. and Med.* **120**, 222–225 (1965).

Gasselin, R. E., Rothstein, A., Miller, G. F., and Berke, H. L.: *J. Pharmacol. Exptl. Therap.* **106**, 180 (1952).

Gazdar, A. F., Stienberg, A. D., Spahn, G. F., and Brown, S.: *Proceedings of the Society of Experimental Biology and Medicine* **139**, 1132 (1972).

Guile, R.: personal communication, 1962.

Hirsch, M. S., Black, P. H., Wood, M. L., and Monaco, A. P.: *Journal of Immunology* **111**, 91 (1973).

Hodnett, E. M., Purdie, N., Dunn, W. J., III, and Harger, J. S.: *J. Med. Chem.* **12**, 1118–1120 (1969).

Kapila, K., Smith, C., and Rubin, A. A.: *Journal, RES* **9**, 447–450 (1971).

Kaplan, A. M., Morahan, P. S., and Munson, J. A.: *Proceedings of the 10th Annual Meeting of the Reticuloendothelial Society*, Williamsburg, Va., 1973.

Kapusta, M. A. and Mendelson, J.: *Arthritis, Rheum.* **12**, 463–471 (1969).

Leavitt, T. J., Merigan, T. C., and Freeman, J. M.: *Am. J. Dis. Child.* **121**, 43–47 (1971).

Mattenheimer, H.: *Z. Physiol. Chem.* **303**, 107 (1956).

Matyasoua, J., Skalka, M., Cejkova, M., and Blazicek, G.: *Folia Biologia (PRAHA)* **17**, 328–339 (1971).

Merigan, T. C. and Regelson, W.: *N.E.J.M.* **277**, 1283–1287 (1967).

Merigan, T. C. and Finkelstein, M. S.: *Virology* **35**, 363–374 (1968).

Miller, G., Berlowitz, L., and Regelson, W.: *Chromasoma* **32**, 251–261 (1971).

Miller, G. J., Berlowitz, L., and Regelson, W.: *Exp. Cell Res.* **71**, 409–421 (1972).

Miller, G. J. and Berlowitz, L.: *Cell Biol.* **59**, 2269 (1973).

Morahan, P. S., Regelson, W., and Munson, A. E.: *Antimicrobial Agents and Chemotherapy* **2**, 16 (1972).

Morahan, P. S., Munson, A. E., Regelson, W., Commerford, S. L., and Hamilton, L. D.: *Proceedings of the National Academy of Science, U.S.A.* **69**, 842 (1972).

Morahan, P. S., Munson, J. A., Baird, L. G., Kaplan, A. M., and Regelson, W.: *Cancer Research* **34**, 506–511 (1974).

Munson, A. E., Regelson, W., and Wooles, W.: *J. Reticuloendothelial Soc.* **7**, 366–374 (1970a).

Munson, A. E., Regelson, W., Lawrence, W., Jr., and Wooles, W. R.: *J. Reticuloendoth. Soc.* **7**, 375–385 (1970b).

Munson, A. E. and Regelson, W.: *Proc. Soc. Exp. Biol. Med.* **137**(2), 553–557 (1971).

Munson, A. E. and Regelson, W.: *Proceedings of the Society of Experimental Biology and Medicine* **137**. 553 (1971).

Munson, A. E., Regelson, W., and Munson, J. A.: *Journal of Toxicology and Applied Pharmacology* **22**, 299 (1972).

Niblick, J. F. and McCreary, M. B.: *Nature New Biology* **233**, No. 36, 52–53 (1971).

Norrby, K.: *Virchows Arch. Abt. B.Z. Hpath.* **9**, 292–319 (1971).

Ottenbrite, R.: personal communication, 1974.

Plescia, O. J. and Braun, W.: *Nucleic Acids and Immunology*, Springer-Verlag, New York, 1968.

Regelson, W. and Holland, J. F.: *Nature* **181**, 46–47 (1958).

Regelson, W., Tunis, M., and Kuhar, S.: *Acta Intern. Contra. Can.* **IVI**, 729–734 (1960a).

Regelson, W., Kuhar, S., Tunis, M., Fields, J., Johnson, J., and Glusenkamp, E.: *Nature* **186**, 778–780 (1960b).

Regelson, W.: *Advances in Exp. Med. Med. Biol.* **1**, 315–332, Plenum Press, 1967.

Regelson, W.: *Advances in Cancer Res.* **11**, 223–287 (1968a).

Regelson, W.: *Advances in Chemother.* **3**, 303–371, Academic Press, N.Y., 1968b.

Regelson, W.: *L'Interferon* **6**, 353–379 (1970a).

Regelson, W., Munson, A. E., Wooles, W. R., Lawrence, W., Jr., and Levy, J.: *L'Interferon* **6**, 381–395 (1970b).

Regelson, W., Munson, A. E., and Wooles, W. R.: *J. Reticuloendoth. Soc.* **7**, 366–374 (1970c).

Regelson, W. and Munson, A. E.: *N.Y. Acad. Sci.* **173**, 831–841 (1970d).

Regelson, W., Munson, A. E., and Wooles, W.: *Symp. Series Immunobiol. Standard.* **14**, 227–236, Karger, Basle, N.Y., 1970e.

Regelson, W. and Munson, A. E.: *Ann. N.Y. Acad. Sci.* **173**, 831–841 (1970f).

Regelson, W.: in W. M. Bikales (ed.), *Water Soluble Polymers* **II**, 161–177, Plenum Press, N.Y. – London, 1973a.

Regelson, W. and Elzay, R.: Work in progress (1973).

Regelson, W., Munson, A. E., and Morahan, P. S.: *Symposium on Polyelectrolytes and their applications*, Pasadena, California, 1973.

Regelson, W.: *J. Med.* **5**, 50–68 (1974a).

Riddle, J. M., Bluhm, G. B., and Barnhart, M. I.: *Proc. Intern Symp. Atherosclerosis Reticuloendothelial System Como.* Sept. 57, Plenum Press, New York, 1967.

Roberts, P. S., Regelson, W., and Kingsbury, B.: *J. of Lab. and Clin. Med.* **22**, 822 (1973).

Rothstein, A. and Meier, R.: *J. Cell Comp. Physiol.* **38**, 245 (1951).

Sandberg, J. and Goldin, A.: *Cancer Chemotherapy Reports* **55**, 233 (1971).

Schuller, G. B., Morahan, P. S., and Snodgrass, M. J.: *Proceedings of the 10th Annual Meeting of the Reticuloendothelial Society*, Williamsburg, Va., 1973.

Shamash, Y. and Alexander, B.: *Biochem. Biophys. Acta* **194**, 449–461 (1969).

Shalka, M.: *Int. Atomic Energy Agency*, Vienna, 1971, 95–103.

Trouet, A.: *Second Workshop, Macrophage Activation Conference Hoechst*, October 25–26, Reisenburg, Germany, 1973.

Tunis, M. and Regelson, W.: *Exp. Cells Res.* **40**, 383–395 (1965).

Vaheri, A.: *Acta Path. Microbiol. Scand.*, 171 (1964).

Van Bekkum, D. W. and Ross, W. M.: in S. Garratini and G. Franchi (eds.), *Int. Symp. on Chemotherapy of Cancer. Dissemenation and Metast.*, Milan, Italy, May 1972, Raven Press, New York, 1973.

HEPARIN AND RELATED POLYELECTROLYTES

L. B. JAQUES

*W. S. Lindsay Professor, Haemostasis-Thrombosis Research Unit, Dept. of Physiology,
College of Medicine, University of Saskatchewan, Saskatoon, Sask., Canada S7N 0W0*

Abstract. Heparin is a sulfated polysaccharide found in certain mammalian tissues in trace amounts. Heparinoids are related sulfated polysaccharides, generally prepared semisynthetically. The sulfated mucopolysaccharides (chondroitin sulfates) are chemically and biologically distinct. The combination of a strong anion and a carbohydrate core make these substances important biological polyelectrolytes which form complexes with many substances. Heparin and heparinoids are the most effective of these in changing the properties of proteins, enzymes, and dyes. The wide range of biological effects thus produced will be indicated, as also the considerable specificity given by the structure of the carbohydrate core. The various biological activities of heparins and heparinoids have been related to various components – accompanying low molecular weight substances, protein or polysaccharide; acid polysaccharide groups, other linkages, molecular weight, combined effect of the major component with a trace material, ability of the main component to provide a clathrate or association complex. The metachromatic color change produced in dyes such as toluidine blue may serve as a model for the association complexes responsible for biological effects produced by heparins and heparinoids.

The polyelectrolyte nature of heparin appears to be responsible for difficulties in its standardization as a drug. The heparin unit is 1/130th of a mg of the International Standard Heparin Preparation. On isolation of heparin from various sources as the crystalline barium salt, identity of gross chemical composition but marked differences in potency by biological assay are evident and the relative assay values are changed in different assay systems. A recent International Collaborative study of the assay of heparin organized by the World Health Organization has shown wide variation in assay values between laboratories due to 'differences in substrate'. A large series of commercial heparins studied extensively over a period of years have shown considerable variations in chemical composition and in potency values by different methods of biological assay. The crystalline barium salt and the pmr spectrum distinguish heparin from mucopolysaccharides. The repeating unit of the main component of commercial heparin is sulfoglucosamine-sulfoiduronic acid (not glucuronic acid as generally given). No correlation has been found in any of the chemical values measured and biological assay values.

Heparin has been used in medicine and surgery for nearly 40 years and in this time has maintained a well-earned reputation as an effective and safe drug [1]. I may claim a share in this as I was a member of the research team at the University of Toronto which developed heparin for clinical use. The major clinical uses of heparin are for the prevention of thrombosis and the prevention of clotting of blood. Thrombosis is the complex plugging of blood vessels which can occur in veins after operation and child-birth and can occur in arteries as the result of diet, age and stress. Clotting of blood is a serious problem in the use of heart-lung machines, artificial kidneys, etc. and the prevention of this by heparin is most important. My presentation reviews points about this drug which indicates its actions are due to its polyelectrolyte nature.

Heparins are mucopolysaccharides isolated from mammalian tissue, which can be isolated and identified as crystalline barium salts with a well-defined chemical composition (12 percent S, 2 percent N, $[\alpha]_D^{20} + 55°$); extremely soluble in water (> 1 gm/ml), insoluble in organic solvents, relatively low viscosity in aqueous solution; with exceptional ability to combine with proteins to alter their physical and biological properties and to produce metachromatic colors with basic dyes.

Heparinoids are sulfated polysaccharides, prepared semisynthetically by sulfuration of partially degraded polysaccharides, or occurring naturally in plant or animal tissue; possessing the properties described above for heparin – high sulfate content, solubility in water, insoluble in alcohol, etc., complexing with proteins, and metachromatically with dyes.

Mucopolysaccharides are naturally occurring carbohydrate polymers containing *glucosamine* and usually a uronic acid; share to some degree the properties listed for heparin and heparinoids; less sulfur, much weaker metachromatic and anticoagulant activities, but show negative optical rotation, differ in solubility characteristics, aqueous solutions show considerable viscosity.

In contrast to the neutral polysaccharides, the carbohydrate skeleton of heparin and mucopolysaccharides is based on an amino sugar uronic acid repeating unit. An important biological difference is that the neutral polysaccharides (glycogen, starch) are metabolic stores, the typical mucopolysaccharides (chondroitin sulfuric acids, keratin sulfate) are important *structural materials* of connective tissue, and the naturally occurring heparins and heparinoids are trace substances and appear to be associated with special cells.

1. Complexing Properties of Heparin, Heparinoids, Mucopolysaccharides

A property shared by all these substances is the ability to form complexes with many other substances [2]. Thus A. Fischer demonstrated complex formation between heparin and casein visually by observing the precipitation of the protein at the isoelectric point. With casein, this occurred at pH 5.0. When heparin is added to casein, no precipitation occurs at 5.0, but a new isoelectric point is observed at 3.0, i.e. the complex has different properties as a protein. Complexing occurs for heparin with small molecules, such as inorganic salts and simple organic bases. Benzidine, cetylpyridinium chloride (C.P.C.), cetyl trimethyl ammonium bromide (Cetavlon) and other long chain amines form insoluble complexes with heparinoids and mucopolysaccharides and are universally used as precipitating agents for these substances.

One of the most distinctive properties of these compounds is metachromasia. Certain dyes change color and this change of color is termed metachromasia. Azure A is blue in dilute solution. Addition of heparin or dextran sulfate to this changes it to red-purple. This color change is suppressed by alcohol, heating and by electrolytes.

Heparins and heparinoids form complexes with proteins and bases, and as shown by the ability to produce metachromasia with submicro quantities are very effective complexing agents in trace amounts. Hence, these substances in trace amounts affect many biological agents such as enzymes, etc. When heparin or a heparinoid is injected in animal or man, an enzyme appears in the blood plasma, lipoprotein lipase. When an oil emulsion is incubated with varying amounts of blood plasma obtained after injection of heparin, the oil is cleared. The heparin has caused the release of this enzyme from tissues to the blood. Heparin and heparinoids have a pronounced action on many enzymes – proteolytic enzymes, carbohydrases, etc. and may inhibit, activate, block active groups of enzyme, of substrate, increase stability or instability and can inhibit enzymic synthesis. Further, for a given enzyme, there are great differences quantitatively and qualitatively in the action of different heparinoids (see [2] for review).

2. Standardization and Variability in Heparin Preparations

For use, drugs must be standardized. This means comparison of batches of the drug with a standard preparation. The standard for heparin is the International Standard Heparin Preparation. This was prepared originally for the League of Nations Commission on Standards and is now administered by the World Health Organization. By definition, a unit of heparin is 1/130 mg of the International Standard Heparin preparation. For most purposes, 0.01 mg of heparin represents one unit.

If a drug can be prepared to constant chemical and biological properties, comparison with the International Standard and hence standardization is relatively easy. Heparin was prepared as a crystalline barium salt by Charles and Scott in 1937. Since crystallization has long been known to be an excellent procedure for obtaining substances with a reproducible degree of purity, one would have thought that this would have become a required procedure for producing heparin for clinical use as is the case for insulin, thyroxin, etc. However, this did not happen. Actually with the barium salts of heparin, it was possible to demonstrate the difficulties in biological standardization for this drug.

TABLE I

Analytical data for crystalline barium salts of lung heparin calculated for the equivalent anhydrous free acid

Source	% S	% N	$[\alpha]_D^{22}$	Potency u/mg[a]			
				MA	Ac	AT	USP
Beef	13.8	2.5	+55°	150	154	153	122
Dog	14.0	2.9	+56°	158	341	225	171
Pork	14.0	2.2	+53°	147	66	75	81
Sheep	14.7	2.6	+53°	166	31	63	59
Dermatan sulfate	5.5	2.6	−52°	39	4	2	2

[a] Directly compared with International Standard Heparin. MA=metachromatic; Ac= =anticoagulant activity on fresh whole blood; AT=inhibition of clotting of blood by thrombin; USP=U.S. pharmacopeia XV assay of inhibition of clotting of sheep plasma. Modified from [3].

The crystalline barium salt of beef heparin proved to have a potency of 100 units/mg. One-third by weight of this salt is base and water and variation in percent of base and of water can be responsible for differences in other values for heparin. Hence it is better to express the values in equivalent anhydrous free acid. When I did this for heparins prepared as the crystalline barium salt from beef, dog, pork and sheep tissue (Table I), there was close agreement for figures for S (14%), N (2%) and optical rotation $[\alpha]_D^{20°} + 55°$. Compare these with the figures for chondroitin sulfate B – S, 5.5% and $[\alpha]_D^{20°}, - 52°$. This was also true for metachromatic potency – mean value of 155 u/mg. However, this was not the case when assays were done for anticoagulant activity. Measured on the clotting of fresh cat whole blood as done by Professor Howell (AC), values for these heparin preparations were from 30 to 350 units/mg. Measured by the prevention of the clotting of ox blood by thrombin (AT) values were from 63 to 153 u/mg. Measured by the prevention of clotting of sheep plasma (USP assay) values were from 60 to 120 u/mg..

Hence, the apparent anticoagulant activities of heparin preparations vary with the type of test used for measuring anticoagulant activity. If potency values are to be assigned to heparin preparations on the basis of a Standard International Heparin prepared from beef heparin it is necessary to specify the assay procedure to be used. Until recently, this was not a problem. Almost all heparin was prepared from beef lung and essentially the same product was made by all manufacturers. However, in the past fifteen years, manufacturers have been changing to pork intestinal mucosa as a cheaper source. This led to W.O.H. organizing an International Collaborative Study of the assay of heparin by 13 laboratories including our own, in which a portion of six coded samples were sent to each laboratory for comparison and determination of relative potency. The raw data was returned to London for examination by the WHO statisticians and reported by them (Bangham and Woodward [4]). All laboratories returned results showing agreement within $\pm 2\%$ for the same heparin preparation compared with itself under code. A wide variation in values was obtained between laboratories even by the same assay procedure when one heparin preparation was compared with another. This was found for all comparisons and pairs and appears to be due to 'differences in substrate' in different laboratories. Hence, this study only proved the limitations of assay methods now used for standardization of heparin.

Our laboratory has conducted extensive studies on a large series of commercial heparins obtained from a number of manufacturers over a period of years and we find that commercial heparins vary considerably in chemical composition [5, 6]. The range of values obtained from the elementary composition of 17 commercial heparins received in the past 15 years are shown in Table II. The range of values were: carbon – 18 to 23%, hydrogen, 3.6 to 5.5%, nitrogen 1.8 to 3.7%, sulfur and sodium 6 to 12%. For each element, values were found throughout the range, i.e. there was no clustering of values. Analytical values for the carbohydrate portion of the heparin preparations show even greater variations. The optical rotation is from $+33$ to $+54$. The uronic acid content determined directly by CO_2 liberation gives values of 23–41% but the carbazole color reaction which is almost universally used gives values up to 51%. Glucosamine determined colorimetrically gave values of 16 to 24% and as indicated two methods of analysis did not give identical values. Acetyl values are found to be from 0 to 5.5%. Cations are $8.7 \rightarrow 16.9\%$. The chief cation is sodium with minor contributions of potassium and ammonium in some preparations. Calculation of molar ratios from this data give values for uronic acid to glucosamine from 0.81 to 1.68 and for sulfur to nitrogen from 1.01 to 2.47. A range of 50% in the ratio S/N would suggest great differences in degree of sulfation. The range for the ratio of cations to anions shows that the heparins as supplied were neutral salts. Anticoagulant activity varied from 101–183 u/mg by an assay using plasma but 14–172 u/mg by an assay using fresh

TABLE II

Range of values of chemical analyses and relative potency by *in vitro* tests of 17 commercial heparin preparations

Analysis	Value (%)	Potency test	Potency (units/mg[b])
Carbon	18.2–24.6	U.S.P.	101–183 u/mg
Hydrogen	3.0– 5.3	Howell	42–172 u/mg
Nitrogen	1.9– 4.5	Metachromatic	63–137 u/mg
Sulphur	6.2–12.1	Microelectrophoresis	59–137 u/mg
Sodium	6.2–13.5		
Glucosamine –		*Microelectrophoresis*	
Elson-Morgan	19.6–24.0	One component	5 preparations
Idole HCl	15.9–21.6	Two components	12 preparations
Uronic acid			
by CO_2	17 –21		
by Carbazole	19.8–40		
Acetyl	none–5.54	*Molar Ratios*[c]	
Cations	8.7–13.8		
Volatile	3.5–23.0	S/N	=0.96–2.36
Total weight assigned[a]	81.1–109.4	CO_2/Carbazole	=0.55–1.06
$[\alpha]_D^{20°} +33.3 \rightarrow +54.1$		Uronic Ac/glucosamine	=0.81–1.68

[a] Total using values for uronic acid from CO_2 liberation, glucosamine by Elson-Morgan reaction, sulphur as SO_3^-.
[b] Relative to the International Standard Preparation of Heparin, No. 2.
[c] Molar ratio: S/N – sulphur nitrogen; CO_2/carbazole – uronic acid value by CO_2 liberation to that by carbazole with glucuronic acid standard; uronic acid/glucosamine – uronic acid to glucosamine. Values taken from [5] and [6].

whole blood. Even the colorimetric tests showed considerable variation. Most of the preparations showed two components on electrophoresis.

The concentration of heparin in tissues from which it is isolated is about 10–50 ppm. Hence, it is a trace substance. Further it combines easily with protein. Hence, its extraction from tissue economically has required considerable ingenuity. Methods for this involve the use of concentrated electrolyte solutions and proteolytic enzymes. Standards set for this compound do not appear to meet the problem presented in identifying this compound and this is responsible for the variability that occurs. This variability is important in standardizing heparin for clinical use [7, 8]. The variations seen in activity values by different tests indicate there are probably similar variations in activity of the preparations for different clinical uses – e.g. prevention of postoperative thrombosis, prevention of clotting in the heart-lung machine, etc.

Fortunately heparin is such a non-toxic drug that it can be given in large enough dosage to compensate for variations in activity.

3. Chemical Structure of Heparin

There has been considerable confusion regarding the chemical structure of heparin. This has been cleared up recently by a combined effort initiated in Saskatoon and involving Drs Peter Dietrich, Lois Kavanagh, Armin Wollin, in our department and Dr Arthur Perlin of the NRC Prairie Regional Laboratory (now at McGill University). Important to this was that we developed simultaneously four technical lines for study of heparin. These were: microelectrophoresis on agarose, nuclear magnetic resonance, bacterial enzymes, and difference spectrophotometry. We have shown [9, 10] that by conducting electrophoresis on agarose-coated microscope slides at 4 °C, it is possible to measure microgram quantities of heparin rapidly and accurately and also to carry out identification tests, etc. on the microscale. This has made possible further chemical and biological studies.

The second procedure which has proved very useful in our hands is nuclear magnetic resonance [11]. Infra-red spectra have long been used in connection with these compounds but give very little differentiation between heparin and mucopolysaccharides. However, proton magnetic resonance spectra have been found to give a finger-print for heparin. In Figure 1 is shown the pmr signals for two heparin prepara-

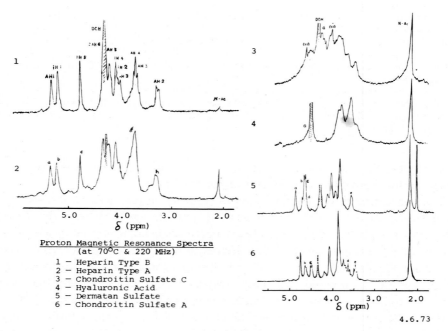

Fig. 1. *Proton magnetic resonance spectra of heparin and mucopolysaccharides.* Recorded at 220 MHz at 70 °C. (Adapted from Perlin *et al.* [12].)

tions compared to those obtained with typical mucopolysaccharides. While, as would be expected, there are some signals in common, the patterns are so different that it is easily possible to distinguish heparins and mucopolysaccharides. The heparin signals are a, b, c, d, e, f, g, h. There is also a small N-acetyl signal in some heparin preparations but not in others.

Flavobacterium heparinum is a microorganism which can be grown on heparin and cause its degradation. C. P. Dietrich [13, 14] in our laboratory showed that as the microorganism became adapted to the substrate, it developed a number of enzymes and that a series of products could be obtained from heparin. The fractions isolated with their constituents are shown in Table III. Heparin shows no reducing power for copper or periodate. Reducing sugar values increase with fractions D_1, C, B_1, B_2, indicating hydrolysis of glycoside bonds. Periodate reduction only appears with removal of sulfate. At the bottom of the table are given for comparison, values for glucosamine, glucuronic acid and glucosamine N–SO$_4$. It is evident that all the products are sulfated. The products correspond to a sulfated polysaccharide, hexasaccharide,

TABLE III

Fractions obtained from heparin by action of flavobacterium heparinum

Fraction	Molar proportions						Activity (i.u./mg)	
	Uronic acid [a]	Gluco-samine	Sul-fate	Re-ducing sugar	Per-iodate reduced	Sacch-aride	Meta-chro-masia	Anti-coagulant
Heparin	1.8	1.1	3.0	0.0	0.0	poly-	114	160
D_1	1.3	0.9	3.0	0.0	–	poly-	136	3.0
D_1	1.3	0.9	3.0	0.1	–	hexa-	130	2.5
C	1.2	1.0	3.0	0.2	–	tetra-	8	0.1
B_1	0.9	1.0	3.0	0.7	0.0	di-	0.4	0.1
B_2	1.0	1.1	2.0	0.8	2.0	di-	0	0.1
A	0.0	1.0	2.0	0.6	3.8	mono-	0	0.1
Gluco-samine	0	1.0	0.0	1.0	4.8	mono-	0	–
Glucu-ronic acid	1.0	0.0	0.0	1.0	5.0	mono-	–	–
Gluco-samine N–SO$_4$	0	1.0	1.0	0.8	5.0	mono-	–	–

[a] Uronic acid by carbazole color, using glucuronic acid standard.
Summary of data from Dietrich [14].

tetrasaccharide, tri-sulfated and disulfated disaccharides, and disulfated glucosamine. The metachromatic activity with toluidine blue is shown fully by the hexasaccharide. However, the anticoagulant activity is completely lost with the initial degradation step.

Dietrich, by using repeated enzymatic hydrolysis and conditions to suppress

152   L. B. JAQUES

sulfatase activity, was able to isolate over 70% of the original heparin as the trisulfated disaccharide. The pmr spectra of this was compared with those of unsaturated oligo- saccharides from chondroitin 4-sulfate and pectin and from some monosaccharide 4-deoxy-hex-4-enopyranosides. In Figure 2 is shown the spectra for the products obtained from heparin through the action of F1. heparinum – A, the 2.6 sulfated

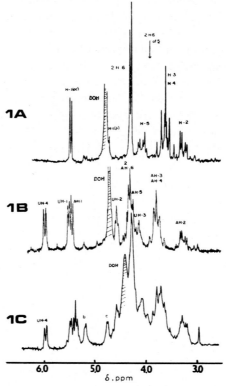

Fig. 2. *Proton magnetic resonance spectra of products obtained from heparin through the action of flavo- bacterium heparinum.* 1A – 2,6 sulfoglucosamine; 1B – trisulfated disaccharide; 1C – oligosaccharide. 100 MHz, 1000-Hz sweep width; A and B at 35°, C at 70°. (From Perlin *et al.* [15].)

glucosamine, B, the sulfated disaccharide from heparin, C, the sulfated oligosaccha- ride. U indicates the uronic acid and A, the amino sugar with the numbers indicating position in the pyranose ring. In the spectrum obtained from 2, 6 sulfoglucosamine (1A) signals are evident which have been identified as H-1, H-6, H-5, H-3, H-4, H-2. Additional signals are observed in the spectrum of the sulfated disaccharide (1B). These are due to the uronic acid and have been identified from type compounds as due to H-1, H-2, H-3 of 2 sulfo-iduronic acid. The heparin-degrading enzyme of F1. heparinum was shown by Linker to be an eliminase which introduced a double bond at C4–C5 of the uronic acid. This is responsible for the signal UH-4, for the vinylic 4-proton. While not in heparin, this double bond and its pmr signal were especially important as it made it possible to establish the configurational assignment of the glycosidic linkage in 1 as the α-linkage. Hence, the disaccharide isolated has been identified as having the structure shown in the upper part of Figure 3. A comparison of the pmr spectra of heparin shown in Figure 1 with those in Figure 3 indicates that all the pmr signals for heparin have been accounted for. The signal *a* can be recognized

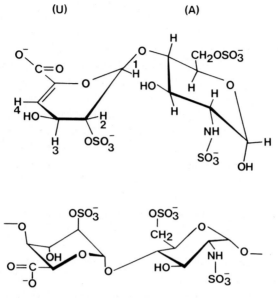

(U)   (A)

Fig. 3.   *Configuration of disaccharide unit in heparin.* Above is disaccharide produced by *flavobacterium heparinum* and below the disaccharide configuration in heparin. (1→4)-linked 4-o-(α-L-idopyranosyluronic acid 2-sulfate)-(2-deoxy-2-sulfoamino-α-D-glucopyranosyl 6-sulfate) biose residue. U = uronic acid, A = amino sugar. (After Perlin *et al.* [15].)

as corresponding to the proton of carbon 1 of the amino sugar, *g* to the protons of carbons 3 and 4, *h* to the proton of carbon 2 of the glucosamine, while *b* corresponds to H-1, *c* to H-5, *d* to H-4, *e* to H-3 and *f* to H-2 of 2 sulfo-iduronic acid. Hence, heparin is a polymer of the disaccharide shown in the lower half of Figure 3 with a repeating sequence of the biose residue (1→4)-linked 4-o-(α-L-iodopyranosyluronic acid 2-sulfate)-(2-deoxy-2-sulfo-amino-α-D-glucopyranosyl 6-sulfate). Further, Perlin and Sanderson [16] demonstrated that iduronic acid was the major uronic acid in heparin by classical degradation methods when they isolated considerable idose from desulfated reduced heparin. The fact that the hexasaccharide showed full metachromatic activity indicates that a chain of only three of these disaccharide residues is sufficient to provide the high metachromatic activity peculiar to heparin.

The array of acid groups on the two sides of the chain of residues of Figure 3 suggests immediately that this will be an effective ion exchange vehicle. Heparin preparations always have considerable accompanying water and base. Wilander [17] found that all the acid groups were titratable. But while heparin is completely dissociated as judged by the titration curve it gives a low osmotic pressure and exhibits the Hammarsten effect i.e. the freezing point of the calcium salt is much less than that indicated for the molarity of the calcium ion. In the determination of the activity coefficient (b/Na$^+$) of sodium ions in heparin using a highly cation selective membrane, Ascoli *et al.* [18] found the values low and nearly invariant with the concentration of heparin.

It has been an essential part of the art that whenever heparin is precipitated by adding organic solvents such as EtOH to aqueous solution, it is necessary to add a pinch of an inorganic electrolyte, usually NaCl. Some of this chloride can be found in some heparin preparations with, of course, a corresponding amount of sodium. Helbert and Marini [19] determined titratable acid groups on a heparin from porcine intestinal mucosa by hydrolysis with Dowex 50 (H$^+$) at 60°. Pretreatment with IRA-400 resulted in a drop of 11 percent in titratable acid groups. The exchangeable acid groups were not dialyzable and after removal could be replaced by adding sulfate. It is significant that higher sulfate values for heparin were reported for earlier preparations and the sulfate content of commercial heparins is now lower. Up to the 1950's the initial extraction of tissues for heparin was made with alkaline ammonium sulfate. Then, the initial extraction was changed to a proteolytic digestion of tissue. Ottoson and Snellman [20] found that when tissue was extracted with potassium thiocyanate with phosphate buffer, the final product had higher antithrombin potency and lower sulfur content than commercial heparin and the difference in sulfate was compensated for by phosphate. This indicates an exchange with inorganic salts can occur in the initial extraction of tissues.

Counterions neutralizing electric charges are distributed throughout the ionic atmosphere with a fraction so close to the polymer skeleton as to be 'bound'. that is a relatively mobile monolayer. The data indicates that heparin probably has a greater fraction of bound univalent cations and also indicates that the functional volume is relatively small. The concentration of cations in the polyelectrolyte phase is much greater in the outside solution. Donnan equilibrium data indicates the co-ions are excluded, but if small the co-ions, penetrate. Heparin can be changed from one 'salt' to another – e.g. barium to ammonium – by double decomposition. However, such double decomposition is not complete usually. Some sites of ion binding are less accessible than others. Some sites are so inaccessible that inorganic ions used in the initial extraction remain bound through the many chemical manipulations involved in purifying the compound. Organic anions and cations can be similarly bound to such sites. It is not surprising from this point of view then to find the considerable variation in commercial heparin preparations previously reported.

The polyelectrolyte nature of heparin has significance for determinations of its molecular weight. We have seen that molecular weight estimations are not practical by oncotic measurements. Measureable viscosity can be due to accompanying protein or polysaccharide. In my experience, viscous polysaccharide can be produced during the extraction procedure and traces of these can be responsible for considerable viscosity. The sites of ion binding and problems of double decomposition means that when organic cations of sufficient chain length are bound to the more inaccessible sites, they can result in the total molecule having a different molecular weight or total net charge. It is, therefore, not surprising that a wide range of molecular weights have been reported – from ultracentrifuge and diffusion studies values of 18 000, 17 000, 16 000, 19 700, 8000–14 000, and from sedimentation and light scattering studies 7600 to 11 800. Fischer reported dialysis of dog heparin through collodion membranes, suggesting a molecular weight of 4000. Problems presented are illustrated by Laurent's study of fractions obtained from a commercial heparin. The results are shown in Table IV. Laurent [21] prepared the CPC (cetylpyridinium chloride) of a commercial

TABLE IV

'Fractionation' of heparin with cetyl pyridinium chloride

(CPC added to heparin in 3.3 N $MgCl_2$. Dilution with water and centrifugation gave fractions at 2.10, 2.00, 1.80, 1.00 $MgCl_2$. Fractions were decomposed with potassium thiocyanate and precipitated with EtOH + NaCl)

	Anticoagulant[a] (u/mg)	N (%)	Hexosamine (%)	Ash (%)	Non-ash (%)	Mol. wt.[b]
Original	110	2.47	22.6	38.3	61.7	8000
Fraction A	138	2.32	23.3	36.5	63.5	11 800
Fraction B	112.5	2.23	20.7	37.3	62.7	9700
Fraction C	112	1.93	18.5	49.2	50.8	8400
Fraction D	96	1.84	17.4	50.8	49.2	7600

	On ash-free basis				S/D^c ($\times 10^7$)	$\bar{V}_{sp}$	Lim. visc. No. (ml/g)
Original	180	(197)	3.77	36.7	1.9	0.42	14.8
Fraction A	217	(223)	3.67	36.7	2.8	0.44	18.8
Fraction B	179	(177)	3.5	33.0	2.3	0.42	14.8
Fraction C	220	(189)	3.80	36.5	2.0	(0.37)	14.0
Fraction D	195	(160)	3.63	36.4	1.8	(0.36)	11.9
Mean	203	(187)	3.66	35.6			$\eta = 1.58$
S.D.	±19	(±27)					$\times 10^{-3}$ M

[a] Anticoagulant activity by method of Studer and Winterstein. Concentration based on nitrogen content (Na heparin $= 44.8 \times N$)
[b] Calculated using 0.42 for value of $\bar{V}_{sp}$.
[c] Archibald method.
Data from Laurent [21] are calculated.

heparin in 3.3 N $MgCl_2$ and collected the fractions precipitating on dilution to 2.10, 2.00, 1.80, 1.00 N $MgCl_2$. These were then decomposed with potassium thiocyanate and precipitated with EtOH plus NaCl. Laurent found the differences shown in the anticoagulant activity of the fractions. From ultracentrifuge data, he calculated that the fractions were of decreasing molecular weight and reported the anticoagulant activity decreased accordingly. However, there is a great difference in the ash content of the fractions (Table IV). If one calculates the anticoagulant activity on an ash-free basis, there is no significant difference in the specific activity of Fractions *A* to *D*. Mean $= 202 \pm 19$. Laurent calculated the concentration of heparin in the solutions measured from the nitrogen content. Correcting for this (values in brackets) does not change the conclusion that there is no significant trend.

In calculating sedimentation and diffusion coefficients and hence molecular weights, Laurent arbitrarily used a value for the partial specific volume of 0.42, arguing that Fractions *C* and *D* were so contaminated with inorganic matter as to render measurements of these fractions invalid. However, in doing so, he ignored that the original material and Fractions *A* and *B* were also contaminated with inorganic material. If the measurements on *C* and *D* are suspect, this must apply for *A* and *B* and to different degrees. The conclusion from Laurent's work must be either that the anticoagulant activity of heparin is not related to molecular weight, or that the calculations of molecular weight of the fractions are not valid and therefore no evidence is provided that anticoagulant activity is related to molecular weight. All physicochemical studies support Laurent's statement that heparin is a relatively small molecule of only 10–20 disaccharide units but do not allow us to determine the molecular weight sufficiently accurately to define chain length more exactly.

We can thus conclude that heparin is a polymer of the disaccharide, 6-sulfoiduronic acid and 2,6-sulfoglucosamine. The fact that the hexasaccharide gives full metachromasia suggests that the basic unit will consist of a chain of three disaccharides and that these can assume a coil shape. What is evident with this structure is that the two sides or planes of the spiral with the sulfate and carboxyl groups have an array which will hold the cloud of counter-ions found experimentally as reported in the literature. The different polar groups of these polymers bind water vapour simultaneously according to steric accessibilities as shown by infrared spectroscopy. With a molecular weight in the range of 4000–12000, the number of coils will be from two to six. While it is possible that heparins occur as a range of molecular weights, it seems to me that this assumption has been made so frequently using very questionable evidence that I propose the alternative assumption should be accepted until proved wrong – that the commercial heparin preparations at least represent a single molecular species. The high negative charge density and flexibility of the heparin molecule will allow matching by basic groups of proteins and complex bases for molecular association and thus explain the marked specificity of effects with wide spectrum of biological activities found [2].

A model [22] of how heparin acts specifically in many biological systems in modifying activities of complex ions may be provided by the metachromatic effect on dyes referred to earlier. The dye, Azure A, shows maximum light absorption at 610 nm. This is decreased when heparin is added, and a new absorption band at 505 mu develops. Heparins and heparinoids are able to produce this color change at very low concentrations and under conditions unfavorable to other metachromatic inducing substances. However, little attention has been paid to the numerous experimental observations reported on metachromasia with heparin and heparinoids, of practical importance to those using this color reaction in studies on heparin and mast cells. In previous publications (cf. [22]), we have demonstrated the following points regarding the phenomenon. The color change is definitely different from those shown by the same dye with acid-base and oxidation-reduction shifts. As is well-known, absorption spectra of the dye in solution show α- and β-bands, characteristic of the dye. Addition of heparin produces a μ-band at a shorter wave-length. The position of the μ-band is characteristic for the substance producing metachromasy, about 120 nm less in the presence of heparin-heparinoids, 95 nm for dermatan sulfate. The reaction can be seen with very low concentrations of heparin and heparinoids (1 μg/ml) in aqueous solution, on paper chromatograms and in agarose gel and shows a simple Mass Law relationship. The μ-band is suppressed by excess heparin but this required a concentration excess of three orders of magnitude. The reaction is suppressed by inorganic electrolytes and increasing temperature but suppression from an increase in ionic strength and hydrogen ion in the medium is much greater with other mucopolysaccharides than with heparin and heparinoids. Dehydration suppressed metachromasy with a shift of light absorption of the α-band. Products obtained from heparin by enzymatic degradation showed that the monosaccharide and disaccharide fragments reacted slightly with the dye with no metachromasy. Full metachromatic activity as compared to heparin, was given by the *heparin hexasaccharide*. Interaction between dye and polyanion which does not lead to metachromasia occurs in the precipitation of dye-polyanion complex from solution. This interaction is not destroyed by ethanol, dehydration or moderate concentration of electrolytes. The limiting concentration of electrolyte depends on the nature of the polyanion.

To explain the metachromatic phenomenon in a manner consistent with these observations, we have proposed combining previous alternative explanations – dye-stacking and pairing concepts – by suggesting that metachromasia is produced by a dye-dimer reacting with the polyanion in a loose manner. The presence of water molecules stabilizes the dimer unit. Due to the resonance in the dye molecules, one would expect a partial charge to form on each end of the dimer and allow loose interaction with the negatively charged heparin molecule. We believe the metachromatic band shift to be the result of this interaction via the amino groups which are part of the dye's chromophore. This interaction partially blocks the N-group in the thiazine dyes from being part of the conjugation system of the dye chromophore and thus reduces π-electron delocalization. The loss in electron mobility causes the dye molecule to absorb light energy at a shorter wavelength. The strength of the binding to the N-group is reflected in the band shift. We explain the exceptional metachromatic activity of heparin as due to the suitable arrangement of anionic sites in the loose coils for interaction with the dye-dimer. The dye-dimer with a partial positive charge on each end interacting with the negatively charged groups of the heparin coil can induce the new absorption band. Special features of the structural arrangement and the charge density associated with the coil are responsible for the high metachromatic activity of heparin and heparinoids observed. The strength of the bonding is responsible for the extent of the band shift. Interaction of dye and heparin, etc. without development of a metachromatic color is conceivably the result of an ion-pairing of dye and polyanion without dye-dye interaction.

Figure 4 illustrates some of these points. Differences were recorded between the absorption spectra of dye and of dye-metachromatic substance. Three types of spectral bands were clearly isolated, which corresponded to those of the absorption spectra of thiazine and related dyes. Michaelis named these α, β and μ-bands and attributed them to the dye monomer, dye dimer and metachromatic form of the dye. Addition of 0.5 M NaCl to a dilute azure A solution (Curve 1) reduced light absorption of the α-band with appearance of light absorption at the α-band, i.e. an increase in the concentration of the dye-dimer at the expense of the dye-monomer. Addition of 0.7 μg/ml of heparin (Curve 2) reduced light absorption for both α- and β-bands with

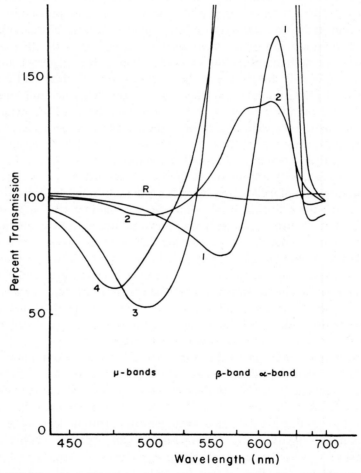

Fig. 4. *Spectral differences in light absorption between azure A and azure A with NaCl or heparin.* Spectra recorded with a double beam spectrophotometer (Beckman DK-2). A deflection below 100% transmission indicates increased light absorption in the sample cell; above 100% transmission, decreased absorption. Reference cell: azure A solution 10 μg/ml. Sample cell: R – azure A solution 10 μg/ml; 1 – plus 3.1 × × 10⁴ μg/ml (0.5 M) NaCl; 2 – plus 0.7 μg/ml heparin; 3 – plus 4.7 μg/ml heparin; 4 – plus 2.2 × 10⁵ μg/ml NaCl (3.8 M). (From Wollin and Jaques [22].)

light absorption now occurring at the μ-band. Increasing heparin concentration to 4.7 μg/ml (Curve 3) decreased further the α- and β-bands, and increased the light absorption in the μ-band with no shift in the metachromatic band. The formation of the μ-bands then occurred at the expense of both the dye-dimer and dye-monomer. It is probable that the dye species were utilized in sequence; i.e. the formation of the μ-band only removed the dye-dimer, and the dye-monomer concentration was reduced to maintain the concentration equilibrium.

Figure 4 shows a μ-band at 475 nm (Curve 4) with the dye in 3 M NaCl. It is evident from this that polyanions are not essential for metachromasy, as is usually thought. We suggest this band shift results from a great number of ions forming a cage around the partial charge of the dye-dimer, thus inducing an effect on the conjugation system similar to that produced by the polyanion. This again supports the idea that the unique ability of heparin to produce the matachromatic phenomenon in very high dilution is due to its unique local charge capacity or polyelectrolyte properties.

We conducted these studies on the metachromatic reaction, not only to find an explanation of the intriguing characteristic color reaction for heparin but also to provide a model of how the polyelectrolyte properties of heparin can affect trace substances and thus explain its marked activities as a trace substance in biological systems. The mucopolysaccharides serve as gross ion-exchangers for connective tissue. The mucopolysaccharides are weaker metachromatic agents, as shown by concentration required for a metachromatic effect, inhibition by electrolytes, and only a moderate absorption band shift. Both heparin and heparinoids are highly active metachromatic agents being effective in trace amounts. They also have marked effects in many biological systems, again in trace amounts. This is due to the higher charge density of heparin and heparinoids, which makes these substances ion-ex-changers for biological substances active in trace amounts (toxic amines, enzymes, constituents of cell membranes) and to be competitive for ATP [23]. The inhibition of blood coagulation is only one example of these powers limited to heparin isolated from pork, beef, dog and whale tissue. The polyelectrolyte nature of heparin is responsible for: (a) The difficulties in preparing a uniform, standard product for clinical use; (b) the difficulties that have been found in determining chemical structure and molecular weight; (c) the ability of heparin to produce in trace amounts many biological effects; (d) the remarkable lack of toxicity and the ability of heparin to act as a detoxifying agent.

As polyelectrolytes, heparin with the related heparinoids provide a unique group of drugs. There has been failure to make use of these drugs due to lack of understanding of their unique nature. A review of the extensive literature [2] has shown that the wide spectrum of biological activities of these compounds is accompanied by a high degree of specificity for individual compounds for individual biological actions useful for therapeutic purposes. Unfortunately, these compounds have been thought of as substitutes for heparin as an anticoagulant. As they are weak anticoagulants, they have been given in massive doses, causing toxic effects. Heparin has been administer-ed to patients for 35 years and has a unique record of freedom from toxic reactions.

This is evidently due to the ability of heparin to complex amines and proteins. Presumably, this will be true of heparinoids when they are used for specific purposes and, therefore, administered in trace amounts to produce the specific biological effect characteristic of the heparinoid used.

Acknowledgements

The experimental studies described in this paper have been supported by grants to the author from the Medical Research Council of Canada.

References

1. Jaques, L. B.: *Anticoagulant Therapy: Pharmacological Principles*, C. C. Thomas, Springfield, Ill., 1965, 174 pp.
2. Jaques, L. B.: '3. The Pharmacology of Heparin and Heparinoids', in G. P. Ellis and G. B. West (eds.), *Progress in Medicinal Chemistry* London, Butterworths, 1967.
3. Jaques, L. B. and Bell, H. J.: 'Determination of Heparin', *Methods of Biochemical Analysis* 7, 253–309 (1959).
4. Bangham, D. R. and Woodward, P. M.: 'A Collaborative Study of Heparins From Different Sources', *Bull. Wld Hlth Org.* 42, 129–149 (1970).
5. Jaques, L. B., Kavanagh, L. W., and Lavallée, A.: 'A Comparison of Biological Activities and Chemical Analyses For Various Heparin Preparations', *Arzneimittel-forschung* 17, 774–778 (1967).
6. Kavanagh, L. W. and Jaques, L. B.: 'A Comparison of Analytical Values for Commercial Heparin', *Arzneimittel-forschung* 23, 605–611 (1973).
7. Jaques, L. B.: 'Heparin From Different Sources', *Lancet Letter* 2, 1262 (1972).
8. Jaques, L. B. and Kavanagh, L. W.: 'Standardization of Heparin for Clinical Use', *Lancet Letter* 2, 1315 (1972).
9. Jaques, L. B., Ballieux, R. E., Dietrich, C. P., and Kavanagh, L. W.: 'A Microelectrophoresis Method for Heparin', *Can. J. Physiol. Pharmacol.* 46, 351–360 (1968).
10. Jaques, L. B. and Wollin, A.: 'Identification and Quantitation of Heparins by Microelectrophoresis on Agarose Gel', *Analytical Biochemistry* 52, 219–233 (1973).
11. Jaques, L. B., Kavanagh, L. W., Mazurek, M., and Perlin, A. S.: 'The Structure of Heparin From Proton Magnetic Resonance Spectral Observations', *Biochem. Biophys. Res. Commun.* 24, 447–451 (1966).
12. Perlin, A. S., Casu, B., and Sanderson, G. R.: '220 MHz Spectra of Heparin, Chondroitins, and Other Mucopolysaccharides', *Can. J. Chemistry* 48, 2260–2268 (1970).
13. Dietrich, C. P.: 'Studies on the Induction of Heparin Degrading Enzymes in *Flavobacterium Heparinum*', *Biochemistry* 8, 3342–3347 (1969).
14. Dietrich, C. P.: 'Novel Heparin Degradation Products. Isolation and Characterization of Novel Disaccharides and Oligosaccharides Produced From Heparin by Bacterial Degradiation', *Biochem. J.* 108, 647–654 (1968).
15. Perlin, A. S., Mackie, D. M., and Dietrich, C. P.: 'Evidence for a (1→4)-linked 4-O-(α-L-idopyranosyluronic acid 2-sulfate)-(2-deoxy-2-sulfo-amino-D-glucopyranosyl 6-sulfate) sequence in heparin. Long-range H–H coupling in 4-deoxy-hex-4-enopyranosides', *Carbohydrate Research* 18, 185–194 (1971).
16. Perlin, A. S. and Sanderson, G. R.: 'L-Iduronic Acid, A Major Constituent of Heparin', *Carbohydrate Research* 12, 183–192 (1970).
17. Wilander, O.: 'Studien Über Heparin', *Skandinavisches Archiev Für Physiologie, Suppl.* 15, 89 pp. (1938).
18. Ascoli, F., Botré, C., and Liquori, A. M.: 'On the Polyelectrolyte Behaviour of Heparin. I. Binding of Sodium Ions', *J. phys. Chem.* 65, 1991–1992 (1961).
19. Helbert, J. R. and Marini, M. A.: 'Structural Studies of Heparin. I. Hydrolytic Cleavage of Sulfates', *Biochemistry* 2, 1101–1106 (1963).
20. Ottoson, R. and Snellman, O.: 'Attempts to Prepare Heparin From Ox Liver Capsules With a Mild Method', *Acta Chem. Scand.* 13, 473–479 (1959).

21. Laurent, T. C.: 'Studies on Fractionated Heparin', *Arch. Bioch. Biophys.* **92**, 224–231 (1961).
22. Wollin, A. and Jaques, L. B.: 'Metachromasy: An Explanation of the Colour Change Produced in Dyes By Heparin and Other Substances', *Thrombosis Research* **2**, 377–382 (1973).
23. Cruz, W. O. and Dietrich, C. P.: 'Antihemostatic Effect of Heparin Counteracted by Adenosine Triphosphate', *Proc. Soc. Exp. Biol. Med.* **126**, 420–426 (1967).

EFFECTS OF IONENES ON NORMAL AND
TRANSFORMED CELLS

R. RAJARAMAN and D. E. ROUNDS

Pasadena Foundation for Medical Research, Pasadena, Calif., U.S.A.

and

S. P. S. YEN and A. REMBAUM

Jet Propulsion Laboratory, California Institute of Technology, Pasadena, Calif., U.S.A.

Abstract. Ionene polymers are polyammonium salts with positive nitrogens in the backbone, resulting from the polycondensation of diamines with dihalides or from the polycondensation of halo-amines. They combine with a variety of molecules of biological importance. The nature of binding and the resulting biological activity depends, to a considerable extent, on the structure and charge density of the polymer.

It has been suggested that malignant cells are more electro-negative than normal cells. If so, malignant cells should demonstrate a greater affinity for the electropositive ionene polymer. A series of studies were carried out to evaluate the degree of cellular adherence and spreading capacity of normal and SV40-transformed human cells (WI38) on normal glass surfaces and on glass which was coated with 3,3 ionene. The transformed cells showed a greater degree of adherence and spreading on the polyionene-coated surfaces than normal cells, thus supporting this basic hypothesis. That cell adhesion even on 3,3 ionene coated glass surface was inhibited by pretreatment of cells with sulfhydryl binding agent N-ethylmaleimide (NEM) indicated the involvement of sulfhydryl groups on cell periphery in the process of cell adhesion.

Different degrees of ionene binding suggested the possibility of a differential toxicity for normal and transformed W138 cells. Coulter counter analyses of 3,3- ionene-treated cells, following 3 days of incubation, showed only moderate differences in the growth or survival rates of the two cell types. However, when the study was repeated, using 6,10 ionene, it was observed that the ionene concentrations up to $4\mu g/ml$ stimulated the growth of normal cells, while this range showed inhibition and death of the transformed cells, possibly by electrostatic cytotoxic interaction. The results of this preliminary study appear to warrant more extensive studies to evaluate this class of polymers as chemotherapeutic agents.

1. Introduction

The presence of positively charged polymeric materials in biological systems is well known. These substances range from the small polycations such as spermine to large proteins like the histones and the DNA-dependent polymerases. The functional and structural roles of such DNA and membrane-bound polymers have been investigated extensively in recent years. The synthetic polymers known as ionenes, which have been well characterized, will contribute as a useful tool for the systematic investigation of these polycation interactions.

1.1. CHEMISTRY OF IONENE POLYMERS

Ionenes, also known as polyionenes, are polyelectrolytes with positively charged nitrogen atoms located in the backbone of polymeric chains. This type of polycation was first prepared by Gibbs *et al.* [1], from dimethylamino-*n*-alkyl halides. The generic name of ionenes was suggested for these salts since the reaction of diamines with dihalides to form polyammonium salts proceeds through ionization of amines [2]. Kern and Brenneisen [3] reported that ionenes are formed by the Menshutkin reaction from ditertiary amines and halides. When the scope of this polycondensation reaction, the effect of concentration and solvent on rates and molecular weights of ionenes and a mechanism was reported [4–7], it was possible under well defined conditions to obtain relatively high molecular weight ionenes. A light scattering study yielded a quantitative relation between viscosity and molecular weight for 3, 4 and 6, 6 ionene [7]. The polymeric reaction products of diamines and dihalides are shown below:

$$\left[\begin{array}{c} CH_3 \\ | \\ -N^+ \underline{Z^-} (CH_2)_x \end{array} \quad \begin{array}{c} CH_3 \\ | \\ -N^+ \underline{Z^-} (CH_2)_y \end{array}\right]_n$$

I

where Z is a halogen, x and y integers and n the degree of polymerization. By controlling the experimental conditions, the values of x and y could be varied up to 16. Thus, it was possible to synthesize a large variety of ionenes including crosslinked ionene networks.

1.2. PREVIOUS STUDIES ON BIOLOGICAL EFFECTS OF IONENES

Linear as well as cross-linked ionenes are endowed with interesting pharmacological properties. The biological effects of the ionene polymers were investigated and compared with the pharmacological properties of monomeric ammonium salts as well as with existing model compounds [8]. Several systems such as DNA, bacteria, fungi, mammalian cells in culture and laboratory animals were used in these studies.

1.2.1. *Interaction with DNA*

An aqueous solution of calf thymus DNA, when mixed with an aqueous solution of 6,10 ionene bromide resulted in a solid fibrous precipitate. This fibrous complex of DNA-ionene polymer exhibited a strong birefringence, which indicated that the helical configuration of the DNA was preserved [8]. The elemental analysis indicated an electrostatic bonding between the negative oxygen of the DNA phosphate groups and the positive nitrogens of the ionene polymer and the elimination of sodium bromide during the complex formation. Although different structures of ionenes yielded similar complexes, an examination of DNA molecular models showed that 6,10 ionene gave the best fit if one assumes that the ionene polymer wraps itself around the DNA double helix. Similar DNA complexes have been obtained with the natural polyamines [9] (spermidine, spermine and putrescine), which are ubiquitous in animal and plant cells and are believed to be associated with a metabolic regulatory role.

1.2.2. *Antimicrobial Activity*

The results of a preliminary screening of the activity of ionenes against a variety of gram positive and gram negative bacteria showed that the low molecular weight ammonium salts at 1000 ppm did not inhibit bacterial growth [8]. However, the various ionenes were very efficient in growth inhibition of these bacteria. The minimum inhibitory concentration of different ionene polymers to the growth of *Staphylococcus aureus* and *Escherichia coli* was generally within a few ppm (Table I). The data shown in Table I indicate that the structure of the ionenes with respect to the

TABLE I

Inhibitory concentrations[a] of different ionene polymers on *Staphylococcus aureus* and *Escherichia coli*

	S. aureus (ppm)	E. coli (ppm)
3,3-ionene bromide	128	128
6,6-ionene bromide	16	16
6,10-ionene bromide	4	4
2,10-ionene bromide	4	8
6,16-ionene bromide	4	32

[a] Similar values of minimum inhibitory concentration were found for *P. aeruginosa*, *B. subtilis*, *C. sporogenes* and *M. smegmatis*.

distribution of the positive charge plays an important role in the antimicrobial activity and that 6,10 ionene bromide appears to have the highest inhibitory effect on *E. coli*.

Insolubilized ionene polymers retain their antimicrobial properties. The insolubilization can be achieved either by exchanging the counterions or by an ionic cross-linking reaction [8]. Thus addition of potassium iodide, iodine aqueous solution to an aqueous solution of ionene chloride yields structure II

$$\left[\begin{array}{c} CH_3 \\ | \\ N^+ \xrightarrow{\ I_3^-\ } (CH_2)_x \end{array} \begin{array}{c} CH_3 \\ | \\ N^+ \xrightarrow{\ I_3^-\ } (CH_2)_y \end{array}\right]_n$$

II

which is insoluble in water and in most organic solvents. The bactericidal activity of the polyammonium salt is enhanced by a slow release of iodine. Thus impregnation of textiles with ionenes followed by treatment with aqueous solution of potassium iodide and iodine leaves under suitable conditions an antibacterial layer on various surfaces, e.g., surgical sutures, cloth etc.

1.2.3. *Toxicity Studies on Whole Animals*

Ionene polymers exhibit toxic effects when injected into mice intraperitoneally (i.p.). The i.p. lethal dose for 50% survival (LD_{50}) is, however, comparable to that of a number of clinically used drugs. Oral administration of ionene polymers was considerably less toxic than i.p. administration. Some of the low molecular weight ammonium salts were also toxic, few of them even at a lower concentration than their large molecular weight polymers [8].

1.2.4. *Ganglionic and Neuro-Muscular Blocking Action*

Some of the ammonium compounds such as tubocurarine, gallamine triethiodide, hexamethonium and decamethonium are clinically used against essential hypertension, the most common form of high blood pressure [10]. These compounds interact with synaptic junctions in ganglia and/or with the motor ends of nerve cells and probably act as inhibitors of the transmitter substance acetyl choline [11]. The activity of the ganglionic blocking agents is strongly dependent on their structure. The ionenes were found to act as more efficient curarizing agents (inhibition at neuromuscular junction) and have longer duration of effects than their monomeric analogues [12]. The less toxic 10,10 ionene when compared with its monomeric analogue showed considerably longer duration of ganglionic blocking action than the monomer. The increase in the duration of the blocking action may be a result of more secure binding to the receptors due to a larger number of positive charges in the polymers. These studies suggest that polymerization of the positive moieties may be an efficient way of increasing the duration of drug action.*

1.2.5. *Interaction with Heparin*

Ionenes form water insoluble complexes with heparin. The stoichiometry of these complexes depend on the structure and the molecular weight of the compounds. Complex formation is probably the reason for the antiheparin activity of ionenes. Investigations with 6,3 ionene bromide in laboratory animals showed that this was more toxic (intravenous (i.v.) LD_{50} 28 mg/kg in mice) than the clinically employed antiheparin agents protamine sulfate (i.v. LD_{50} 44 mg/kg) and toluidine blue (i.v. LD_{50} 45 mg/kg). However, cumulative i.v. doses of 6,3 ionene bromide up to 5 mg/kg as 1% solution could be given rapidly to anesthetize dogs without markedly affecting either respiration or circulation. Heparin was found to offer a protective action in neutralizing the toxicity of 6,3 ionene bromide in both mice and dogs. Thus pretreatment of mice with heparin enabled them to survive doses three times the LD_{50} values with only mild toxicity symptoms [13, 14].

2. Current Studies on Mammalian Cells in Culture

In the present studies, we have used mammalian cells (human and mouse cells) in culture with 3,3 and 6,10 polyionenes as molecular probes to understand the cell surface phenomena of cell adhesion and cell spreading on the substratum and evaluated the toxic effects of these two polycations on normal and transformed mammalian cells in culture.

* See also paper by J. Schmidt and M. A. Raftery, this book, p. 175.

The mammalian and avian cells have a net negative charge on their surfaces and this negative charge is known to be involved in several cell surface phenomena such as cell aggregation, cell motility, cell adhesion, membrane integrity, cell recognition, nerve excitability, cation transport and antigenic properties. Surface membrane sialic acid appears to be the major source of surface negativity. Several altered membrane characteristics like the presence of new antigens, embryonic antigens, agglutinability, altered response to physiologic regulators, lack of contact inhibition, impaired cell communication, altered permeability and change in electrophoretic mobilities are most noticeable in transformed cells [15]. This results in an impairment of the social interactions of cells evolving at cell surfaces inherent in malignancy. As a general rule, the net surface negativity increases in a transformed cell by about 20% in comparison with its normal progenitor (Table II). The difference

TABLE II

A survey of electrophoretic mobilities of normal and neoplastic cells

Cells	Mobility	References
Clone c13 Normal	-1.02 ± 0.06	16
Clone P Transformed type 1	-1.26 ± 0.06	
Clone Q Transformed type 1	-1.29 ± 0.05	
Clone S Transformed type 1	-1.25 ± 0.05	
Clone V Transformed type 1	-1.27 ± 0.06	
Clone X Transformed type 1	-1.23 ± 0.06	
Clone Y Transformed type 1	-1.28 ± 0.09	
Clone N Transformed type 1	-1.30 ± 0.06	
Adult mouse fibroblasts	-0.85 ± 0.03	17
Ehrlich mouse ascites carcinoma	-1.05 ± 0.05	
Ehrlich-Landschutz mouse ascites strain	-1.00 ± 0.02	
Ehrlich ascites	-1.14 ± 0.01	18
Sarcoma 37 ascites	-1.28 ± 0.04	
Mouse liver	-1.03 ± 0.02	

in the surface charge results in the differential cell-cell aggregation and also in differential cell-substratum adhesion, since the substratum, e.g., glass, also has a net negative charge.

Using 3,3 and 6,10 ionenes, three different kinds of experiments were carried out: (a) stimulation of cell adhesion and spreading, (b) inhibition of cell adhesion and spreading, and (c) toxicity studies with normal and transformed human cells in culture.

2.1. STIMULATION OF CELL ADHESION AND SPREADING

The rationale of this series of experiments was to employ the high positive charge density of ionenes to increase the rate of cellular adhesion and spreading onto a glass surface. Since the polyionenes have about 100 times more positive charges than the glass surface, they would bind to the glass surface and would still provide enough positive charges to interact with the negative charges in the cell surface. Thus a shift of charge in favor of positivity in the ionene polymer coated glass surface would accelerate cell adhesion and spreading. Furthermore, since the transformed cells have about 20% more negative charge than their normal counterparts, the transformed cells may adhere and spread faster than the normal cells to polycation-coated glass surfaces.

Normal human diploid WI-38 cells of passage 17–22 (kindly supplied by Dr Hayflick of Stanford University) and its SV40 transformed derivative WI-38VA132RA (SV40-WI-38), of passage 109–210 (kindly supplied by Dr Girardi of the Wistar Institute) and SV40 transformed 3T3 mouse cells (supplied by Dr Paul of the Salk Institute) were used in these experiments. Adhesion of cells was routinely measured in multiplates (Lux Scientific Co.) in which Gold Seal coverglasses were used as a substratum for cell adhesion. Two of the coverglasses were soaked for 30 minutes in Ca^{++} and Mg^{++} free Hank's solution containing 2 μg/ml 3,3 ionene bromide, rinsed twice with Ca^{++} and Mg^{++} free Hank's solution (pH 7), while the other two served as controls. Cells from 3–4 day old cultures were trypsinized with 0.25% trypsin; then rinsed three times in serum-free Eagle's medium. Four ml aliquots of 5×10^5 cells in serum-free medium (pH 7.2–7.4) were inoculated in each chamber and incubated at 34 °C. At different intervals the multiplates were subjected to gyratory shaking for 3 minutes in a shaker at 220 rpm; the weakly attached and floating cells were drained; the cover slips were rinsed in saline and fixed in 2% glutaraldehyde in saline.

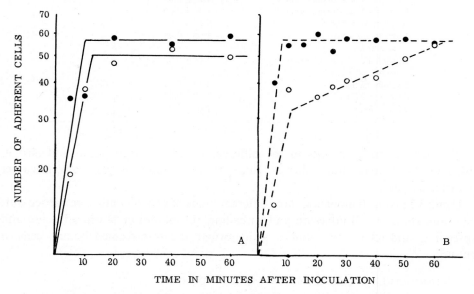

Fig. 1. The number of adherent cells on 3,3 ionene-coated and control coverglasses at different time periods after inoculation. (A) The adherence rate of normal WI38 cells on polyionene-coated (solid circles) and uncoated (open circles) coverglasses. (B) The adherence rate of SV40-WI38 cells on polyionene-coated (solid circles) and uncoated (open circles) coverglasses.

The cover slips were examined with a phase-contrast microscope, the cells were counted in 8 different randomly selected areas and the mean number of cells per unit area calculated. The number of cells which adhered, as a function of time, is shown in Figure 1. The data indicate that the number of cells attached to the 3,3 ionene treated coverglass is greater than that for the untreated coverglass. The increase in the rate of adhesion of cells in the polycation treated coverglass is more pronounced in the SV40 WI-38 cells than in the normal WI-38 cells. This would probably mean that more adhesive sites are established in the presence of the polycation coating, thus increasing the strength of adhesion within the given time.

Cell flattening or spreading was also accelerated in both normal and transformed WI-38 cells. This was studied by randomly selecting photographs at a uniform magnification of cell samples from the previous experiment. An average of 30–40 cell images for each control and experimental condition was carefully cut out and weighed to give a relative area measurement. The cellular areas on the glass surfaces are shown in arbitrary units in Figure 2 as a function of time. In spite of the limitations of light

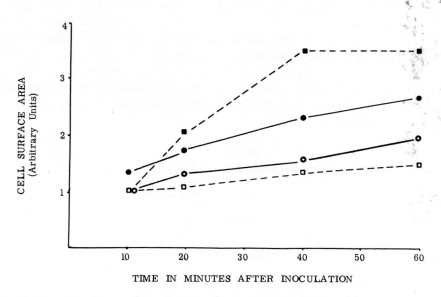

TIME IN MINUTES AFTER INOCULATION

Fig. 2.   Relative cell surface areas observed at various time intervals after adherence to 3,3 ionene-coated and control coverglasses. Normal WI38 cells are represented by circles connected by solid lines. SV40-WI38 cells are represented by squares connected by dashed lines. The lower lines (open symbols) are cells on untreated glass surfaces. The upper lines (closed symbols) are cells on 3,3 ionene-coated coverglasses.

microscope resolution, it was obvious that the rate of cell spreading was accelerated in using normal WI-38 cells in the presence of 3,3 ionene coating, and was markedly pronounced in the SV40 WI-38 cells. This is compatible with the assumptions that polycation-coated glass surfaces have more positive charges than untreated glass and the transformed cell membrane has a higher negative charge than normal cells.

SV40 3T3 mouse cells behaved similarly to SV40 WI-38 cells on the polyionene treated glass surface.

2.2. INHIBITION OF CELL ADHESION AND SPREADING

Although surface membrane sialic acid appears to be the major source of negative charges, other sources such as charged side chains of membrane proteins, membrane bound RNA and membrane lipids also contribute to the net cell surface charge. Increase in negative charges in the transformed cells is usually attributed to increased concentration of surface sialic acid. Altered sialic acid, aminosugar and glycoprotein compositions in the plasma membrane of transformed cells have been reported [15]. According to Wallach, even studies showing alterations of membrane sialic acid in neoplastic conversion cannot readily be related to surface potential, because the electrokinetic expression of ionic groups on membrane depends, among more complex variables, upon (a) the effective radius of the charge bearing site, (b) other charges and (c) the depth of the charge in the membrane, all of which are unknown. For example, the leukemic mouse cells have a lower anodic mobility because the increased charge contribution of sialic acid is outweighed by a rise in surface cationic groups. Sialic acid may also influence cell-cell interaction, (sialic acid removal impairs normal embryonic cell aggregation) and antigenic expression of tumor cells. Apart from sialic acid, thiol groups may contribute to the adhesion of cells to negative surfaces [16]. In order to study the involvement of free —SH groups in the processes of cell adhesion and spreading, we have studied the effect of sulfhydryl binding agents on cell adhesion and spreading.

TABLE III

Effect of N-ethylmaleimide on cell adhesion

Concentration of NEM	Percent of cells attached expressed in relation to respective controls	
	at 20 min	at 60 min
1.0 mM		
Untreated glass surface	0	0
3,3 Ionene treated glass surface	0	0
0.5 mM		
Untreated glass surface	0	0
3,3 Ionene treated glass surface	0	0
0.1 mM		
Untreated glass surface	14	18
3,3 Ionene treated glass surface	28	36

N-ethylmaleimide (NEM) binds preferentially and irreversibly with free —SH groups; while p-chloromercuribenzoate (PCMB) binds preferentially and reversibly with the free —SH groups of the membrane. Suspension of WI-38 cells for 10 minutes in Ca^{++} and Mg^{++} free Hank's solution containing 1.0 or 0.5 mM NEM before seeding on multiplates containing coverglasses totally inhibited cell adhesion irrespective of the presence or absence of polyionene coating on the coverglasses (Table III). There was a significant reduction in the number of adherent cells, even with 0.1 mM NEM pretreatment. Similarly, pretreatment with PCMB also inhibited the adhesion of WI-38 cells for the duration of the experiment (60 min). When the cells were allowed to adhere for 2 minutes and then were placed in the NEM or PCMB-containing Hank's solution, all spreading was arrested. These studies imply that thiol groups are involved in the processes of cell adhesion and spreading, and hence, probably also in cellular locomotion. The exact mechanism of inhibition is not known. Probable explanations are:

(1) permeability of cell membrane to cations is related to the —SH groups in the membrane proteins and the —SH groups may form a bond with the polycation directly,

(2) —SH groups may form an essential bond with other interposed molecules that bind with the substratum, and

(3) —SH groups may be part of an active site in an enzyme involved in cell adhesion and spreading.

2.3. TOXICITY STUDIES WITH NORMAL AND TRANSFORMED CELLS

Toxicity of 3,3 and 6,10 polyionenes to normal WI-38 cells and SV40 WI-38 cells in monolayer cultures was studied by the standard 3-day growth response. Thirty ml Falcon plastic culture flasks were seeded with 0.25×10^6 cells/flask in 5 ml of Eagle's medium $+10\%$ fetal calf serum. After 24 hours, the medium was replaced with the medium containing different concentrations (0–10 μg/ml of 3,3 or 6,10 ionene polymer and the cells were allowed to grow for an additional three days. On the fourth day the cells were trypsinized and the number of cells per ml was calculated with the use of a Coulter particle counter. Three replicate flask cultures were analyzed at each concentration and the experiments were repeated twice. The growth factor (GF) for each treatment was calculated as the ratio of the number of cells/ml on day 4 to the number of cells/ml on day 1 (on the day of initiation of the treatments) and was expressed as percent of the control. The toxicity was plotted as a function of concentration as shown in Figures 3 and 4.

The 3,3 ionene polymer was slightly more toxic to the SV40 WI-38 cells at concentrations up to 4 μg/ml; but above this concentration the normal WI-38 cells were more sensitive to 3,3 ionene polymer than the SV40 WI-38. The nature of the toxicity response of SV40 WI-38 cells to 3,3 ionene polymer indicates that a fairly large proportion of the transformed cell population was resistant to this compound, while the remainder was very sensitive. However, the transformed cells were extremely sensitive to 6,10 ionene polymer even at the lowest concentration tested (2 μg/ml). The

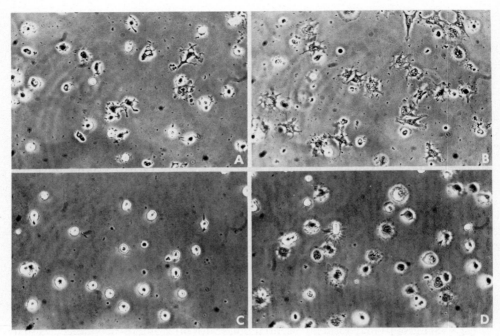

Fig. 3. Appearance of cells 40 min after inoculation on coverglasses. (A) WI38 cells on uncoated glass. (B) WI38 cells on 3,3 ionene-coated coverglass. (C) SV40-WI38 cells on uncoated glass. (D) SV40-W138 cells on 3,3 ionene-coated coverglass.

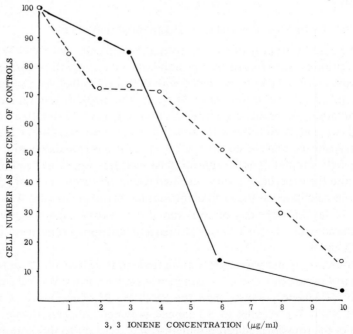

Fig. 4. Cell number, relative to untreated control populations following a 3 day incubation in varying concentrations of 3,3 ionene Cl. The solid line represents WI38 cells, while the dashed line represents SV40-WI38 cells.

exact mechanism of specific toxicity to transformed cells by the 6,10 polymer is not known. It is possible that the surface changes in the membrane were conducive to a more efficient binding of the 6,10 ionene polymer, whose charge density is lower than that of the 3,3 ionene polymer. It is probable that (1) electrokinetic binding to nega-

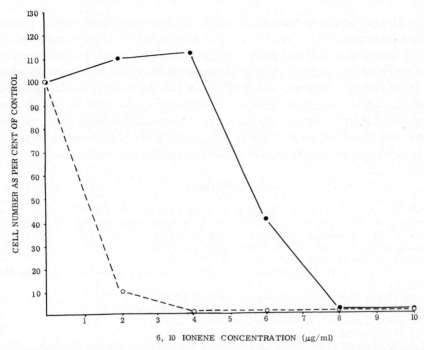

Fig. 5.   Cell number, as a percent of control populations, following a 3 day incubation in varying concentrations of 6,10 ionene Br. The solid line represents WI38 cells, while the dashed line represents SV40-WI38 cells.

tively charged surface, causing charge neutralization might result in a cytotoxic effect, or (2) by entering the cells by pinocytosis, the polycation may bind to nuclear material. It would appear that the charge distribution in the 6,10 polymer backbone results in higher toxicity to cells than the charge distribution in the 3,3 polymer.

3. Summary and Conclusions

Polyionenes and their low molecular weight analogues constitute a unique model system for a molecular probe of the living cell machinery, because their structure, their positive charge densities, counterions and their molecular weights can be varied systematically. These considerations apply not only to the study of toxicity or antimicrobial activity, but also to the understanding of the interaction of the polyionenes with DNA or as molecular probes to elucidate the properties of cell membranes.

The high charge density of polyionenes is responsible for their bactericidal and fungicidal activities, for the prolonged duration of the curarizing action and for the formation of complexes with DNA and heparin. The differences in the biological properties or stability of the complexes may be explained on the basis of electrostatic association between the negative and the positive moieties of the interacting molecules.

Since the normal and neoplastic cells show an array of different surface properties including an increase in the anodic mobility after transformation, some polyionenes (e.g., 6,10 ionene bromide) may preferentially bind to cancer cells and inactivate them. This differential binding and toxicity to transformed cells is in conformity with the topological changes of membrane binding sites in the transformed cells. It would appear that these polyelectrolytes may be uniquely favorable by virtue of their polycationic nature for certain studies in experimental cell biology *in vitro*, since one can alter the social behavior of the cells by coating the substratum with the polyionenes or by suspending the cells in polyionene-containing medium.

References

1. Gibbs, C. F., Littman, E. R., and Marvel, C. S.: *J. Am. Chem. Soc.* **55**, 753 (1933).
2. Rembaum, A., Baumgartner, W., and Eisenberg, A.: *J. Polymer Sci.*, Part B 6, 159 (1968).
3. Kern, W. and Brenneisen, E.: *J. Prakt. Chem.* **159**, 194 (1941).
4. Noguchi, H. and Rembaum, A.: *Macromolecules* **5**, 253 (1972).
5. Rembaum, A. and Noguchi, H.: *Macromolecules* **5**, 261 (1972).
6. Yen, S. P. S., Casson, D., and Rembaum, A.: in N. M. Bikales (ed.), *Water Soluble Polymers*, Plenum Press, N.Y., 1973.
7. Casson, D. and Rembaum, A.: *Macromolecules* **5**, 75 (1972).
8. Rembaum, A.: *Appl. Polymer Symp.*, J. Wiley & Sons, No. 22, 299 (1973).
9. Bacharach, U.: *Ann. Rev. Microbiol.* **24**, 109 (1970).
10. Paton, W. and Zaimis, E. J.: *Pharmacol. Rev.* **4**, 219 (1952).
11. Burger, A. and Dekker, M.: *Drugs Affecting the Peripheral Nervous System*, N.Y., 1967.
12. Schueler, F. W. and Keasling, H. H.: *J. Am. Pharm. Assoc.* **45**, 792 (1956); see also chapter: 'Interactions of Polyammonium Compounds with Acetyl Choline Receptors', this volume, J. Schmidt and M. A. Raftery, p. 175.
13. Kimura, E. T., Young, P. R., Stein, R. J., and Richards, R. K.: *Toxicol. Appl. Pharmacol.* **1**, 185 (1959).
14. Preston, F. W.: *Proc. Central Soc. Clinic Res.* **25**, 63 (1952).
15. Wallach, D. F. H.: *The Plasma Membrane: Dynamic Perspectives, Genetics and Pathology*, Springer-Verlag, 1972.
16. Grinnell, F., Milan, M., and Srere, P. A.: *J. Cell Biol.* **56**, 659 (1973).
17. Forrester, J. A., Ambrose, E. J., and Stoker, M. G. P.: *Nature* **201**, 945 (1964).
18. Simon-Reuss, I., Cook, G. M. W., Seaman, G. V. F., and Heard, D. H.: *Cancer Res.* **24**, 2038 (1964).
19. Mayhew, E. and Nordling, S. J.: *J. Cell Physiol.* **68**, 75 (1966).

INTERACTIONS OF POLYAMMONIUM COMPOUNDS
WITH ISOLATED ACETYLCHOLINE RECEPTORS

J. SCHMIDT and M. A. RAFTERY

California Institute of Technology, Pasadena, Calif., U.S.A.

Abstract. The acetylcholine receptor from the electric organ of *Torpedo californica* was solubilized with the nonionic detergent Triton X-100 and purified by chromatography on an affinity resin containing an acetylcholine analogue. The drug binding properties were investigated by two independent methods: (a) equilibrium dialysis using radioactive ligands, and (b) inhibition of bungarotoxin binding; the latter approach permits the determination of receptor affinities for a wide range of non-radioactive compounds. A simple bungarotoxin binding assay was developed and employed to estimate dissociation constants for a variety of cholinomimetic compounds. Ionenes are polymeric forms of the cholinergically active methonium drugs. It was found that the monomers such as hexamethonium and decamethonium are less efficient than acetylcholine in protecting the receptor against bungarotoxin binding; the ionenes, on the other hand, bind much more tightly than their monomeric counterparts, some of them displaying affinities much higher than those of acetylcholine and potent curariform drugs like *d*-tubocurarine and flaxedil. Binding efficiency was observed to depend on charge density, with a 6,10 ionene being most efficient and the 3,4 ionene binding least strongly.

1. Introduction

Most nerve cells communicate by means of chemical signals which are transmitted and received in specialized structures called 'synapses'. Much of our present knowledge of synaptic function is derived from studies of the vertebrate neuromuscular junction which connects the ending of a motor nerve to a muscle fiber, and uses acetylcholine (AcCh) as a transmitter substance. When an impulse arrives at the nerve terminal, acetylcholine is released and reaches the muscle membrane which responds with an increase in cation permeability. The resulting influx of sodium ions into the muscle cell is accompanied by a change in membrane potential which initiates a sequence of events that eventually lead to a contraction of the muscle fiber [1]. The membrane components which interact with the neurotransmitter are termed 'receptors'. Because of their central role in chemical signalling neurotransmitter receptors have figured prominently in attempts to reach a better understanding of synaptic processes on a molecular level.

From work on nerve-muscle preparations it is known that compounds other than AcCh bind to the receptor. Substances such as carbamylcholine and decamethonium mimic the effect of the natural transmitter ('agonists'), whereas the curariform drugs such as *d*-tubocurarine or gallamine triethiodide prevent the action of AcCh by blocking receptor sites ('antagonists'). Numerous experiments have been conducted to elucidate the structure-activity relationships of such cholinergic agents (for a review see [2]), and thereby gain insight, albeit indirect, into the nature of the acetylcholine receptor (AcChR). Recent progress in the isolation of AcChR molecules [3–9], has opened the door to a more direct, biochemical investigation of receptor structure and receptor-ligand interactions. In this laboratory a large-scale purifica-

Alan Rembaum and Eric Sélégny (eds.), Polyelectrolytes and Their Applications, 175–185. All Rights Reserved.
Copyright © 1975 by D. Reidel Publishing Company, Dordrecht-Holland.

tion procedure for the AcChR from the electric organ of *Torpedo californica* has been developed [10]; sufficient quantities of purified receptor have thereby become available to carry out a variety of detailed binding experiments.

The techniques used include equilibrium dialysis [11] and fluorescence procedures [12]. Another method is based on the observation that the rate of binding of α-bungarotoxin (αBgt) to the receptor is affected by cholinergic ligands [13] and permits rapid estimation of the receptor affinities of a wide spectrum of nonradioactive, nondialyzable drugs.

It is well established that bis- (or poly-) onium compounds, such as dimethyl *d*-tubocurarine and flaxedil are among the most active blocking agents [2]; Lewis and his collaborators have investigated the pharmacological properties of linear hydrocarbons containing up to six intra-chain quaternary ammonium functions, and, depending on the inter-onium distance, have obtained polymers as powerful as, or even more potent than, curare [14]. Polycations of this type of high molecular weight are easily synthesized and have been termed 'ionenes' [15]. In this report we wish to describe the effect of ionenes on the toxin-bonding activity of purified Torpedo AcChR.

2. Materials and Methods

Torpedo AcChR was purified by an affinity chromatography procedure [11]; DEAE-cellulose chromatography of the purified material, followed by concentration on a PM 10 dia-flow membrane yielded a product with a specific activity of 10 nmoles αBgt binding sites per mg protein (or 1 binding site per 100000 daltons of protein).

α-Bungarotoxin was isolated from lyophilized *B. multicinctus* venom (Miami serpentarium) [16] and labeled with ^{125}I following the method of Hunter and Greenwood [17]; Specific radioactivity of the toxin preparations varied from 2 to 4×10^6 cpm/μg depending on batch and age. In order to reduce background in the disk assay (see below), the iodinated toxin was passed through a small column of DEAE-cellulose before use.

Binding of iodotoxin to AcChR was monitored using a DEAE-cellulose disk technique described elsewhere [13] with the following modifications: receptor concentration was reduced to 3×10^{-9} M; concentration of toxin to 6×10^{-9} M; all experiments were carried out at 0° in 0.25 ml 2 mM sodium phosphate, pH 7.4, 0.1% Triton X-100. Receptor was incubated with ligand for an hour; formation of receptor-toxin complex was started by the addition of toxin and (unless otherwise stated) terminated after 0.20 min by applying an 0.1 ml aliquot to the DEAE-cellulose disk and immersing it in the wash fluid. Controls are obtained with receptor denatured by heating to 100° for 3 min.

Ionenes were kindly provided by Dr A. Rembaum; the number of repeating bis-onium unit segments is of the order of 50 to 100, unless stated otherwise.

Ionenes of the structure

$$\left[\begin{array}{cc} CH_3Z^- & CH_3Z^- \\ | & | \\ -N^+ \!-\!(CH_2)_x\!-\!N^+\!-\!(CH_2)_y\!- \\ | & | \\ CH_3 & CH_3 \end{array} \right]_n$$

are abbreviated x–y–Z (n in most cases is ca. 100).

3. Results

The binding of iodinated αBgt to AcChR is specific and irreversible [13]. Formation of the receptor-toxin complex, under the conditions described in the methods section, is a very fast process: the reaction goes to completion in approximately one minute (Figure 1). Contrary to enzyme kinetics, reaction conditions cannot be chosen such as

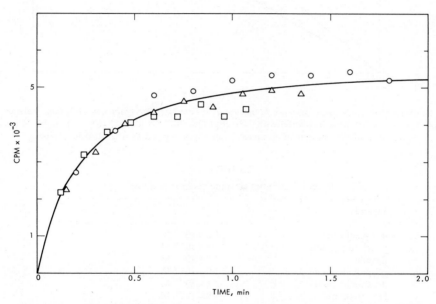

Fig. 1.   *Rate of binding of* ^{125}I-αBgt *to AcChR*. Receptor (3×10^{-9} M) and toxin (6×10^{-9} M) were allowed to react at $0°$ in 2 mM sodium phosphate pH 7.4; 0.1-ml aliquots were monitored for complex formation, as described in the Methods section, after the intervals indicated. Background-corrected data from three separate experiments ($\bigcirc$, $\triangle$, $\square$) are presented.

to hold the velocity constant even over a very short period of time, since the receptor is used up in the reaction. Nevertheless it can be shown that complex formation rate, for receptor concentrations of 3×10^{-9} M and short incubation times, is linearly dependent on the concentration of receptor (Figure 2). Therefore, under the conditions outlined in Methods total complex formed represents a measure of the amount of active receptor present. This amount can be reduced, by either lowering the quantity of

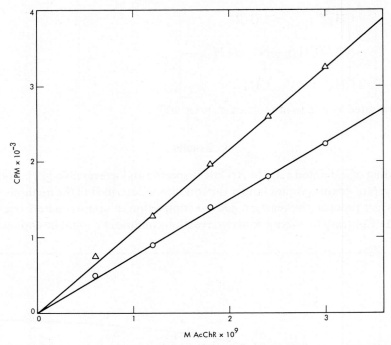

Fig. 2. *Rate of receptor-toxin complex formation as a function of receptor concentration.* Various concentrations of AcChR were incubated with ^{125}I-αBgt (2×10^{-9} M) for 0.15 (○—○), and 0.30 (△—△) minutes and the amount of complex formed was assayed as described in Methods.

TABLE I

Receptor affinities of some cholinergic drugs

Ligand	I_{50}
d-tubocurarine	1.2×10^{-7} M
benzoquinonium	1.8×10^{-7} M
flaxedil	2.0×10^{-7} M
propionylcholine	7.0×10^{-7} M
decamethonium	2.1×10^{-6} M
hexamethonium	2.8×10^{-6} M
acetylcholine	$(3.13 \pm 1.32) \times 10^{-6}$ M ($n = 10$)
atropine	6.5×10^{-5} M
carbamylcholine	7.7×10^{-5} M
nicotine	9.2×10^{-5} M
phenyltrimethylammonium	1.4×10^{-4} M
eserine	1.5×10^{-4} M
pilocarpine	4.5×10^{-4} M
choline	1.75×10^{-3} M

The ability of each drug to affect the rate of binding of ^{125}I-αBgt to purified Torpedo AcChR was tested over a wide range of concentrations, as described in Methods. Concentrations required to reduce the toxin binding rate to 50% (I_{50}) were found by interpolation, as described for AcCh in Figure 3.

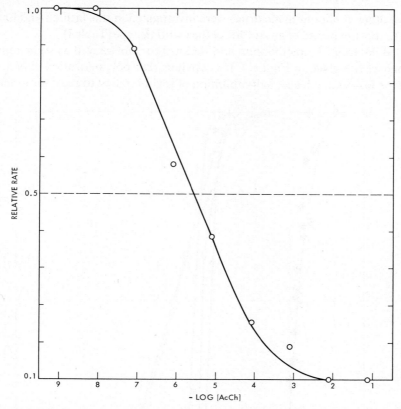

Fig. 3. *Effect of AcCh on toxin binding rate.* The rate of toxin-receptor complex formation is measured as outlined in the Methods section, in the presence of the indicated concentrations of AcCh. Rates are normalized to the binding velocity in the absence of ligand. The broken line indicates half-maximal rate.

receptor added to the incubation mixture or by adding ligands which block the toxin binding site. If one assumes identity or complete overlap of the toxin binding site with the ligand binding site, the reduction of the toxin binding rate caused by a reversibly binding ligand can be used to estimate its affinity for the receptor. The dissociation constant K_D is frequently used to quantitate affinity; it is defined as that ligand concentration at which a ligand occupies 50% of the binding sites of some binding macromolecule. Accordingly, at this concentration only half of the binding sites of AcChR will be available for interaction with ^{125}I αBgt, and toxin binding therefore will proceed at half the original rate.

Results of such an experiment are given in Figure 3. Plotting the rate of toxin binding as a function of AcCh concentration, one observes virtually complete inhibition above 10^{-3} M, while below 10^{-8} M no effect of the transmitter on the toxin binding rate can be seen. The reaction rate is reduced to 50% at an AcCh concentration of 2.5×10^{-6} M. This K_D value is in good agreement with the dissociation constant as determined by equilibrium dialysis (2.3×10^{-6} M, Reference 11). The affinities of a variety of other cholinomimetic drugs were investigated in this fashion and found to be close to those expected of an AcChR of the nicotinic type (Table I).

Detailed data on hexamethonium and decamethonium as well as some representative ionenes are given in Figure 4. It is obvious that polymerization enhances the effect of the bis-onium drugs; half-inhibition of toxin binding to receptor is achieved

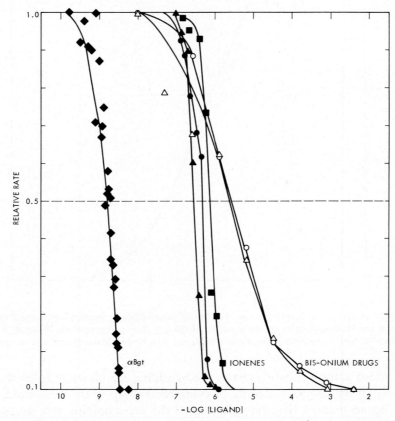

Fig. 4. *Comparison of the effect of bis-onium and poly-onium compounds and of αBgt on the rate of* [125]*I-αBgt binding to AcChR.* The rate of complex formation is measured, as described in Methods, in the presence of the indicated concentrations of hexamethonium (○—○), decamethonium (△—△), 6–6–Cl (●—●), 6–10–Br (▲—▲), and 3–3–Br (■—■). Concentrations of the ionenes are based on the repeating unit segment. For comparison, the effect of native αBgt on the binding of radioactive toxin is included in the graph (◆—◆).

with less than micromolar concentrations of the poly-onium compounds. The I_{50} values compare with the inhibition constants of the powerful cholinergic blocking agents dimethyl-*d*-tubocurarine and gallamine triethiodide. All ionenes tested proved to be potent inhibitors of the toxin binding site of Torpedo AcChR; the data, summarized in Figure 5, seem to indicate a correlation between inhibitory potency and length of the unit segment.

Interestingly, the inhibition curves for all ionenes are much steeper than those for hexamethonium or decamethonium and resemble that for αBgt. The slope of the inhibition curve for αBgt is a consequence of the irreversible binding between receptor and toxin, i.e., free receptor is poisoned gradually by addition of toxin, until at 3×10^{-9} M αBgt the binding sites are saturated and no active receptor is left that could react with the radioactive toxin added in the assay. Receptor in equilibrium

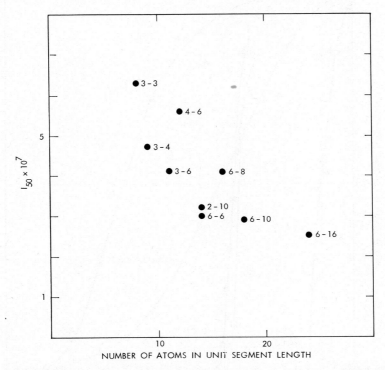

Fig. 5. *Correlation of ionene structure and inhibitory potency.* Concentrations required to reduce toxin binding rate to 50% (I_{50}) were determined, as described in the legend to Figure 4, for a variety of ionene bromides, and plotted as a function of the length of the unit segment.

with a reversible ligand, on the other hand, will always contain a fraction of un-occupied binding sites which will result in limited receptor – ^{125}I αBgt complex formation even at fairly high inhibitor concentrations, thereby causing inhibition curves to appear 'shallower'. Thus ionenes seem to bind with low reversibility, a suggestion that can be tested in yet another way. All reversible ligands are eventually displaced by bungarotoxin, if incubation is continued long enough. This effect can be demonstrated by a shift of the inhibition curve to higher ligand concentrations, i.e. by an apparent decrease in ligand affinity. Irreversibly binding ligands, on the other hand cannot be dislodged from the binding site, and their binding curves are not shifted. Such an experiment on the binding characteristics of a well-known reversible ligand (hexamethonium), an established irreversible agent (αBgt), and a representative ionene (6–6–Cl) is shown in Figure 6. Clearly prolonged incubation does not result in the displacement of native αBgt by the iodotoxin. The midpoint of the hexamethonium curve, on the other hand, shifts to much higher concentrations indicating gradual reversal of the block exerted by the bis-onium compound. In the case of

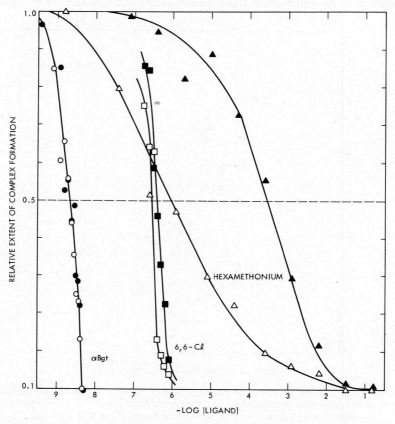

Fig. 6.   *Inhibition of receptor-toxin complex formation by several ligands as a function of incubation time.* The effect of various concentrations of native αBgt (circles), hexamethonium (triangles) and 6–6–Cl (squares) on toxin binding rate was measured as described in the Methods section. Formation of complex was monitored after 0.2 min (open symbols) and again after 20 min (filled symbols).

6–6–Cl the two inhibition curves almost superimpose suggesting that the ionene dissociates rather slowly from the toxin binding site of the AcChR.

It should be pointed out here that the half inhibition concentration of an irreversible ligand such as αBgt or cobrotoxin, is not an inhibition constant, but a variable depending on the concentration of binding sites in the incubation mixture. As the line between irreversible binding and slowly reversible binding can in practice not always be drawn very accurately, similar considerations may hold for the ionenes and the I_{50} values presented in Figure 5, though reflecting affinities in some qualitative manner, are not necessarily identical with inhibition constants.

In order to investigate whether affinity is greatly affected by the overall length of the polymer, two fractions of 6–10–Br of different molecular weight were compared. As can be seen in Figure 7, the two inhibition curves superimpose. Similar results

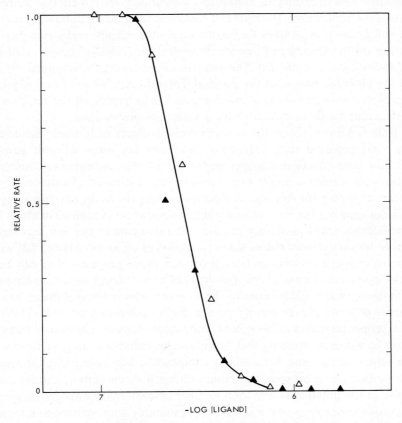

Fig. 7.   *Comparison of the inhibitory potency of ionenes of different lengths.* A low MW (△—△) and a high MW (▲—▲) fraction of the ionene 6–10–Br were compared for their effects on toxin binding rate as described in the text.

were obtained with three fractions of 6–6–Cl, ranging in molecular weight from 6000 to 25 000. Obviously only a relatively small number of monomeric units have to be linked together to achieve maximum affinity; further extension of the molecule then has no additional effect.

4.   Discussion

Only few studies have been conducted so far on the binding properties of purified receptors. The affinities of isolated AcChR from *Electrophorus electricus* and *Torpedo californica* for a small number of cholinergic ligands have been described [11, 18]. The equilibrium dialysis experiments with purified Torpedo receptor have yielded data on the affinities of AcCh, dimethyl-*d*-tubocurarine, and decamethonium that are in good agreement with results of the present studies. Quantitative differences are presumably due to modification of the receptor binding characteristics by the salt concentrations necessary to minimize Donnan equilibrium effects during equilibrium dialysis procedures. We thus feel confident that the toxin binding rate assay permits estimation of K_D values for readily reversible ligands, and yields qualitative information on the affinities of ligands in general (provided, of course, that ligand and toxin binding sites coincide). The assay has been used in a simplified version to confirm the nicotinic nature of the purified Torpedo AcChR [11] and is presently being employed in establishing a more detailed drug profile of the receptor [19]; among the compounds investigated are a variety of polycations.

Very little is known about the neurobiological effects of ionenes. Schueler and Kaesling [20] reported that polyammonium salts are more efficient ganglionic blockers and their effects last longer compared to their monomeric counterparts. Their observation finds a simple explanation in the increased affinity of polymeric methonium drugs for the AcChR, as documented by the toxin binding assay. This affinity is not simply a function of molecular weight, since variations in chain length do not noticeably affect inhibitory potency. The structure of the unit segment does not seem to be very critical either, since I_{50} values of all ionenes tested fall within a very narrow range. Nevertheless it appears that those polymers that can be considered as polymeric forms of the potent cholinergic drugs hexamethonium and decamethonium bind slightly better to the receptor. Alternatively it might be argued that ionenes of lower charge density possess higher inhibitory potency. The importance of a proper geometry of bis-onium compounds is a well established fact [3, 14] and it may be worth mentioning that Lewis and his collaborators [21], from a study of linear tetra-, penta-, and hexa-onium compounds, concluded that the length of the hydrocarbon chain between quaternary nitrogen atoms is more crucial than the overall size of the polymer or the total number of quaternary nitrogen atoms.

The neuropharmacology of the ionenes is presumably not restricted to interactions with AcChR. Choline re-uptake, release of transmitter, and esterase activity may also be affected by this class of polymers, since in all cases recognition sites for quaternary nitrogen functions are involved. Thus the present paper is no more than a beginning in the application of ionenes to the pharmacological and biochemical study of the various macromolecular structures functioning in cholinergic transmission.

References

1. Katz, B.: *Nerve, Muscle, and Synapse*, McGraw-Hill, 1966.
2. Triggle, D. J.: *Neurotransmitter-Receptor Interactions*, Academic Press, London and New York, 1971.
3. Schmidt, J. and Raftery, M. A.: *B.B.R.C.* **49**, 572 (1972).
4. Karlsson, E., Heilbronn, E., and Widlund, L.: *FEBS Letters* **28**, 107 (1972).
5. Olsen, R. W., Meunier, J. C., and Changeux, J. P.: *FEBS Letters* **28**, 96 (1972).

6. Klett, R. P., Fulpius, B. W., Cooper, D., Smith, M., Reich, E., and Possani, L.: *J. Biol. Chem.* **248**, 6841 (1973).
7. Lindström, J. and Patrick, J.: in M. V. L. Bennett (ed.), *Synaptic Transmission and Neuronal Interaction.* Raven Press, New York, in press (1974).
8. Karlin, A.: in *1972 Intra-Science Symposium on Mechanisms of Drug Action*, in press (1973).
9. Eldefrawi, M. E. and Eldefrawi, A. T.: *Arch. Biochem. Biophys.* **159**, 362 (1973).
10. Schmidt, J. and Raftery, M. A.: *Biochemistry* **12**, 852 (1973).
11. Moody, T., Schmidt, J., and Raftery, M. A.: *BBRC* **53**, 761 (1973).
12. Martinez-Carrion, M. and Raftery, M. A.: *BBRC* **55**, 1156 (1973).
13. Schmidt, J. and Raftery, M. A.: *Anal. Biochem.* **52**, 349 (1973).
14. Lewis, J. J. and Muir, T. C.: in A. Burger (ed.), *Drugs Affecting the Peripheral Nervous System*, Vol. 1, M. Dekker, Inc., N.Y., 1967, p. 327.
15. Casson, D. and Rembaum, A.: *Macromolecules* **5**, 75 (1972).
16. Clark, D. G., Macmurchie, D. D., Elliott, E., Wolcott, R. G., Landel, A. M., and Raftery, M. A.: *Biochemistry* **11**, 1663 (1972).
17. Hunter, W. M. and Greenwood, F. C.: *Nature* **194**, 495 (1962).
18. Meunier, J. C. and Changeux, J. P.: *FEBS Letters* **32**, 143 (1973).
19. Schmidt, J. and Raftery, M. A.: *J. Neurochem.*, in press (1974).
20. Schueler, F. W. and Kaesling, H. H.: *J. Am. Pharm. Assoc.* **45**, 192 (1956).
21. Edwards, D., Lewis, J. J., Stenlake, J. B., and Stothers, F.: *J. Pharm. Pharmacol. Suppl.* **11**, 87T (1959).

BIOMEDICAL APPLICATIONS OF POLYCATIONS

HAROLD MOROSON and M. ROTMAN

Dept. of Radiology, New York Medical College, New York, N.Y., U.S.A.

Abstract. Biomedical applications of polycation follow from their electrostatic attraction to negatively charged surfaces in silicosis therapy, immunochemistry, and as immune adjuvant. Polycation treated mouse tumor cells show a reduction in negative charge, increased cell aggregation, and increased cell lysis. Injecting polycations into tumor-bearing mice at dosages non-lethal to the host inhibits the growth of several types of tumor. Injecting polycations before challenging mice with tumor cells also increases their survival.

The mechanism of the polycation inhibition of tumor growth appears to be due to two separate effects: direct cytotoxicity and immune adjuvant activity.

Polycation molecules may have low or high molecular weight, may be linear or branched, and may have their charged quaternary group, (ammonium, sulfonium, or phosphonium), integral or pendant to the chain, with resulting biological properties extremely dependent upon these parameters. They are all water soluble and many are commercially available.

There are several biomedical applications of polycations stemming from the property of this class of polyelectrolyte to bind to surfaces with negative electrostatic charge. A theoretical understanding of the biological interactions of acidic and basic polyelectrolytes and effects on enzymatic processes in cells has been provided by A. Katchalsky [1].

Each type of cell exhibits its own characteristic electronegativity i.e. lymph node and spleen cells are more negative than thymocytes and macrophages. Cell membrane acid anions are the major cause of excess surface negative charge. Early work with cell electrophoresis suggested that neoplastic, proliferating, and embryonal cells have a higher electro-negative surface charge than other types of cells [2, 3]. More recently, a uniformity of electrokinetic pattern between normal and leukemic cells has been reported [4]. The malignant cell lines HeLa (cervical cancer), HEp2 (laryngeal carcinoma), 256 (rat sarcoma), normal lymphocytes, and normal and malignant trophoblastic cells were found to have a common cell surface coating of an ionic nature not generally found for other normal cell types tested [5].

Electron microscopic studies have also shown that malignant cells have an increased surface thickness of mucopolysaccharide material, containing terminal groups of negatively charged sialic acid moieties [6].

A difficulty in attempting to exploit electrostatic phenomenon in therapy of malignant disease stems from the lack of specificity for tumor cells by polycations and also the constantly changing surface chemistry of cell membranes. Nevertheless, the possibility of employing polycations for preferential growth inhibition of malignant cells has been attempted, with some promising results reported [7–10].

Before discussing this type of effect, we would like to briefly review some other applications of polycations in biomedicine.

1. Silicosis Therapy

Pneumoconiosis is an interstitial pneunomia caused by irritation of the lungs during occupational exposure to dust particles. Silicosis is the most serious of these diseases and is contracted by inhalation of fine particles of silica over long periods of time in gold mines, tin mines, stone quarries, and during sand blasting.

Poly vinyl pyridine N-oxide (PVNO) in aerosal spray form is used to counteract the pathogenic effects produced by inhaled quartz or silica particles in mines. This polycation is a potent inhibitor of silica or quartz hemolysis of red blood cells. Protection is presumably due to electrostatic binding of PVNO to the electronegative surface of silica particles in the lung. It is also possible that macrophages may be stimulated by PVNO to engulf the silica particles, thus protecting surrounding stroma.

2. Immunochemistry

Poly-l-lysine finds application in an assay for hemolytic plaque-forming cells, important in immunology. Poly-styrene plastic dishes are treated with a dilute solution of poly-l-lysine, which neutralizes the negative surface charge on the plastic, enabling a uniform layer of blood cells to form directly on the plastic dish, eliminating need for an agar-gel intermediate layer. The agar-gel layer has anticomplement activity thus reducing sensitivity of the assay [11].

Diethylaminoethyl Dextran (DEAD) is employed in column chromatography to separate protein molecules, due to its binding affinity to antibody or antigen bearing different net electronegative charges.

DEAD also increases the cellular uptake of viral RNA by cells in tissue culture by a factor of up to 10^5 [12]. Thus DEAD can be employed to increase the transfer of immunity by means of immune RNA to non-immune cells and prolong the survival of animals [13]. The ability of a cancer virus to cause tumors in mice can be increased up to 10-fold following administration of DEAD. Hence as a research tool, DEAD may be of use in the detection of viruses with low cancer-causing activity in mice and other species. The recently reported carcinogenicity of DEAD in mice may be due to this mode of action, therefore care should be exercised in the handling of this compound.

3. Immune Adjuvant

In 1965, Johnson et al. [14] showed histones could enhance the antibody response of mice to bovine γ-globulin. Moroson [15] suggested some polycations had immune adjuvant properties in the non-specific rejection of experimental tumors in mice. In 1972, Gall et al. reported that several synthetic polycations (Primafloc C-7, (poly-vinylimidazoline) C-5, and C-3) showed strong adjuvant activity with diphtheria and tetanus toxoids in both guinea pigs and mice [16]. The relative potency of Primafloc C-7 appeared to be 100 times greater than that of endotoxin, and 3000 times greater than poly-*l*-lysine. These workers conclude that polycation adjuvants act as potentiators of antigen uptake by cells, whereas polyanions do not, and that polycations and polyanions enhance the immune response in different ways.

4. Growth Inhibition of Tumor Cells

Several classes of polycations, primarily quaternary polyethylene imine (PEI), polypropylene imine (PPI), polyvinylimidazoline (PVA) [15], and diethylaminoethyl dextran (DEAD) [9] have been found to inhibit the growth of allogeneic and syngeneic tumors in mice at non-toxic levels to the host. Anti-tumor activity appears related to the polycation type, molecular weight, zeta potential and route of injection. Table I lists the polycations found to have tumor growth inhibiting properties, and the zeta potential of several polycations is shown in Table II. Ultraviolet absorption spectra of the polycations PEI and PPI are shown in Figures 1 and 2. These agents are cytotoxic, but without some specificity for tumor cells, polycations would appear

TABLE I

Polycations found to have tumor growth inhibiting effect in mice

AGENT	STRUCTURE OF MONOMERIC UNIT	MOLECULAR WEIGHT	SOURCE
POLYETHYLENEIMINE (PEI)	$\{CH_2-CH_2-N-CH_2CH_2-NH\}$ CH_2 CH_2 $-N-$	600 1,800 60,000 100,000	DOW CHEMICAL CO. (MIDLAND, MICH.)
POLYPROPYLENEIMINE (PPI)	CH_3 CH_3 $\{CH_2-CH-NHCH_2\ CHNH\}$	1,000-1,500	INTERCHEMICAL CORP. (NEW YORK, N.Y.)
POLYAMINE (C3)	$\{(CH_2)_n-NH_2-(CH_2)_n-NH_2\}$	30,000-50,000	ROHM AND HAAS CO. (PHILADELPHIA, PA.)
POLYVINYL IMIDAZOLINE (PVA)	$\{(CH_2-CH)\}$ C HN NH CH_2-CH_2	$1 \times 10^6 - 3 \times 10^6$	ROHM AND HAAS CO. (PHILADELPHIA, PA.)
POLY 4 VINYL PYRIDINE N-OXIDE (PVNO)	$\{(CH_2-CH)\}$	CA. 10^5	POLYSCIENCES INC. (RYDAL, PA)
DIETHYLAMINOETHYL DEXTRAN (DEAD)	$(C_2H_5)_2\ N-CH_2CH_2-DEXTRAN$	500,000	PHARMACIA (SWEDEN)
DIETHYLAMINOETHYL CELLULOSE (DEAC)			UNION CARBIDE CORP. (TARRYTOWN, N.Y.)
DIETHYLAMINOETHYL POLYACRYLATE (DEAP)			NATIONAL STARCH CORP. (NEW YORK, N.Y.)

TABLE II

Zeta potential of several polycations
in saline phosphate buffer

Polycation	Zeta potential (mV)
PEI	+ 33.0
PPI	+ 18.5
PVA	+ 16.0
DEAD	+ 7.0

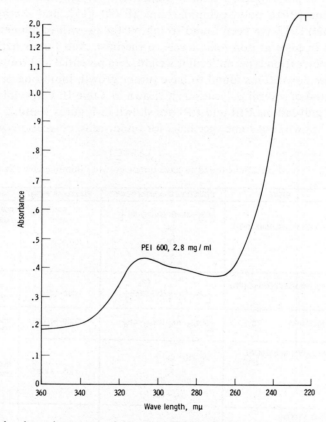

Fig. 1.   Ultraviolet absorption spectra of the polycation polyethylenimine, molecular weight 6×10^4, in saline phosphate buffer, pH 7.0.

to have little to offer. However, a specificity of binding to mouse tumor cells compared to non-tumor cells is suggested by some precipitation experiments, the results of which are summarized in Table III. Ehrlich ascites cells were withdrawn from Swiss white mice intraperitoneally inoculated 5 days prior with 1×10^6 Ehrlich ascites cells, and diluted to 3×10^6 cells/ml with physiological saline. When 6 or 180 μg/ml of polycation was added these cells precipitated, while untreated cells remained in

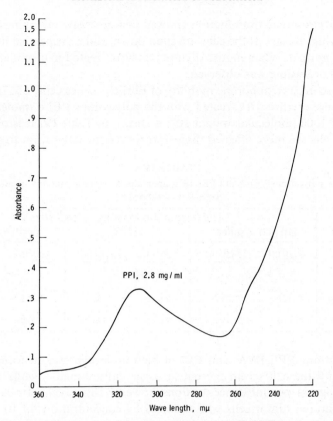

Fig. 2.   Ultraviolet absorption spectra of the polycation polypropylenimine, molecular weight 1.5×10^3, in saline phosphate buffer, pH 7.0.

TABLE III

Comparison of tumor and non-tumor cell precipitation by polycation treatment in tris-maleate buffer in pH 7.0

Polycation 6 µg/ml and 180 µg/ml	Ehrlich Ascites Cells (3×10^6/ml)			Mouse Thymocytes (6×10^6/ml)		
	10 min 0°	30 min 0°	10 min 37°	10 min 0°	30 min 0°	10 min 37°
Control	−	−	−	−	−	−
PEI (10^5)	+++	+++	+++	−	−	−
PEI (6×10^4)	+++	+++	+++	−	−	−
PEI (1.8×10^3)	+	+++	+++	−	−	+
PEI (1.2×10^3)	−	+++	+++	−	−	+
PEI (6×10^2)	−	+++	+++	−	−	+
PPI (1.5×10^3)	+	+++	+++	−	+	+++
POLYVINYLIMIDAZOLINIUM PVA (1×10^6)	+++	+++	+++	−	−	+
DEA DEXTRAN (5×10^5)	−	+++	+++	−	−	−

− = no settling   + = slight   +++ = complete settling

suspension. Microscopic examination showed characteristic clumping, with grape-bunch appearing clusters. If the cells are spun down, and resuspended in saline, they again appear normal. When mouse thymocytes were treated in a similar fashion, far less clumping or settling was observed.

The decrease in electrophoretic mobility of Ehrlich ascites cells, L5178Y leukemia cells, and mouse erythrocytes treated with the polycations PEI 6 (molecular weight 600) and PEI 1000 (molecular weight 10^5) is shown in Table IV. It is observed that these tumor cells are more affected than erythrocytes by polycation treatment.

TABLE IV

Effect of the polycations PEI 6 and PEI 1000 upon electrophoretic mobility of mouse tumor cells and erythrocytes

Polycation	Electrophoretic Mobility, μ/sec/V/cm.					
	Ehrlich Ascites		L5178Y		Erythrocytes	
	3 µg/ml	15 µg/ml	3 µg/ml	15 µg/ml	3 µg/ml	15µg/ml
Control	−2.4		−2.2		−3.6	
PEI 6	−1.7	−1.7	−2.0	−1.9	−4.1	−4.1
PEI 1000	−1.5	−1.1	−1.8	−1.8	−3.6	−3.4

The polycations PPI, PVA, and PEI of high molecular weight form a colloidal precipitate with the calf serum present in tissue culture media, while DEAD does not. At microgram per ml concentrations however this precipitate is not visible. Growth of cultured tumor cells was inhibited by concentrations of 10–20 μg/ml of high molecular weight PEI, PPI, or PVA in the medium.

A dose of 4–5 mg/kg body weight injected i.p. into Ehrlich ascites tumor-bearing mice was inhibiting to growth of this tumor *in vivo*, as evidenced by increased survival employing PEI 1000 but not PEI 6, (Table V). Both polycations were equally

TABLE V

Survival of Swiss mice 40 days after i.p. inoculation of 10^5 Ehrlich ascites cells, treated one day later with poly-cation injected i.p. at a dosage of 4–5 mg/kg

Treatment	% Survival
Saline control	0
Pei 6	0
PEI 1000	40

effective in increasing survival if injected 3 days, but not 8 days, prior to the tumor cell inoculation (Table VI).

Preliminary results with a transplanted methyl cholanthrene induced fibrosarcoma,

TABLE VI

Survival of Swiss mice at 88, 100, and 150 days after s.c. inoculation of 10^5 Ehrlich ascites cells.
Prior to tumor cell passage mice were treated with polycation according to
schedule indicated on chart

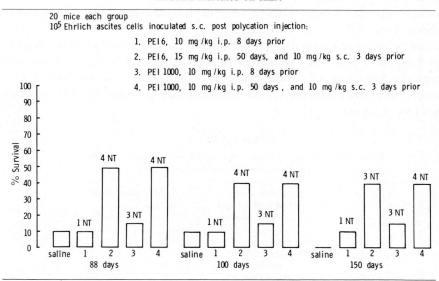

20 mice each group
10^5 Ehrlich ascites cells inoculated s.c. post polycation injection:

1. PEI 6, 10 mg/kg i.p. 8 days prior
2. PEI 6, 15 mg/kg i.p. 50 days, and 10 mg/kg s.c. 3 days prior
3. PEI 1000, 10 mg/kg i.p. 8 days prior
4. PEI 1000, 10 mg/kg i.p. 50 days, and 10 mg/kg s.c. 3 days prior

TABLE VII

PEI Stimulation of protein and DNA synthesis in mouse spleen and inhibition in
Ehrlich ascites tumor

Sample	2 hrs. ^{14}C amino acid (10μCi)			3 hrs. ^{3}H Thymidine 50μCi	
	EA cells/mouse x 10^{-7}	cpm/10^6 cells x 10^{-4}	cpm/mg spleen x 10^{-5}	EA cells/mouse x 10^{-7}	cpm/10^6 cells x 10^{-3}
Saline control	2.5	7.6	86	1.9	15.6
PEI 600 20 mg/kg	0.40	2.1	157	0.67	27.8

termed MC-2, which grows rapidly in syngeneic C57/B mice from an inoculum of
10^5 cells, and which kills the host within 2 months, indicate that this tumor, like the
Ehrlich ascites carcinoma, is growth-inhibited if polycations are injected 4 days prior
to s.c. transplant of 2×10^5 MC-2 tumor cells.

Peritoneal macrophages removed from mice previously injected with PPI (i.p.,

0.045 mg/mouse) show increased activation by microscopic observation of giemsa stained smears. The polycation PEI 600 was found stimulating to mouse spleen cells as measured by increased tritium-thymidine and Carbon-14 incorporation while simultaneously the Ehrlich ascites tumor cells showed a decreased protein and DNA synthesis (Table VII).

L1210 mouse leukemia cells show a reduction in migration out of capillary tubes when 1 μg/ml of polycation is present in the surrounding medium, in comparison to controls. Migration of mouse spleen cells under the same conditions was not affected (Table VIII). The polycation treatment was non-toxic to cells as shown by trypan blue stain exclusion.

TABLE VIII

Effect of polycation on migration from capillary tubes of L1210 tumor cells and normal mouse spleen cells

Cell type	Incubated with MEM + 20% Fetal calf serum plus:	% Migration after 24 hours
L1210	control	100
L1210	1 μg/ml PPI	20
L1210	1 μg/ml PEI	20
L1210	1 μg/ml PVA	25
spleen	control	100
spleen	1 μg/ml PPI	90
spleen	1 μg/ml PEI	90
spleen	1 μg/ml PVA	100

Most proteins at pH 7–8 have a net negative charge, and measurements of cell electrophoretic mobility are fairly constant during the cell cycle. Hence cell interactions with polycations should be independent of the cell cycle position. Polycations are potent cytotoxic agents, and they may find at extremely low concentrations, useful application in inhibition of tumor cells.

Acknowledgements

This work was supported in part by Public Health Service research grant CA 14374 from the National Cancer Institute.

References

1. Katchalsky, A.: 'Polyelectrolytes and Their Biological Interactions', *Biophys. J.* **4**, 9–41 (1964).
2. Vassar, P. S.: 'The Electric Charge-Density of Human Tumor Cell Surfaces', *Laboratory Investigation* **12**, 1072–1077 (1963).
3. Cook, G. M. W., Heard, D. H., and Seaman, G. V. F.: 'The Electrokinetic Characterization of the Ehrlich Ascites Carcinoma Cell', *Exptl. Cell Res.* **28**, 27–39 (1962).
4. Patinken, D., Schlesinger, M., and Doljanski, F. A.: 'Study of Surface Ionogenic Groups of Different Types of Normal and Leukemic Cells', *Cancer Research* **30**, 489–497 (1970).

5. Hause, L. L., Pattillo, R. A., Sances, A., and Mattingly, R. F.: 'Cell Surface Coatings and Membrane Potentials of Malignant and Nonmalignant Cells', *Science* **169**, 601–603 (1970).
6. Martinez-Paloma, A. and Brailovsky, C.: in *The Year Book of Cancer*, Year Book, Chicago, 1969, pp. 494–496.
7. Ambrose, E. J., Easty, D. M., and Jones, P. C. T.: 'Specific Reactions of Polyelectrolytes with the Surfaces of Normal and Tumor Cells', *Brit. J. Cancer* **12**, 439–447 (1958).
8. Moroson, H.: 'Tumor Growth Inhibiting Effects of the Polycations PEI, PPI, and PVA', in S. Garrattini and G. Franchi (eds.), *Chemotherapy of Cancer Dissemination and Metastasis*, Raven Press, N.Y., 1973.
9. Thorling, E. B. and Larsen, B.: 'Inhibitory Effect of DEAE-Dextran on Tumour Growth', *Acta Pathol. Microbiol. Scand.* **75**, 237–246 (1969).
10. Sirica, A. E. and Woodman, R. J.: 'Selective Aggregation of L1210 Leukemia Cells by the Polycation Chitosan', *J. Nat. Cancer Inst.* **47**, 377–388 (1971).
11. Kennedy, J. C. and Axelrad, M. A.: 'An Improved Assay for Haemolytic Plaque-Forming Cells', *Immunology* **20**, 253–257 (1971).
12. Hull, R.: 'The Effect of DEAE-Dextran on the Nucleic Acids of Two Plant Viruses', *J. Gen. Virol.* **11**, 111–117 (1971).
13. Rigby, P. G.: 'Prolongation of Survival of Tumor Bearing Animals by Transfer of Immune RNA with DEA-Dextran', *Nature* **221**, 968–969 (1969).
14. Johnson, A. G., Hoekstra, G., and Merritt, K.: 'Enhancement of Early Antibody Synthesis by Heterologous Nucleic Acids and Histones', *Fed. Proc.* **24**, 178–183 (1965).
15. Moroson, H.: 'Polycation Treated Tumor Cells *in Vivo* and *in Vitro*', *Cancer Research* **31**, 373–380 (1971).
16. Gall, D., Knight, P. A., and Hampson, F.: 'Adjuvant Activity of Polyelectrolytes', *Immunology* **23**, 569–575 (1972).

DNA AS A POLYELECTROLYTE: RECENT INVESTIGATIONS ON THE Na–DNA SYSTEM

DANE VASILESCU, HENRI GRASSI, and MARIE-AGNÈS RIX-MONTEL

Laboratoire de Biophysique – UER-DM – Université de Nice,
Parc Valrose – 06034, Nice-Cedex, France

Abstract. The behaviour of DNA as a polyelectrolyte has been considered in the simplest case, i.e. DNA dissolved in an aqueous 1.1 electrolyte and compensated by monovalent metallic counter-ions.

In this study we will first present an outline of the principal theoretical models and try to rationalise the notation for the parameters describing the electrolyte which acts as solvent for the DNA. We will thus see that one of the fundamental parameters is the Landau length:

$$l = \mathscr{L}/\varepsilon_r' \, T \quad \text{where} \quad \mathscr{L} = \frac{e^2}{4\pi\varepsilon_0 k} \text{ in MKSA units.}$$

The model proposed by Manning, which makes use particularly of the charge parameter $\xi = l/b$ (where b is the distance between two phosphate sites projected on the axis of the double helix), has been chosen as a basis of discussion for our experimental results. Manning's laws allow us to relate in a simple way the measurable activity coefficients γ_1 and γ_2 of the counter and co-ions to the charge parameter ξ and the number $X = n_e/n_s$. This latter is the ratio between the number of ionised phosphate sites and the concentration of the electrolyte solvent.

The activity coefficients γ_1 and γ_2 have been measured using electrodes selectively sensitive to Na$^+$ and Cl$^-$ respectively for the system Na$^+$–DNA dissolved in NaCl which is in excess compared to the concentration n_e of ionised phosphate sites. In the range of concentrations $X \in [0.2 \to 1.2]$ the values obtained for γ_{Na^+} and γ_{Cl^-} are compatible with Manning's theoretical values.

A study of the thermal transconformation of DNA has allowed us to follow the evolution of the charge parameter ξ and the length b as functions of the temperature.

If we suppose that the helix configuration is that of DNA form B$\xi_h = 4.2$), we can deduce that for the coil configuration, the value of b_{coil} is smaller than the theoretical value of 7 Å corresponding to state where each nucleotidic thread is completely unwound and stretched out.

Lastly we have been able to determine a value for the change of enthalpy as the melting point (T_m) is crossed.

List of Symbols

Latin Letters

a	Radius of the DNA double helix or radius of the cylindrical model
a_i	Ionic activity of ion i
b	Distance between two charge sites projected onto the cylindrical axis
B	Bjerrum parameter
c	in superscript = coil state of DNA
e	Absolute value of electronic charge
E	Nernst potential
E_{pot}	Potential energy
f^*	Degree of binding
h	in superscript = helix state of DNA Helix parameter
H	Enthalpy
k	Boltzmann constant
l	Landau length
$\mathscr{L}$	The constant $e^2/4\pi\varepsilon_0 k$
L	Length of the rod-like cylinder equivalent to DNA macromolecule
m	Concentration of DNA in gram litre^{-1}
m_{M^+}	Molar concentration of counter-ions

Alan Rembaum and Eric Sélégny (eds.), Polyelectrolytes and Their Applications, 197–216. All Rights Reserved.
Copyright © 1975 by D. Reidel Publishing Company, Dordrecht-Holland.

n	Number of cations (or anions) per m^3 in a 1.1. aqueous electrolyte
n_1	Molar concentration of counter-ions
n_2	Molar concentration of co-ions
n_e	Molar concentration of phosphate sites onto DNA
n_S	Molar concentration of the added salt
P	Number of monovalents sites on DNA
P_{M^+}	Colog of counter ions concentration
r_{12}	Distance between two ions 1 and 2
r_D	Debye length
R	Distance of a point from the axis of the cylinder
$\mathscr{R}$	Ideal gaz constant
T	Absolute temperature
T_m	Melting point of DNA
$X = n_e/n_S$	

Greek Letters

γ_1	Activity coefficient of counter-ions
γ_2	Activity coefficient of co-ions
ε_0	Permittivity in free space
ε_r'	Real part of the relative permittivity of water.
η	$\eta = \xi_c^{-1} - \xi_h^{-1}$
ξ	– Generalised Bjerrum parameter
	– Charge parameter of DNA
ϖ	Potential energy in kT units
ϕ	Angular parameter of DNA helix (in cylindrical coordinates)
χ^{-1}	Debye length
χ_+	$\chi/\sqrt{2}$
ψ	Electrostatic potential
	Extent of binding

1. Introduction

The purpose of this work is to present a synthesis of the important facts known about DNA when considered as a polyelectrolyte. Amongst the numerous physico-chemical properties of this fundamental bio-molecule, its behavior in aqueous salt solution is particulary important since it reflects closely the conditions *in vivo*. We will center the discussion on the interaction between the polyion and its surrounding ionic atmosphere, leaving apart the dielectric properties of the solution. This latter point has already been largely discussed (see References 1–5).

In an attempt to clarify the situation, we have endeavoured to present the main theoretical models using the same notation and the same parameters whilst retaining the liaison with the parameters which characterise an electrolyte 1.1 in aqueous solution. This procedure has allowed us to remark a certain singularity concerning the macromolecule DNA. A series of experimental results obtained in our laboratory on the system Na^+–DNA are then presented. These results concern the activity of the counter-ion Na^+ and the co-ion Cl^-, both during the thermal transconformation and in the two conformal extremes: helix and coil. The study of the activity of the counter-ion during the process of denaturation of DNA is important because it allows us to show up a mechanism of expulsion of this ion during the destruction of the double-helix [6]. This duality – 'ionic interaction – change of conformation' is then discussed on the basis of Manning's model [7, 8].

2. The Known Conformation and the Existing Models

The polyelectrolyte DNA consists of two parts: a macromolecular negative polyion, and an electrolyte in aqueous solution. It is known that the exact nature of the counter-ion (metallic cation) controls the structure of the DNA [9]. In solution there exist two principal conformations: form B for counter-ions such as Na^+ and form C for the counter-ion Li^+. A bivalent counter-ion such as Mg^{++} does not seem to deform the form B, but it is known, however, that Cu^{++} can destroy the structure by penetrating the molecule to reach the prefered fixation sites at the level of the coupled bases G–C. [10–13]. Thus Na^+ is a counter-ion whose action is purely external (electrostatic interaction with the phosphate sites moderated by thermal agitation) and which maintains the DNA in form B, stable and with cylindrical symmetry.

2.1. THE MACROION

At this point it is desirable that we should explain exactly what we mean by a phosphate site. We will base the discussion on model 3 of form B, described by Langridge et al. [14], where we can give the average position of a phosphate site in terms of cylindrical coordinates (see Figure 1).

If we suppose that an electron is exchanged between the oxygen atoms O_3 and O_2 bound to a phosphorous atom (O_1 and O_4 participate directly in the chain ribose-phosphate) it is possible to trace an average helix passing through all the atoms O_2 and O_3 of a nucleotidic chain. This helix contains, by definition, the points having a charge $-e$ and which we will call phosphate sites. If we suppose that the

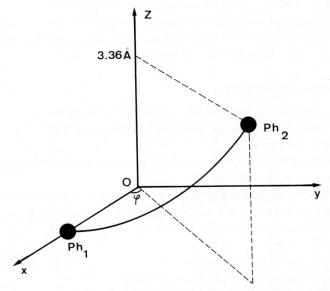

Fig. 1.   Mean helix arc between two neighbour phosphate sites plotted in cylindrical coordinates for the B form of DNA (model 3 of Langridge et al.).

thread of the helix is equal to 33.6 Å per turn with 10 phosphate sites per turn and if the average radius of the helix containing these phosphate sites is given by

$$\langle a \rangle = \frac{r_{O_3} + r_{O_2}}{2} = \frac{10.33 + 9.14}{2}$$

$$\langle a \rangle = 9.73 \text{ Å}$$

then the parametric equation of this helix is

$$x = a \cos \phi = 9.73 \cos \phi$$

$$y = a \sin \phi = 9.73 \sin \phi$$

$$z = h\phi = 5.347\phi \qquad \left(h = \frac{33.6}{2\pi} \text{ Å rad}^{-1} \right)$$

where x, y and z are expressed in Å and ϕ in radians. If we place one of the phosphate sites Ph_1 on the x axis ($x = 9.73$ Å; $y = z = 0$), the coordinates of the next site Ph_2 are $x = 7.87$ Å; $y = 5.719$ Å; $z = 3.36$ Å ($\phi = 2\pi/10$). The length of the arc $Ph_1 - Ph_2$ is

$$Ph_1 - Ph_2 = \frac{36\pi}{180} ((9.73)^2 + (5.347)^2)^{1/2} =$$

$$= 6.975 \text{ Å}.$$

The length of the arc is thus close to 7 Å. If we now consider an anti-parallel double helix structure, then the projection of all the phosphate sites onto the z axis will yield $(3.36/2)$ Å $= 1.68$ Å $\simeq 1.7$ Å between consecutive projections.

It is thus possible to consider two types of model: the DNA is supposed to be a cylinder of radius a and length L with a uniform charge density on its surface; or alternatively a rod of length L with a discrete charge $-e$ every 1.7 Å.

2.2. THE ELECTROLYTE SOLVENT

If we confine ourselves to the case of a symmetrical electrolyte 1.1 in aqueous solution, it is possible to describe the solution, in normalised MKSA units, in terms of its characteristic parameters.

In the absence of an external electric field, we need consider only the thermal noise of the ions and the water [15].

We suppose that the electrolyte contains n cations (or anions) per m³, that the minimum possible distance between ions is d (hard sphere model) and that $\langle r_{12} \rangle$ is the average distance between two ions 1 and 2. In a volume $V = 1$ m³ we then have $\langle r_{12} \rangle = (V/2n)^{1/3}$.

If ε_0 is the permittivity of free space,

ε_r' is the real part of the permittivity of water,

k is Boltzmann's constant, and

T is the absolute temperature, we can introduce;

the universal constant:

$$\mathscr{L} = \frac{e^2}{4\pi\varepsilon_0 k} = 1.669 \times 10^{-5}\ \text{m} \times (^\circ\text{K})$$

the Landau length:

$$l = \frac{\mathscr{L}}{\varepsilon_r' T}$$

the generalised Bjerrum parameter:

$$\xi = \frac{\mathscr{L}}{\varepsilon_r' T \langle r_{12} \rangle} = \frac{l}{\langle r_{12} \rangle}$$

the Bjerrum parameter:

$$B = \frac{\mathscr{L}}{\varepsilon_r' T d} = \frac{l}{d} \quad \text{and}$$

the Debye length:

$$\chi^{-1} = r_D = (8\pi n l)^{-1/2}.$$

The electrostatic interactions between ions are not important if

$$l \ll \langle r_{12} \rangle$$

and the electrolyte may be considered as a plasma. Now, at 25 °C, $\varepsilon_r' = 78.33$ and $l = 7.155$ Å. For molar concentrations of 10^{-4} M, 10^{-2} M and 10^{-1} M we have $\langle r_{12} \rangle = 203$ Å, 43.6 Å and 20.2 Å respectively. We may thus assimilate an electrolyte 1.1 to a plasma for concentrations of less than 10^{-2} M.

3. The Models

We will recall in this paragraph, in our rationalised notation, a non-exhaustive list of the most important results arising from the principal theoretical models which have been proposed and which may be adapted to the case of DNA. A critical comparison of these various models has already been published by Manning [16]. All the models suppose, as a starting point, that DNA is rigid rod with cylindrical symmetry and proceed, in general, by a resolution of the Poisson-Boltzmann equation for the system polyion plus electrolyte (first treated by Alfrey et al. [17] and Fuoss et al. [18]).

In order to fix our ideas let us recall, following Oosawa [19], that for a cylinder of length L and radius a containing P monovalent sites (charge e) distributed uniformly over the surface of the cylinder, the potential energy drop E_{pot} of a monovalent ion situated at a distance R from the axis is:

$$|E_{pot}| = \frac{Pe^2}{2\pi\varepsilon_0\varepsilon_r' L} \log\left(\frac{R}{a}\right) \qquad \text{(MKSA units)}$$

If we reduce this energy to unities of kT, we find:

$$\varpi = \frac{|E_{\text{pot}}|}{kT} = \frac{e^2}{4\pi\varepsilon_0 k} \frac{2P}{\varepsilon_r' TL} \log\left(\frac{R}{a}\right) = 2\frac{\mathscr{L}}{\varepsilon_r' T} \frac{P}{L} \log\frac{R}{a}$$

$$\varpi = 2\frac{l}{b} \log\left(\frac{R}{a}\right),$$

where $b = L/P$ is the distance between two charge sites (projected onto the z axis).

For DNA form B we have already seen that $b = 1.7$ Å and $a = 9.73$ Å. At 25 °C, $l = 7.15$ Å so that at a distance $R = 100$ Å

$$\varpi \simeq 19.$$

In the model developed by Kotin and Nagasawa [20] the 'degree of binding' is defined by the expression:

$$f^* = 1 - \left(\frac{2Pe^2}{\varepsilon kTL}\right)^{-1} \qquad \text{(u.e.s.c.g.s.)}.$$

In MKSA units and our notation this becomes

$$f^* = 1 - \frac{1}{2P}\frac{L}{l} = 1 - \frac{1}{2}\frac{b}{l}$$

introducing the charge parameter for DNA

$$\xi = \frac{l}{b}$$
$$f^* = 1 - \tfrac{1}{2}\xi^{-1}.$$

For DNA in the helix conformation ($b \simeq 1.7$ Å and $\xi_h = 4.7$) we thus find

$$f^* = 0.88.$$

According to Katchalsky et al. [21] the resolution of the Poisson-Boltzmann equation in the presence of added salt leads to a definition of a minimum distance of approach, R_0, for a free counter-ion such that:

$$\left|\frac{e\psi(R_0)}{kT}\right| = 1,$$

where $\psi(R_0)$ is the electrostatic potential created by the cylinder at a distance R from its axis.

If $|e\psi(R)| > kT$, the counter-ions are immobilised
If $|e\psi(R)| < kT$, the counter-ions are free.

The thickness of the atmosphere of 'immobilised' counter-ions is given by

$$\chi_+ R_0 = 0.7 \pm 0.3,$$

where $\chi_+^2 = 4\pi n l$ in our notation. (n is the number of cations per m³.)

If we consider the case of an electrolyte 1.1 at a concentration of 10^{-3} M, we have

$$n = 6.10^{23} \text{ cations/m}^3$$
$$r_D^+ = \chi_+^{-1} = 136 \text{ Å}$$

so

$$R_0 \simeq 0.7 \times 136 \simeq 95 \text{ Å}.$$

This sheath of immobilised counter-ions is considerably reduced if we increase the concentration of the electrolyte, for a concentration of 10^{-1} M

$$R_0 = 9.5 \text{ Å}$$

Under these conditions the interaction between polyions can be appreciable. It seems then that the range of salt concentrations 10^{-3} M to 10^{-2} M is particularly interesting because it is compatible both with the notion of the electrolyte as a plasma and with the non-agregation of the polyions.

Lastly we point out, that for this series of models, Daune [22] has extended the Lifson-Katchalsky potential for a single helix to the case of the double helix of DNA and concludes that for concentrations of added (monovalent) salt less than 5×10^{-2} M, there should be an ejection of counter-ion M^+ of magnitude

$$\Delta P_{M^+} = 0.52,$$

where $P_{M^+} = \text{colog}[M^+]$.

Manning [7, 8] does not make a clear distinction between these 'free' and 'immobilised' counter-ions. In his 'Limiting Laws' which he has developped for polyelectrolytes, he has succeeded in linking the activity of the counter and co-ions with the concentrations of DNA and the added salt. Moreover the phenomenon of ejection of the counter-ion during the transconformation helix-coil may be accounted for with the help of the charge parameter ξ. We take as a starting point the definition of this funcdamental parameter:

$$\xi = \frac{e^2}{4\pi\varepsilon_0\varepsilon_r' kTb} = \frac{l}{b}.$$

If $\xi > 1$ the system is unstable and the counter-ions will condense on to until the limiting value of $\xi = 1$ is reached. It is interesting to note that the critical point obtained corresponds, at 25 °C to $b = 7.15$ Å ($l = 7.15$ Å). Thus b corresponds almost exactly to the length of the arc of the helix Ph_1-Ph_2 calculated above. The Landau length varies little in the temperature range 20–80 °C and this assertion is thus valid for all the conditions under which DNA in solution is usually studied (see Table VI).

Let n_1 and n_2 be the molar concentrations of the counter and co-ions respectively.

$$n_1 = n_e + n_s$$
$$n_2 = n_s,$$

where n_e is the molar concentration of phosphate sites and n_s is the molar concentration of the added salt.

The activities of the counter and co-ions are:

$$a_1 = \gamma_1 n_1$$
$$a_2 = \gamma_2 n_2,$$

where the γ_1, γ_2 are the activity coefficients.

Manning calculates the following expressions:

$$\left\{ \begin{array}{l} \gamma_1 = \dfrac{(\xi^{-1}X + 1)}{(X+1)} \exp\left(-\dfrac{1}{2} \dfrac{\xi^{-1}X}{(\xi^{-1}X+2)} \right) \\[3mm] \gamma_2 = \exp\left(-\dfrac{1}{2} \dfrac{\xi^{-1}X}{(\xi^{-1}X+2)} \right), \end{array} \right.$$

where $X = n_e/n_s$.

For the two limiting phases of DNA (helix and coil denoted by subscripts h and c respectively), the important parameters are:

$$\left\{ \begin{array}{l} \eta = \xi_c^{-1} - \xi_h^{-1} \\[2mm] \dfrac{\gamma_1^h}{\gamma_1^c} = \dfrac{(1 + \xi_h^{-1}X)}{(1 + \xi_c^{-1}X)} \exp \dfrac{\eta X}{(\xi_c^{-1}X + 2)(\xi_h^{-1}X + 2)} \\[3mm] \dfrac{\gamma_2^c}{\gamma_2^h} = \exp \dfrac{-\eta X}{(\xi_c^{-1}X + 2)(\xi_h^{-1}X + 2)} \end{array} \right.$$

We add, lastly, that Manning gives an expression for the change in enthalpy at T_m [8] (melting point of DNA)

$$(\Delta H)_{T_m} = 1.15 \, \mathscr{R} T_m^2 \left(\frac{dT_m}{d\log m_{M^+}} \right)^{-1} \times \eta,$$

where R is the ideal gas constant and m_{M^+} the concentration of the counter-ions.

We have conducted a series of experiments in our laboratory in order to explore the limits of validity of Manning's model and his 'Limiting Laws' as applied to DNA form B in the helix and coil phases.

4. Experimental Study of the Activities of the Counter and Co-ions Associated with DNA in a Solution of NaCl

It is important to realise that a measurement of the activities a_{Na^+} and a_{Cl^-} for the helix and coil states of DNA is sufficient to test the validity of Manning's model

[23]. In order to obtain the values of these ionic activities experimentally, we have developed a technique using glass-membrane or single cristal electrodes which are selectively sensitive to Na^+ or Cl^- ions [24, 25].

The potentiometric measurements were performed with a sensibility threshold fixed at 0.1 mV, the thermal noise of the electrodes is estimated to be less than the threshold. The reproductibility of our measurements indicates that the precision with which we can determine the Nernst potential of the electrode is approximately 0.5 mV [26].

The Nernst potential of the selective electrode in thermal equilibrium is linked to the activity of the corresponding ion by the expression

$$E = E_0 + \frac{kT}{e} \log a,$$

where E is the potential difference between a calomel reference electrode and the selective electrode. E_0 is a constant for the pair of electrodes and for the same type of ion, and a the activity of the ion under investigation.

The electrodes are calibrated, using solutions of known activity, before every measurement following the procedures described in References 24 and 25.

Our experimental measurements were performed within a limited range of concentrations, both of the salt solution in which the DNA is dissolved and for the DNA itself.

The salt solutions were confined to the range of concentrations from 10^{-3} M to 10^{-2} M. This is because we must have $n_s < 10^{-2}$ M to be able to consider the solution as a plasma and $n_s > 10^{-3}$ M in order to have a good signal from the electrode. As for the DNA, if m is the concentration in grams per liter, the molar concentration of ionised sites for DNA dissolved in NaCl is given by

$$n_e = \frac{m}{331}$$

331 being the average molecular mass of a nucleotidic unity compensated by Na^+.

We have used concentrations in the range 0.3 g/l $< m <$ 0.8 g/l. If the solution is too concentrated, agregates are formed and have a masking effect on the electrode, if the solution is too weak it is no longer possible to detect the ejection of the counterions. Our sources of DNA were from calf-thymus and from E.-Coli.

During the thermal transconformation of DNA, the activities a_i of the surrounding ions change. From the ratio of the ionic activities a_h and a_c measured for the helix and coil phases, it is possible to deduce the ratio $(\gamma_h/\gamma_c)_{exp}$.

This ratio may be compared directly with the theoretical value proposed by Manning. Table I shows the results thus obtained for the counter-ion sodium.

We see that the experimental results agree with theory to within 15%; however, for solutions too concentrated in DNA (e.g. 0.8 g/l DNA in 2 10^{-3} M NaCl) or solutions too weak in added salt, Manning's model seems to be inapplicable. We have

TABLE I

Theoretical and experimental ratio of activity coefficients in helix and coil states $\gamma^h_{Na+}/\gamma^c_{Na+}$

DNA	DNA Conc.	n_e	n_s	X	a^h_{Na+}	a^c_{Na+}	$\dfrac{\gamma_h}{\gamma_c}$ exp	$\dfrac{\gamma_h}{\gamma_c}$ th
V. 378	0.375 g/l	1.133×10^{-3}	5×10^{-3}	0.227	4.7×10^{-3}	5.2×10^{-3}	0.904	0.892
					2.9×10^{-3}	3.6×10^{-3}	0.805	
					2.9×10^{-3}	3.7×10^{-3}	0.784	
V. 11	0.325 g/l	0.982×10^{-3}	3×10^{-3}	0.327				0.855
					3×10^{-3}	3.6×10^{-3}	0.833	
					3×10^{-3}	3.7×10^{-3}	0.812	
						2.45×10^{-3}	0.857	
E. Coli	0.343 g/l	1.036×10^{-3}	2×10^{-3}	0.518	2.1×10^{-3}			0.796
						2.5×10^{-3}	0.84	
					4.7×10^{-3}	5.1×10^{-3}	0.922	
V. 11	0.8 g/l	2.417×10^{-3}	4.5×10^{-3}	0.537				0.796
					4.8×10^{-3}	5.1×10^{-3}	0.941	
V. 378	0.4 g/l	1.2×10^{-3}	2×10^{-3}	0.6	2×10^{-3}	2.5×10^{-3}	0.80	0.775
						2.6×10^{-3}	0.846	
E. Coli 6	0.47 g/l	1.42×10^{-3}	2×10^{-3}	0.71	2.1×10^{-3}			0.749
						2.65×10^{-3}	0.830	
					3.6×10^{-3}	5×10^{-3}	0.72	
V. 11	0.77 g/l	2.326×10^{-3}	3×10^{-3}	0.775				0.735
					3.7×10^{-3}	5×10^{-3}	0.74	
					2.2×10^{-3}		0.786	
E. Coli 6	0.62 g/l	1.873×10^{-3}	2×10^{-3}	0.936		2.8×10^{-3}		0.704
					2.3×10^{-3}		0.821	
						1.65×10^{-3}	0.727	
V. 7	0.375 g/l	1.133×10^{-3}	10^{-3}	1.133	1.2×10^{-3}			0.672
						1.7×10^{-3}	0.796	
E. Coli 6	0.4 g/l	1.2×10^{-3}	10^{-3}	1.2	1.2×10^{-3}	1.76×10^{-3}	0.68	0.662
V. 378	0.8 g/l	2.417×10^{-3}	2×10^{-3}	1.208	2.3×10^{-3}	2.9×10^{-3}	0.793	0.661
V. 7	0.73 g/l	2.205×10^{-3}	10^{-3}	2.205	1.49×10^{-3}	2.02×10^{-3}	0.737	0.557

n_e	molar concentration of DNA ionized sites.
n_s	molar concentration of monovalent added salt.
$X =$	n_e/n_s.
a^h_{Na+}	activity of free sodium ions in native solution at 20 °C, DNA being assumed to exist exclusively in the helix form.
a^c_{Na+}	activity of free sodium ions in thermally denatured solutions at 70 °C, the DNA being assumed to exist exclusively in the coil form; measurements taken when hot.
$(\gamma_h/\gamma_c)^{exp}$	Ratio of activity coefficients of sodium corresponding to the helix and coil states of DNA, obtained from the corresponding ratio of activities measured by selective electrodes.
$(\gamma_h/\gamma_c)_{exp}$	Ratio of activity coefficients of sodium corresponding to the helix and coil states of DNA, obtained from the corresponding ratio of activities measured by selective electrodes.
$(\gamma_h/\gamma_c)_{th}$	Same ratio as the previous one, obtained from Manning's theoretical model.

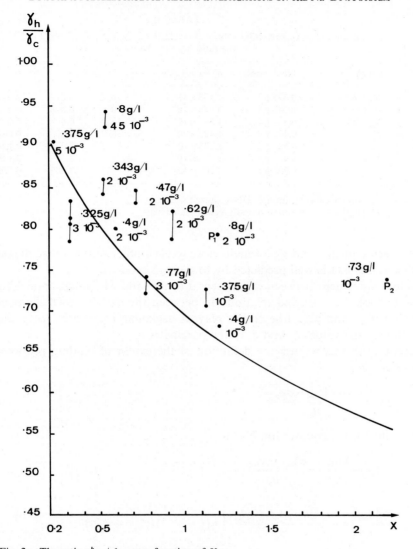

Fig. 2.   The ratio $\gamma_{Na^+}^h/\gamma_{Na^+}^c$ as a function of X.
· The solid curve represents the theoretical ratio given by Manning's model.
· The dots represent the experimental values obtained by means of the electrodes.

carried out an analogous study of the co-ion Cl^- [27]. For this ion the ratio of the activity coefficients may be written:

$$\frac{\gamma_{Cl^-}^c}{\gamma_{Cl^-}^h} = 1 - \varepsilon \quad \text{with} \quad \varepsilon = \frac{\eta X}{(2 + \xi_c^{-1}X)(2 + \xi_h^{-1}X)}.$$

The values of ε, calculated for a number of experimental points, are shown in Table II.

TABLE II

Calculation of experimental values of $\varepsilon = [(\xi_c^{-1} - \xi_h^{-1})\,X]/[(2 + \xi_c^{-1}X)\,(2 + \xi_h^{-1}X)]$ in the ratio $(\gamma_{Cl^-}^c / \gamma_{Cl^-}^h) = 1 - \varepsilon$

DNA	DNA conc.	n_e	n_s	X	ε
V. 378	0.375 g/l	1.133×10^{-3}	5×10^{-3}	0.277	0.038
E. Coli 6	0.343 g/l	1.036×10^{-3}	2×10^{-3}	0.518	0.073
V. 378	0.4 g/l	1.2×10^{-3}	2×10^{-3}	0.6	0.082
E. Coli 6	0.47 g/l	1.42×10^{-3}	2×10^{-3}	0.71	0.092
E. Coli 6	0.62 g/l	1.873×10^{-3}	2×10^{-3}	0.936	0.109
E. Coli 6	0.4 g/l	1.2×10^{-3}	10^{-3}	1.2	0.125
V. 7	0.73 g/l	2.205×10^{-3}	10^{-3}	2.205	0.158

n_e molar concentration of DNA ionized sites.
n^s molar concentration of monovalent added salt.
X n_e/n_s.

It can be seen that $\gamma_{Cl^-}^c$ and $\gamma_{Cl^-}^h$ remain close to each other in this range of concentrations and this fact is well predicted by Manning's model.

The close agreement between experimental results and Manning's model has incited us to look for a second relationship between the activity coefficients of the counter-ion $\gamma_{Na^+}^h$ and $\gamma_{Na^+}^c$ (the co-ion plays a negligible role) so as to be able to obtain directly the values ξ_h and ξ_c of the parameter ξ.

To accomplish this we use the definition of the extent of binding proposed by Archer et al. [28]

$$\psi = \frac{[Na^+]_{\text{total}} - a_{Na^+}}{n_e}$$

we may write, for the counter-ion Na^+:

$$\Delta\psi = \frac{\Delta a_{Na^+}}{n_e} = \frac{a_{Na^+}^h - a_{Na^+}^c}{n_e}$$

or

$$\frac{\Delta a_{Na^+}}{n_e} = \frac{(n_e + n_s)\,(\gamma_{Na^+}^c - \gamma_{Na^+}^h)}{n_e} = (1 - X^{-1})\,(\gamma_{Na^+}^c - \gamma_{Na^+}^h)$$

we thus have

$$\gamma_{Na^+}^c - \gamma_{Na^+}^h = \frac{\Delta a_{Na^+}}{n_e} \cdot \frac{1}{1 + X^{-1}}.$$

The experimental values of $\gamma_{Na^+}^c$ and $\gamma_{Na^+}^h$ thus calculated are given in Table III.

Using Manning's expression for the activity of the counter-ion

$$\gamma_{Na^+} = (\xi^{-1}X + 1)(X + 1)^{-1}\,\exp{-\frac{1}{2}\frac{\xi^{-1}X}{(\xi^{-1}X + 2)}}$$

we can thus calculate independantly of each other the values of ξ_h and ξ_c correspond-

TABLE III

Activity coefficients in helix and coil states ($\gamma^h_{Na^+}$ and $\gamma^c_{Na^+}$) obtained from our experiments

DNA	Δa_{Na^+}	n_e	$\Delta\psi$	X^{-1}	$(\gamma^c_{Na^+} - \gamma^h_{Na^+})_{exp}$	$(\gamma^c_{Na^+})$	$(\gamma^h_{Na^+})$
V 378	0.5×10^{-3}	1.133×10^{-3}	0.442	4.405	0.0817	0.851	0.769
V 11	0.7×10^{-3}		0.713		0.1757	0.901	0.725
	0.8×10^{-3}	0.982×10^{-3}	0.814	3.058	0.2006	0.929	0.728
	0.6×10^{-3}		0.611		0.1505	0.901	0.750
	0.7×10^{-3}		0.713		0.1757	0.934	0.759
E. Coli 6	0.35×10^{-3}	1.036×10^{-3}	0.338	1.930	0.1153	0.812	0.696
	0.4×10^{-3}		0.386		0.1317	0.823	0.691
V 11	0.4×10^{-3}	2.417×10^{-3}	0.165	1.862	0.0576	0.738	0.681
	0.3×10^{-3}		0.124		0.0433	0.734	0.691
V 378	0.5×10^{-3}	1.2×10^{-3}	0.416	1.667	0.156	0.78	0.624
E. Coli 6	0.4×10^{-3}	1.42×10^{-3}	0.282	1.408	0.117	0.759	0.643
	0.45×10^{-3}		0.317		0.1316	0.774	0.642
V 11	1.4×10^{-3}	2.326×10^{-3}	0.602	1.290	0.263	0.939	0.676
	1.3×10^{-3}		0.559		0.2441	0.939	0.695
E. Ccoli 6	0.6×10^{-3}	1.873×10^{-3}	0.320	1.068	0.1547	0.723	0.568
	0.5×10^{-3}		0.267		0.1291	0.721	0.592
V 7	0.45×10^{-3}	1.133×10^{-3}	0.397	0.883	0.2108	0.772	0.561
	0.5×10^{-3}		0.441		0.2342	0.797	0.562
E. Coli 6	0.56×10^{-3}	1.2×10^{-3}	0.466	0.833	0.2542	0.794	0.540
V 378	0.6×10^{-3}	2.417×10^{-3}	0.248	0.828	0.1357	0.656	0.520
V 7	0.53×10^{-3}	2.205×10^{-3}	0.240	0.453	0.1651	0.627	0.463

n_e molar concentration of DNA ionized sites.
n_s molar concentration of monovalent added salt.
a_{Na^+} activity of free sodium ions.
$X = n_e/n_s$.
$\Delta\psi = \Delta a_{Na^+}/n_e$.

ing to the activity coefficients γ_{Na^+} listed in Table III. The resulting values are shown in Table IV.

The average values found for the charge parameter for the ensemble of our measurements are:

$$\langle \xi_h \rangle = 5.3$$
$$\langle \xi_c \rangle = 1.6$$

from which

$$\eta = 0.43$$

DANE VASILESCU ET AL.

TABLE IV

ξ_h and ξ_c parameters obtained from some of our experiments

DNA	X	$\gamma^h_{Na^+}$	ξ_h^{-1}	ξ_h	$\gamma^c_{Na^+}$	ξ_c^{-1}	ξ_c
E. Coli 6	0.518	0.969	0.1544	6.475	0.812	0.618	1.619
		0.691	0.1255	7.969	0.823	0.676	1.48
V 11	0.537	0.681	0.112	8.95	0.738	0.335	2.983
		0.691	0.149	6.712	0.734	0.326	3.068
E. Coli 6	0.71	0.643	0.190	5.259	0.759	0.547	1.829
		0.642	0.183	5.461	0.774	0.634	1.578
V II	0.775	0.676	0.316	3.163	0.939	1.238	0.807
		0.695	0.413	2.422	0.939	1.238	0.807
E. Coli 6	0.936	0.568	0.139	7.20	0.723	0.598	1.671
		0.593	0.214	4.68	0.721	0.598	1.671
V 7	1.133	0.561	0.238	4.196	0.772	0.821	1.219
		0.562	0.238	4.196	0.797	0.891	1.121
E. Coli 6	1.2	0.540	0.217	4.616	0.794	0.9	1.11
V 378	1.208	0.520	0.166	6.04	0.656	0.521	1.917
V 7	2.205	0.463	0.308	3.243	0.627	0.676	1.480

n_e molar concentration of DNA ionized sites.
n_s molar concentration of monovalent added salt.
$\gamma^h_{Na^+}$ activity coefficient of Na$^+$ in helix state.
$\gamma^c_{Na^+}$ activity coefficient of Na$^+$ in coil state.

It is now interesting to deduce the values of Manning's conformation parameter, b, which describes the distance between two phosphate sites in the two phases helix and coil.

$$\langle b_h \rangle = 1.3 \text{ Å}$$
$$\langle b_c \rangle = 4.7 \text{ Å}.$$

5. Discussion

It may be remarked that the average value $\langle b_c \rangle$ is rather far from the value of 7 Å chosen by Manning and which corresponds to the distance between two phosphate sites, measured along the arc of a helix.

This disagreement may be explained since Manning supposed that in the 'coil' phase, the two polynucleotidic strands are completely separated and stretched out straight. We conclude that after thermal denaturation, the molecule of DNA is in a form which is only partly unwound. This change is accompanied by an ejection of the counter-ions which is revealed in our measurements by Δa_{Na^+}. This fact has been confirmed by dielectric measurements [29] noise measurements [15] and also by recent investigations by Schmitz and Schurr using the technique of inelastic light scattering [30].

It is not possible, however, to accept the value $\langle b_h \rangle = 1.3$ Å for the double helix structure since, as we have already seen, the minimum value for DNA form B is 1.7 Å. It is not possible to compress to a greater extent the successive plateaux formed by the pairs of bases without decreasing appreciably the accepted radii of the Van der Waals' forces.

It is now interesting to reverse our reasoning and to recalculate the value of $\langle \xi_c \rangle$ using for ξ_h the more likely value of 4.2 corresponding to $b = 1.7$ Å. The only experimental value we use is the ratio $(\gamma^h_{Na^+}/\gamma^c_{Na^+})_{exp}$ given theoretically by the expression

$$\frac{\gamma^h_{Na^+}}{\gamma^c_{Na^+}} = \frac{(1 + 0.24X)}{(1 + \xi_c^{-1}X)} \exp \frac{(\xi_c^{-1} - 0.24)\,X}{(2 + \xi_c^{-1}X)(2 + 0.24X)}.$$

The values thus obtained for ξ_c are listed in Table IV. The average value for ξ_c in the coil phase is

$$\langle \xi_c \rangle = 1.26$$

and

$$\langle b_c \rangle = 6.08 \text{ Å}$$

which implies

$$\eta = 0.55$$

The new value for $\langle b_c \rangle$ is still less than that chosen by Manning.

TABLE V

ξ_c calculated from Manning's expression:

$$\frac{\gamma^h_{Na^+}}{\gamma^c_{Na^+}} = \frac{1 + \xi_h^{-1}X}{1 + \xi_c^{-1}X} \exp \frac{(\xi_c^{-1} - \xi_h^{-1})\,X}{(2 + \xi_c^{-1}X)(2 + \xi_h^{-1}X)}$$

DNA	X	$(\gamma^h_{Na^+}/\gamma^c_{Na^+})_{exp}$	ξ_c^{-1}	ξ_c
V 378	0.277	0.904	0.915	1.092
V 11	0.327	0.805	1.36	0.735
		0.784	1.49	0.671
		0.833	1.16	0.862
		0.812	1.31	0.763
E. Coli 6	0.518	0.857	0.745	1.342
		0.84	0.818	1.222
V 11	0.537	0.922	0.485	2.062
		0.941	0.42	2.38
V 378	0.6	0.80	0.915	1.092
E. Coli 6	0.71	0.846	0.665	1.5037
		0.830	0.73	1.370

Table V (Continued)

DNA	X	$(\gamma^h_{Na^+}/\gamma^c_{Na^+})_{exp}$	ξ_c^{-1}	ξ_c
V 11	0.775	0.72 0.74	1.098 1.01	0.910 0.990
E. Coli 6	0.936	0.786 0.821	0.75 0.645	1.33 1.55
V 7	1.133	0.727 0.706	0.855 0.925	1.170 1.081
E. Coli 6	1.2	0.68	0.985	1.015
V 378	1.208	0.793	0.647	1.54
V 7	2.205	0.737	0.61	1.639

$\gamma^h_{Na^+}/\gamma^c_{Na^+}$ is given by our experimental results.
$\xi_h = 4.2$.

TABLE VI
ξ_c and b_c as functions of temperature from:

$$\frac{\gamma^h_{Na^+}}{\gamma_{Na^+}} = \frac{1+\xi_h^{-1}X}{1+\xi^{-1}X} \exp\frac{(\xi^{-1}-\xi_h^{-1})\,X}{(2+\xi^{-1}X)(2+\xi_h^{-1}X)}$$

DNA	X	$T(°C)$	γ_h/γ	ξ	DO	ε_r'	l	b
V 378	0.227	20°C→40°C	1	4.2	0.75	78.33 73.15	25°:7.15 40°:7.29	1.702 1.736
		50°C	0.995	4.0	0.77	69.89	7.39	1.85
		60°C	0.953	1.818	0.88	66.78	7.51	4.13
		65°C	0.928	1.333	0.96	65.28	7.56	5.67
		70°C	0.911	1.163	1.0	63.82	7.62	6.55
		>70°C	0.904	1.092	1.04	62.38 60.98	75°:7.69 80°:7.75	7.04 7.09
V 7	1.133	20°C→45°C	1	4.2	0.72	78.33 73.15	25°:7.15 40°:7.29	1.70 1.736
		50°C	0.923	2.564	0.84	69.89	7.39	2.88
		55°C	0.857	1.96	0.93	68.32	7.45	3.80
		60°C	0.774	1.235	0.97	0.97	66.78	6.08
		≥70°C	0.706	1.081	1.01	63.82 60.98	70°:7.62 80°:7.75	7.05 7.17
V 7	2.205	20°C→40°C	1	4.2	1.46	78.33 73.15	25°:7.15 40°:7.29	1.702 1.736
		45°C	0.898	2.778	1.55	71.50	7.34	2.64
		50°C	0.814	2.105	1.70	69.89	7.39	3.51
		55°C	0.772	1.852	1.82	68.32	7.45	4.02
		60°C	0.749	1.695	1.87	66.78	7.51	4.43
		≥70°C	0.737	1.639	1.90	63.82 60.98	70°:7.62 80°:7.75	4.65 4.73

$\gamma^h_{Na^+}/\gamma_{Na^+}$ is given by our experimental results at different temperatures.
$\xi_h = 4.2$.
$b = l\xi^{-1}$, where l is Landau length $1.669 \times 10^{-5}/\varepsilon_r'(T)\,T$ (in meters).
ε_r' relative permittivity of solvent.

Keeping the same assumption ($\xi_h = 4.2$) we have calculated the change in the charge parameter ξ during thermal transconformation, as a function of temperature (Table VI and Figure 3).

Figures 4 and 5 show respectively the change in the distance between successive phosphate sites as a function of temperature, $b(T)$, and the optical density of the same solutions. It is interesting to note the correspondance between these last two figures.

This semi-empirical calculation of the parameter η based on the assumption of an only partly unwound coil ($\langle b_c \rangle = 6$ Å, $\eta = 0.55$) now permits us to readjust the value

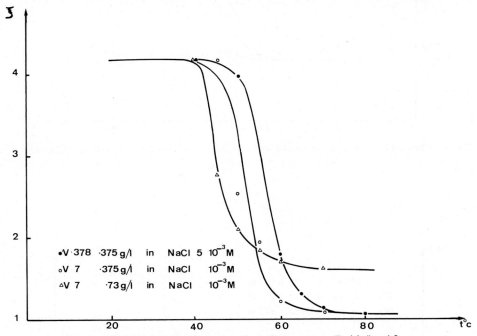

Fig. 3.   Charge parameter ξ as a function of temperature T with $\xi_h = 4.2$.

of the change in enthalpy at T_m calculated by Manning [23]. If we take the theoretical value $\eta = 0.76$ corresponding to $\xi_c = 1$ and $\xi_h = 4.2$ and taking $T_m = 325$ K and $\mathrm{d}\, T_m / \mathrm{d}(\log m_{\mathrm{Na}^+}) = 18$ we find

$$(\Delta H)_{325\,\mathrm{K}} = 10.2 \text{ kcal (mole base)}^{-1}.$$

However, if we take our new value $\eta = 0.55$

$$(\Delta H)_{325\,\mathrm{K}} = 7.4 \text{ kcal (mole base)}^{-1}.$$

Privalov *et al.* [31] have performed calorimetric measurements on DNA solutions with concentrations of salt and DNA different from those used here. They find

$$(\Delta H)_{T_m} = 4.5 \text{ kcal (mole base)}^{-1}.$$

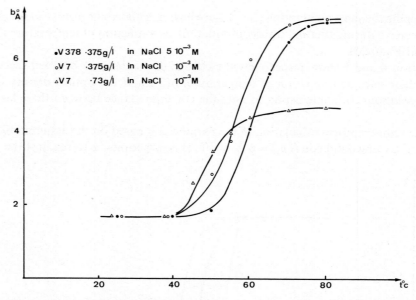

Fig. 4.   Distance b between two phosphate sites as a function of temperature T with $\xi_h = 4.2$ and $b = \xi^{-1}l$.

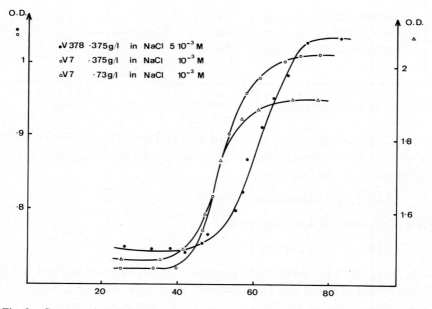

Fig. 5.   Correspondant optical density as a function of temperature T. Measured at 2600 Å.

It may be seen that the large discrepancy between this experimental value and the value of ΔH calculated from a theoretical value for the parameter η is reduced when the semi-empirical value of η is used.

6. Conclusion

DNA is a polyelectrolyte for which the positive counter-ion plays a dominant role. In this respect, if we consider the Landau length for an electrolyte 1.1. supporting the counter-ion, it is remarkable that this latter corresponds almost exactly to the length of the arc of the average helix passing through two consecutive phosphate sites. This fact assures that the charge parameter ξ will have its theoretical value of 1 for the stable states of DNA.

A good model for a polyelectrolyte may be applied to the case of DNA provided that it takes into account.
- the activity of the ions forming the ionic atmosphere around the polyion.
- the ejection of the compensating ions during the transition helix – coil.

On the experimental background which we have developped, Manning's model is the most adept for a description of these phenomena.

Nevertheless it is necessary to make a semi-empirical adjustment of the parameter $\eta = \xi_c^{-1} - \xi_h^{-1}$ in order to agree with the known conformation of DNA in the helix state.

A creditable hypothesis is that of a 'coil' form in which the two strands of the double helix are only partially unwound. This hypothesis leads to a lowering of the calculated value of the enthalpy of fusion.

It remains to be seen if this procedure which we have adopted for the counter-ion sodium remains valid for other mono and bi-valent compensating ions having a purely external action.

Acknowledgement

We express our thanks to G. M. Searby for his help in translating this manuscript.

References

1. Vasilescu, D.: 'Some Electrical Properties of Nucleic Acids and Components', in J. Duchesne (ed.), *Physico-Chemical Properties of Nucleic Acids*, Academic Press, New-York, 1973, T. I, p. 31.
2. Mandel, M.: *Mol. Phys.* **4**, 489 (1961).
3. Hanss, M.: Doctoral Thesis, Paris, 1965.
4. Takashima, S.: *Biopolymers* **5**, 899 (1967).
5. Grassi, H.: Thèse de Spécialité, Marseille, 1969.
6. Vasilescu, D., Fiancette, C., and Mesnard, G.: *Biochim. Biophys. Acta* **129**, 417 (1966).
7. Manning, G. S.: *J. Chem. Phys.* **51**, 924 (1969).
8. Manning, G. S.: *J. Chem. Phys.* **51**, 934 (1969).
9. Eichhorn, G. L.: *Nature (G.B.)* **194**, 474 (1962).
10. Daune, M. and Chambron, J.: *J. Chim. Phys.* **65**, 72 (1968).
11. Zimmer, C. and Venner, H.: *Eur. J. Biochem.* **15**, 40 (1970).
12. Liebe, D. C. and Stuehr, J. E.: *Biopolymers* **11**, 145 (1972).

13. Liebe, D. C. and Stuehr, J. E.: *Biopolymers* **11**, 167 (1972).
14. Langridge, R., Marvin, D. A., Seeds, W. E., Wilson, H. R., Hooper, C. W., Wilkins, M. H. F., and Hamilton, L. D.: *J. Mol. Biol.* **2**, 38 (1960).
15. Vasilescu, D., Teboul, M., Kranck, H., and Gutmann, F.: *Electrochimica Acta* **19**, 181 (1974).
16. Manning, G. S.: *Annual Review of Physical Chemistry*, p. 117, 1972.
17. Alfrey, T., Berg, P. W., and Moravetz, H.: *J. Polym. Sci.*, **7**, 543 (1951).
18. Fuoss, R. M., Katchalsky, A., and Lifson, S.: *Proc. Natl. Acad. Sci.*, **37**, 579 (1951).
19. Oosawa, F.: *Polyelectrolytes*, Marcel Dekker, New-York, 1971.
20. Kotin, L. and Nagasawa, M.: *J. Chem. Phys.*, **36**, 873 (1962).
21. Katchalsky, A., Alexandrowicz, Z., and Kedem, O.: in B. E. Conway and R. G. Barradas (eds.), J. Wiley, New-York, 1966, p. 295.
22. Daune, M.: *Biopolymers* **7**, 659 (1969).
23. Manning, G. S.: *Biopolymers* **11**, 937 (1972).
24. Rix-Montel, M. A.: Thèse de Spécialité, Marseille, 1970.
25. Vasilescu, D. and Rix-Montel, M. A.: *Biochim. Biophys. Acta* **199**, 553 (1970).
26. Rix-Montel, M. A., Grassi, H., and Vasilescu, D.: *Biophysical Chemistry*, in press (1974).
27. Vasilescu, D., Rix-Montel, M. A., and Grassi, H.: *Compt. Rend. Acad. Sci. Paris* **D273**, 1154 (1971).
28. Archer, B. G., Craney, C. L., and Krakauer, H.: *Biopolymers* **11**, 781 (1972).
29. Grassi, H. and Vasilescu, D.: *Biopolymers* **10**, 1543 (1971).
30. Schmitz, K. S. and Schurr, J. M.: *Biopolymers* **12**, 1543 (1973).
31. Privalov, P. L., Ptitsyn, O. B., and Birshtein, T. M.: *Biopolymers* **8**, 559 (1969).

SOME RECENT STUDIES OF DNA IN AQUEOUS SOLUTION: INTERACTIONS WITH CERTAIN DYES AND ANTIBIOTICS

V. CRESCENZI and F. QUADRIFOGLIO

Laboratory of Macromolecular Chemistry, Institute of Chemistry, University of Trieste, Trieste, Italy

Abstract. A review is given of the results of studies recently carried out or in progress in this laboratory aiming at a thermodynamic characterization of the 'strong-binding' by DNA (calf thymus) of the synthetic dyes proflavine and ethidium bromide, of the natural dye zoanthoxanthin, and of the antibiotic daunomycin and actinomycin, respectively, in dilute aqueous solution.

Confirmatory evidence is afforded that calorimetry may be a particularly useful tool in these investigations.

1. Introduction

A number of papers have appeared in the literature during the last thirty years dealing with certain peculiar features exhibited by a variety of ionic dyes in aqueous solutions of polyelectrolytes. In particular dye binding by biopolymers has been studied extensively as it provides an example of small molecule-macromolecule interactions, which are of major importance in biochemistry [1]. Moreover, certain dyes are known to possess antibacterial and mutagenic activities [2].

From the standpoint of the physical chemistry of aqueous solutions many dyes, e.g. the acridines, are also of interest by themselves as they can form dimers, trimers and higher aggregates by 'vertical-stacking' even in very dilute solutions with concomitant, often dramatic, changes in spectral properties [3]. The interplay of driving forces governing such aggregation phenomena has not, however, been clearly understood as yet.

Marked changes in spectral properties of a given dye generally also occur upon binding to an oppositely charged polyelectrolyte in aqueous solution, and this effect is usually advantageous to quantitatively assess and to draw indications on the mechanism of binding by comparison with spectral perturbations brought about by self-aggregation of the dye in water.

With synthetic polyelectrolytes, to summarize the matter in a very qualitative fashion, one generally speaks in terms of dye bound by electrostatic forces in monomeric form by the macroions (for very low dye to polymer concentration ratios) and of bound dye 'stacking' into the macroions. It is worth mention, however, that the tightly coiled globular chains of poly(methacrylic acid) in the *unionized* state in aqueous solution are capable of binding different dyes [4]. In these cases, having to neglect electrostatic interactions one might invoke a sort of dye molecules 'solubilization' within the globular polymer chains [5], a phenomenon reminiscent of that taking place in colloidal electrolyte solutions (viz. dye solubilization into the 'micelles').

With biopolymers, in addition to all binding mechanisms schematically indicated above, very selective interactions may take place using certain dyes through mecha-

Alan Rembaum and Eric Sélégny (eds.), Polyelectrolytes and Their Applications, 217–230. All Rights Reserved.
Copyright © 1975 by D. Reidel Publishing Company, Dordrecht-Holland.

nisms involving *specific sites* along the biopolymer chains and leading in the special case of DNA to 'intercalation' of dye molecules [6]. A typical example in this connection is afforded by the dye ethidium bromide [7].

Recently, the interaction in vitro of antibiotic molecules, like actinomycin, daunomycin, etc·, with DNA has received ever increasing attention in view of the great importance of these systems [8]. Such natural compounds are generally of quite complex molecular structure so that, as a consequence, the mechanism through which they can bind to DNA, not to speak of the mechanism of their biological action, may be a complicated one.

To simplify the matter in a crude way it is convenient for our scope to divide the antibiotics that exert the first stage of their biological activity by interacting with DNA, into two classes: those which would 'intercalate' their aromatic portions into the DNA double-helix, at least under appropriate experimental conditions, (e.g. daunomycin), and those which are linked only 'on-the-surface' of DNA (e.g. distamycin) [9]. Let us emphasize that this classification is made only in the attempt of formulating, on the basis of physico-chemical data, possible criteria to establish whether or not a given antibiotic does 'intercalate' without any presumption of disclosing the real biological implications of the different modes of binding.

As mentioned above, a wealth of spectroscopic, hydrodynamic, etc. type of information is already available for the various systems considered, i.e. synthetic polyelectrolyte-dye as well as biopolymer-drug systems. Of these, however, a comparative detailed thermodynamic characterization is still lacking. For instance, very few calorimetric data which give unambiguous enthalpy of interaction values, and hence of entropy figures with the aid of classical binding constants measurements, have been reported in the literature [10]. The same applies to the case of dyes or antibiotics self-aggregation in water, of which a better thermodynamic picture is also of relevance to our main subject.

The results of the work carried out in the last two years in our laboratory, based essentially on microcalorimetric experiments and which we wish to summarize in what follows, are obviously not meant to fill these gaps. Our research is in fact a first step toward a more complete description of equilibrium properties of aqueous solution of a few dyes and antibiotics in water and in aqueous solution of DNA, to ascertain if this type of approach can actually help in discriminating selective interaction mechanisms *via* what we may call a 'calorimetric-criterion'.

More specifically, our data following the order of exposition from relatively simple to more complex cases, mainly concern the energetics of: (a) (i) the dimerization in aqueous solution of the dyes ethidium bromide (EB), proflavine (PF), and acriflavine (AF); (ii) the interaction of EB with synthetic polycarboxylates. (b) The dimerization of the antibiotics Daunomycin (D), Adriamycin (A), and Actinomycin (Act), and of the interaction of Act with deoxyguanosine-5'-phosphate (dGMP) in aqueous solution. (c) The interaction of the dyes and of the antibiotics indicated above with DNA (calf-thymus). (d) The interaction with DNA of a natural dye (and of one simple derivative) belonging to a novel class of compounds from colonial anthozoans (namely: tetraazacyclopentazulene dyes) having a selective inhibitory power on DNA synthesis.

2. Summary of the Results

2.1.1. THERMODYNAMICS OF EB, PF, AND AF DIMERIZATION IN WATER

2.1.1.1. For the dye EB in water we have evaluated both the dimerization constant, K_D, defined for the process $2D \rightleftarrows D_2$, and the enthalpy, ΔH_D at 25° by means of calorimetry [11]. The results are:

$$K_D = 41 \text{ M}^{-1} \left(\Delta G_D^\circ = -2.2 \text{ kcal mole}^{-1}\right)$$
$$\Delta H_D = -7.6 \pm 0.3 \text{ kcal mole}^{-1}; \; \Delta S_D = -18 \text{ e.u.}$$

The absorption spectra of EB in monomeric and in dimeric form in water are given in Figure 1. It is interesting to mention that the spectrum of EB 'intercalated' into

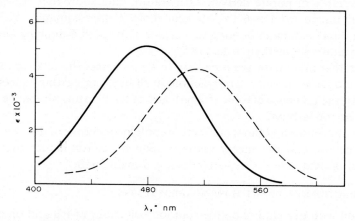

Fig. 1.   Spectrum of ethidium bromide in water. – Continuous curve; monomer spectrum. Dotted curve; dimer spectrum. (From Reference 11.)

DNA also exhibits a maximum at 518 nm with a molar extinction coefficient of about 3800 [1d].

2.1.1.2. Working with PF the following figures have been obtained by means of both calorimetric and spectroscopic measurements [12]*.

$$H_2O \; (25°); \; K_D = 477 \text{ M}^{-1} \left(\Delta G_D = -3.64\right) \Delta H_D =$$
$$= -6.3 \pm 0.6 \text{ kcal mole}^{-1} \text{ spectroscopy}$$
$$H_2O \; (45°); \; K_D = 200 \left(\Delta G_D = -3.34\right) \Delta S_D = -8.9 \quad \text{e.u.}$$
$$0.5 \text{ M KCl} \; (25°); \; K_D = 1840 \text{ M}^{-1} \left(\Delta G_D = -4.44 \text{ kcal mole}^{-1}\right)$$
$$\Delta H_D = -8.5 \pm 0.5 \text{ kcal mole}^{-1} \text{ calorimetry}$$
$$\Delta S_D = -13.6 \text{ e.u.}$$

* Manipulation of spectral data for PF, AF, and Adriamycin respectively to derive K_D values has been made according to the procedure described by Schwarz [13]. All ΔH_D values are in kcal per mole of dimer.

For AF (N-methyl PF) we have so far calculated van 't Hoff enthalpies of dimerization from spectroscopic K_D values at three temperatures in water. The figures are:

$$K_D (25°) = 848 \text{ M}^{-1}; \, K_D (35°) = 554 \text{ M}^{-1}; \, K_D(45°) = 376 \text{ M}^{-1}.$$

The resulting approximate ΔH_D value is -7.4 kcal mole^{-1} at, nominally, 25°. $\Delta S = -11.5$ e.u.

Inspection of the data reported above and examination of the literature reveals that dye dimerization enthalpies at near 25° cluster around a mean value of about -6.7 kcal mole^{-1}. As a matter of fact individual ΔH_D figures for six different dyes do not differ by more than about 15% from the average figure given above [13–17]. Furthermore, entropies of dimerization result systematically negative although quite different in absolute value passing from one dye to another. Speaking of dimerization processes by 'vertical-stacking' in aqueous solution, it may be interesting to recall that also for a number of purine derivatives literature data show that the enthalpy of dimerization ranges from -6 to -9 kcal mole^{-1} approximately with negative entropies of dimerization amounting to -22.4 e.u. for both 6-dimethylaminopurine and 6(methylamino)-9 methylpurine [18].

It appears that more data are necessary to try to assess the relative relevance of hydrophobic, coulombic and π-π interactions in all such aggregating aqueous systems (in particular, the influence of ionic strength and of temperature should be studied in a more systematic fashion).

Although our results and those of others are still inconclusive in this respect, we are inclined to believe that the energy released upon π electron clouds overlap in the dimerization of dyes and purine derivatives, is a main factor.

2.1.2. INTERACTIONS OF EB WITH POLYCARBOXYLATES

In connection with our studies on the solution behaviour of EB and on its interactions with DNA it was considered worthwhile to investigate also a few essential aspects of the binding of EB by synthetic polyelectrolytes in dilute aqueous solution [14]. Three maleic acid (MA) copolymers have been examined, namely: MA-isobutene (MAiB), MA-propylene (MAP), and MA-ethylene (MAE). The polyelectrolytes were half-neutralized with $(CH_3)_4$ NOH. In essence, the results of such a study which may be of some relevance to the main theme of this review, are:

2.1.2.1. UV data indicate that for P/D ratios up to near unity (P = polymer concentration in monomole l^{-1}, and D = dye stoichiometric concentration in mole l^{-1}) the dye spectrum becomes hypochromic and is progressively red shifted. For P/D $\simeq$ 1 the spectrum of EB resembles that reported in Figure 1 for the dimeric dye in water. Under these conditions, at least a large fraction of bound EB molecules should be stacked to form dimers (or higher aggregates) along the macroions chains. For P/D $\gg$ 1, the dye spectrum slowly reverts to that characteristic of the free monomeric form in water. This indicates, in our opinion, that EB molecules in the presence of an excess of the polycarboxylates would be bound essentially in monomeric form exhibiting a spectrum very similar to that of free monomers. In all cases considered, the spectra show a clear isosbestic point at 510 nm, i.e. at the same wavelength as for the EB-DNA system [1d].

2.1.2.2. Calorimetric measurements carried out under experimental conditions similar to those of the spectral runs, have allowed collection of heat of interaction data which, with the aid of a few equilibrium dialysis data on the extent of EB binding by the polycarboxylates, finally yielded the following enthalpy of binding ΔH_B values (in the case of MAiB):

$$\Delta H_B = -0.2 \text{ kcal mole}^{-1} \quad \text{for} \quad P/D \gg 1$$
$$\Delta H_B = -3.5 \text{ kcal mole}^{-1} \quad \text{for} \quad P/D \simeq 1$$

We consider these results as evidence in favour of a stacking of most EB molecules bound by half-neutralized MAiB for P/D near to unity. As a matter of fact, if we normalize the observed heat of interaction per mole of pairs of bound EB molecules we quite obviously get: $\Delta H_B = -7 \text{ kcal mole}^{-1}$ of pairs; this value happens to be close to the enthalpy of dimerization of EB in water (see above). The ΔH_B value estimated for $P/D \gg 1$ might then be taken as the enthalpy of EB binding in monomeric form to the polycarboxylate chains. The limits of the above qualitative conclusions are apparent, however, once it is recalled that dye bound wholly in monomeric form $(P/D \gg 1)$ or only in aggregated form $(P/D \simeq 1)$ are necessarily limiting, quantitatively unrealistic cases.

2.2. On the solution behaviour of daunomycin, adriamycin and of actinomycin

2.2.1. Little is known on the solution behaviour of daunomycin (D) and of adriamycin (A) (Figure 2) in water, because, quite naturally most studies have dealt directly with the more important problem of their interaction *in vitro* and *in vivo* with DNA. In our opinion, however, certain features of the solution behaviour of D and of

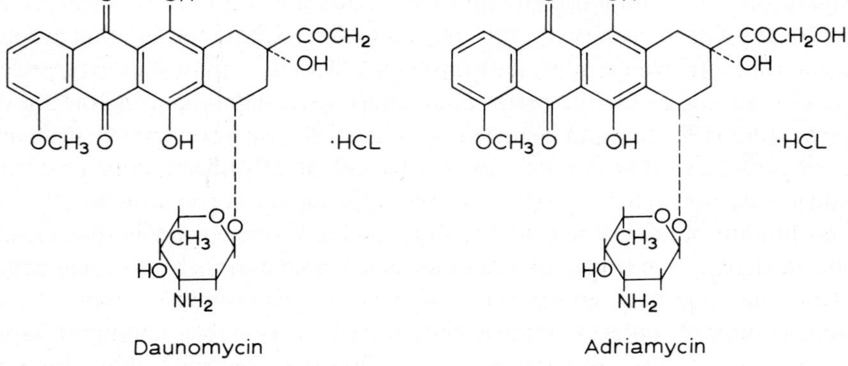

Fig. 2.   Structures of daunomycin and adriamycin.

A may have a bearing on the latter problem. We have found by means of UV and circular dichroism (CD) measurements that spectral properties of both A and D change significantly with concentration and have interpreted these data in terms of dimerization.

In the case of D, the estimated dimerization constant values are: $K_D(25°) = 3000 \pm 200$; $K_D(35°) = 1830 \pm 200$; $K_D(45°) = 1160 \pm 50$ M^{-1} (in 0.01 M phosphate buffer, pH = 7.0). From these values one evaluates: $\Delta H_D = -8.9 \pm 1$ kcal mole^{-1} and $\Delta S_D = -14.0$ e.u. at nominally, 25°. Direct calorimetric measurements at 25° have finally yielded $\Delta H_D = -9.1 \pm 0.3$ kcal mole^{-1}, an average value of seven independent experiments, in good agreement with the van 't Hoff enthalpy.

Working with adriamycin (A), both K_D and ΔH_D could be directly evaluated with the aid of calorimetry, handling a series of data on heat of dilution to give the following figures: $K_D = 1100 \pm 300$ M^{-1}; $\Delta H_D = -9.6 \pm 0.6$ kcal mole^{-1}; $\Delta S_D = -18.3$ e.u. (always at 25°).

The dimerization process of adriamycin and of daunomycin in aqueous solution certainly involves stacking of a pair of molecules with overlap of their fused ring systems, similar to what happens in the case of simple dye molecules like PF and EB. This stacking process is seen to involve an enthalpy decrease of 8–9 kcal mole^{-1}, again not dissimilar from that observed for other aggregating systems mentioned in Section 2.1 above.

In the case of actinomycin (Figure 3) we have not yet completed our measurements. At this stage it may be thus safer to quote literature data [19] and simply report that ΔH_D for actinomycin dimerization would be -15 kcal mole^{-1}, with $\Delta S_D = -38$ e.u. In view of the fact that actinomycin is uncharged at neutral pH, the large negative entropy of dimerization is really surprising. We suspect that on the basis of our preliminary calorimetric data, the ΔH_D figure given above is somewhat in error; a less negative ΔH_D would obviously lead, among other things, to a more acceptable ΔS_D figure.

2.2.2. It is known on the basis of spectral data that actinomycin (Act) interacts with deoxyguanosine-5'-monophosphate (dGMP) in aqueous solution to form 1:2 complexes [20, 21]. The interest in this type of studies derives from the fact that according to some authors Act would intercalate into DNA, with its actinocyl moiety, preferentially at G-C sites [22]. On the basis of our calorimetric data and using the K_D value for the Act-(dGMP)$_2$ complex reported by others [20] we have arrived at the mean value $\Delta H_D = -17.4 \pm 4$ kcal mole^{-1} of Act bound, at 25°, (mean value of seven independent measurements; in each case the appropriate corrections for the small heats of dilution of dGMP and of Act were made). If one considers that in such a complex the fenoxazine ring of one Act molecule would establish two equal contacts with the purine rings between which it is sandwiched, the assumption that the energy involved per mole of contact is about 8.5 kcal appears reasonable. This figure happens to be quite close to the enthalpy of dimerization data reported above for simple dyes as well as for adriamycin and daunomycin. It also indirectly supports the as-

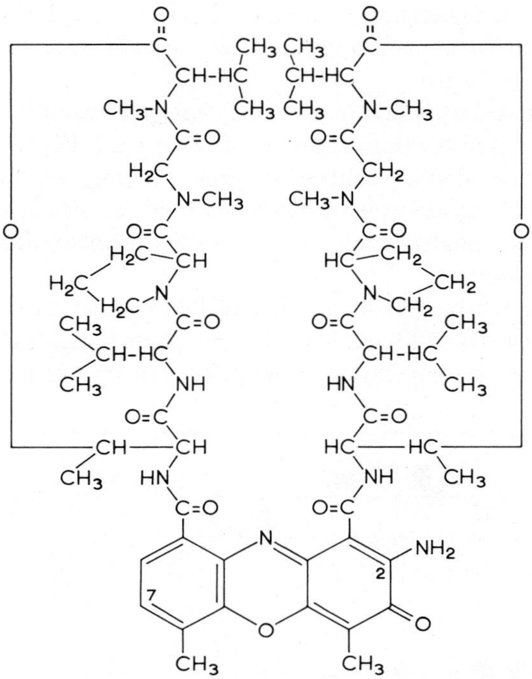

Fig. 3.   Chemical structure of actinomycin C_1 (D). – Abbreviations: MeVal, methyl-valine; sar, sarcosine; Pro, proline; Val, valine; Thr, threonine.

sumed stoichiometry of complexation, despite some uncertainty due to the approximate K_D value and to the possible residual self-aggregation of dGMP molecules in the final solutions.

It is finally interesting to mention that no heat exchange has been detected upon mixing Act with dAMP under experimental conditions similar to those employed with dGMP.

2.3. As mentioned at the outset, different authors have reported spectroscopic, viscometric and ultracentrifugation data pointing out convincingly that both EB and PF when 'strongly' bound by native DNA are actually *intercalated* between adjacent base pairs along the biopolymer double helical chains. This would be also true for a number of other dyes of suitable molecular structure with fused aromatic rings. The 'strong' binding, characterized by association constants greater than 10^5–10^6 M^{-1}, is generally limited however to molar ratios, r, of bound dye per DNA nucleotide smaller than about 0.15–0.20 in dilute solutions of relatively low ionic strengths. At higher r values, where all sites available for intercalation appear to be saturated, a weaker type of binding prevails. Additional dye molecules would then interact only with the 'surface' of the rodlike macroions of DNA. Qualitatively speaking a similar type of behaviour has been found by others studying the interaction of DNA with the two antibiotics actinomycin (Act) and Daunomycin (D). For the latter two it is also commonly assumed that for $r < 0.1$, approximately, intercalation of the aromatic moieties of Act or D takes place.

In the case of Act, intercalation involves dG-dC sites preferentially, with the cyclic pentapeptide rings located in the minor groove of DNA [22]. For all systems mentioned above while one can find a wealth of apparent binding constants determinations, little effort has been made toward a better thermodynamic characterization of intercalative and non-intercalative binding. Our results seem to add something of interest from this standpoint.

Calorimetric data on the heat of interaction of DNA (native, calf-thymus) with PF and EB, respectively, in 0.015 M and in 0.1 M phosphate buffer (pH = 7) at 25° are reported in Figure 4. Despite the scatter in the points, partly due to difficulties en-

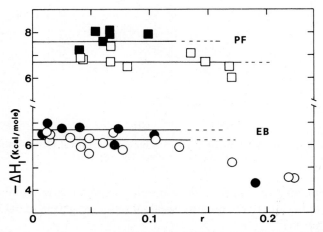

Fig. 4. Enthalpy of interaction between proflavine (PF) and ethidium bromide (EB) with calf thymus DNA as a function of stoichiometric molar ratio dye/DNA, r. – Open symbols: ionic strength $I = 0.015$ (phosphate buffer pH = 7); full symbols: ionic strength $I = 0.1$ (phosphate buffer pH + KCl). (From Reference 12.)

countered in the calorimetric experiments [12], one can conclude that in the range of r values within which intercalation would take place one finds constant (average) ΔH_B values of -6.2 and -6.7 ± 0.5 kcal mole^{-1} dye bound for EB and of -6.7 and -7.6 ± 0.6 kcal mole^{-1} dye bound for PF, depending on the ionic strength. (The fraction of dye bound was derived in each case from equilibrium dialysis experiments.)

The results obtained working with daunomycin and actinomycin are reported in Figure 5. For the antibiotic D it is again observed that in the range of r values for which D would be intercalated into DNA a constant negative enthalpy of interaction, amounting to $-6.5 + 0.5$ kcal mole^{-1}, results [23].

It is interesting to recall the enthalpy of dimerization data for PF, EB and D reported in Sections 2.1 and 2.2 and to compare them with the enthalpies of intercalation given above.

Despite inaccuracies inherent to each set of measurements, we think that our data clearly point out that self-association of PF, EB and D and their 'strong interaction' with DNA are characterized by quite similar heat effects. We believe this is not mere coincidence but, on the contrary, original evidence that strong binding of the three species considered with DNA is synonymous with intercalation. Intercalation of an

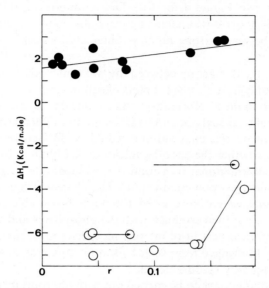

Fig. 5. Enthalpy of interaction of native calf-thymus DNA with: (a) actinomycin D (full symbols), (b) daunomycin hydrochloride (open symbols). – Solvent: 0.01 M phosphate buffer, pH 7.0. r is the final stoichiometric ratio between the molar concentration of antibiotic and the molar (phosphorous base) concentration of DNA. $T = 25\,°C$. In two experiments with daunomycin, solutions of DNA already containing a given amount of the drug were added of more daunomycin to span the r indicated by the arrows. (From Reference 23.)

aromatic polycyclic molecule like, for instance, PF into the *native* structure of DNA must involve interruption of one base-pair contact (requiring 5–6 kcal mole^{-1} of contacts, approximately) and the establishment of two new contacts between the intercalated molecule and the previously stacked bases. Our data would then indicate that intercalation leads to dye-bases interactions energetically as favourable as base-base or dye-dye ring interactions per actual surface of contact. It is important also to point out that, always according to our data, *intercalation is a process* which, as distinct from dye (or antibiotic) dimerization, *is favoured also by the entropy* (e.g. about +10 e.u. for EB intercalation [12]). This should be attributed in large part to the release of water molecules from the hydration sheaths of interacting species (also intuitively consistent with an intercalation process) which would more than compensate the loss in enthalpy due to the concomitant increase in DNA chains stiffness and to the cratic term.

Passing finally to the case of Act, two main aspects of our calorimetric results deserve some comments: (1) the enthalpy of Act-DNA interaction, ΔH_B, is positive (around $+2$ kcal mole^{-1}). Therefore a large positive entropy change strongly favours this interaction (Act is uncharged at neutral pH in water). (2) The enthalpy of Act-dGMP complexation is on the contrary negative and quite large (approximately -8.5 kcal mole^{-1} of Act-dGMP contact; see Subsection 2.2.2).

Moreover no discontinuity is evident in the ΔH_B vs. r plot for Act as opposed to the case of D (but see also Figure 4 for EB). Discontinuities in the trend of spectral properties of D, PF and EB with changing r are also more clearly evidentiable and mark a boundary between 'strong-binding' (intercalation) and 'weak-binding' r regions.

In the case of Act the distinction between the two r intervals would however result, to our knowledge, only from Scatchard plots of equilibrium dialysis data. We cannot conclude from our limited information that, contrary to the widespread belief, actinomycin molecules strongly bound to DNA are not intercalated but at least cast doubts on this point. In this connection it would be, in fact, very important to be able to understand whether the endothermicity of Act-binding by DNA may result from an alleged conformational transition the pentapeptide rings of the antibiotic molecules should undergo upon binding [24]. This change in conformation would be energetically uncompensated however by the many favourable contacts each actinomycin molecule might then establish with the phosphates and the bases of DNA. Anyway, this geometrical best fit of interacting species must maximize the number of water molecules disengaged from their hydration layers, as this ultimately makes the binding process a very favourable one.

Evidently more studies have to be carried out with appropriate model compounds and different actinomycin derivatives to see whether our doubts can find further support.

2.4. Let us finally consider the case of other two recently identified natural compounds, two nitrogenous pigments of unusual structure, named zoanthoxanthin (I) and 3-norzoanthoxanthin (II).

A most interesting feature of their structure is the previously unknown planar tetraazacyclopentazulene ring system which bears a close relationship with that of some synthetic tropoids [25]. This feature has prompted an investigation on their possible biological activity and a parallel study of their mode of interaction with DNA. For the former inportant aspect a group of research of the Italian National Cancer Institute in Milan, with which we now work in collaboration has found that both compounds I and II selectively inhibit DNA synthesis in vitro. Both I and II moreover have greater inhibitory effect against rat liver high molecular weight DNA polymerase than E. Coli DNA polymerase I [26].

On our side, encouraged by these findings we have carried out a few physico-chemical measurements to try to understand how molecules I and II do actually interact with DNA (calf-thymus) in dilute aqueous solution [26]. We have observed that the absorption spectra of I and II in the presence of an excess of DNA (in 0.01 M acetate buffer, pH = 5.0 so that species I and II bear a positive charge) are red-shifted and hypochromic with respect to free dye spectra. These results are a conclusive evidence of dye 'binding'. More revealing are the results of viscosity measurements reported in Figure 6, from which it clearly appears that the intrinsic viscosity of DNA

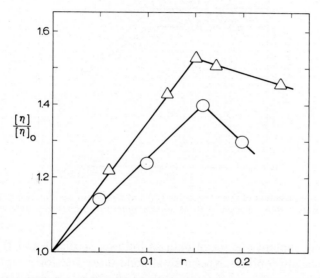

Fig. 6. Intrinsic viscosity of dye-DNA complexes relative to the intrinsic viscosity of DNA alone as a function of the molar ratio dye/DNA. – Solvent: 0.01M acetate buffer (pH 5.0). △ zoanthoxanthin; ○ norzoanthoxanthin. (From Reference 26.)

increases with increasing r, the molar ratio dye/DNA and, in the case of zoantho-xanthin (I), it reaches for $r = 0.15$ a value 50% greater than that measured in the absence of dye (always in acetate buffer, and 25°). Beyond this critical r value the intrinsic viscosity of DNA exhibits the tendency to decrease.

There is general agreement that an increase in viscosity of DNA consequent to dye or drug binding is very strong support of 'intercalation'. Incidentally, the effect found by us working with compounds I and II is qualitatively similar to, but less marked than that already reported by others for proflavine, ethidium bromide and daunomycin [27–29]. To estimate the association constant, K_B, of I and II with DNA we have performed a series of spectrophotometric experiments. Scatchard plots of the results of these experiments indicate that, for both I and II, K_B is about $3 \times 10^5 \, \text{M}^{-1}$

and that the number of binding sites per phosphate group of DNA is 0.15 approximately. It is interesting to point out that the discontinuity in the viscosity plots is observed just for $r \simeq 0.15$; the K_B value above should thus refer to the intercalative binding of our dyes.

Finally, calorimetric measurements (Figure 7) have shown that for $r < 0.15$ the enthalpy of binding ΔH_B of I and II by DNA is nearly a constant equal to -3.5 kcal mole^{-1} for I and -3.1 kcal mole^{-1} for II, approximately. For $r > 0.15$ the enthalpy drops to lower absolute values.

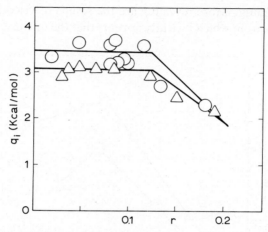

Fig. 7.   Enthalpy of interaction of zoanthoxanthin ($\triangle$) and norzoanthoxanthin ($\bigcirc$) as a function of the molar ratio dye/DNA – Solvent: 0.01 M acetate buffer (pH 5.0). (From Reference 26.)

We tentatively explain our set of data as follows: species I and II intercalating into DNA under appropriate experimental conditions and for $r < 0.1$ would do it either 'partially' or with a different 'stacking-efficiency' in comparison to other typical intercalating agents (e.g. proflavine, ethidium bromide, daunomycin) thus leading to a relatively small enthalpy change (nearly half, in absolute value, of that found for the three compounds cited). 'Partial intercalation' might be attributed to the slight deviation from planarity of molecules (I) and (II) and/or to other steric effects. On the other hand intercalation might be complete, or at least comparable to that taking place with other dyes, but the interactions between the bases and the azacyclopentazulene rings would be not so efficient in view of the particular electronic structure of these dyes. In the zoanthoxanthin molecule in fact, the π-electronic current is essentially equivalent to that of an azulene system thus possibly leading to an interaction with DNA bases like that of a bicyclic system. Relevant to this point is the fact that the observed differences in the enthalpy of interaction parallel a significantly lower inhibitory action exhibited by I and II against nucleic acids polymerizing enzymes when compared to daunomycin (a potent inhibitor) under similar essay conditions.

This, together with the observation the ΔH_B is consistently negative for all compounds tested so far and for which there are unambiguous evidences in favour of their intercalation into DNA, makes the calorimetric criterion a useful one in these type of studies. A natural, interesting extension of our research consists in passing to a thermodynamic characterization of the interaction of DNA with different antibiotic molecules which *do not* give rise to intercalation. Work is in progress along these lines in our laboratory.

Equilibrium data shall be supplemented also by kinetic data as the latter appear very important in elucidating the specificity and mechanisms of interaction between DNA and biologically active compounds.

Acknowledgement

Studies summarized in this review have been carried out with financial support of the Italy Consiglio Nazionale delle Ricerche.

References

This list contains references to a few relevant articles or reviews and is by no means a comprehensive one.

1. (a) von Hippel, P. H. and McGhee, J. D.: *Ann. Rev. Biochem.* **41**, 231 (1972).
   (b) Stone, A. L. and Bradley, D. F.: *J. Am. Chem. Soc.* **83**, 3627 (1961).
   (c) Blake, A. and Peacocke, A. R.: *Biopolymers* **6**, 1225 (1968).
   (d) Waring, M. J.: *J. Mol. Biol.* **13**, 269 (1965).
2. (a) Brenner, S., Barnett, L., Crick, F. H. C., and Orgel, A.: *J. Mol. Biol.* **3**, 121 (1961).
   (b) Albert, A.: *Selective Toxicity*, 2nd ed., Methuen, London, 1968.
   (c) Tomchick, R. and Mandel, H. G.: *J. Gen. Microbiol.* **36**, 225 (1964).
   (d) Waring, M. J.: *Mol. Pharmacol.* **1**, 1 (1965).
3. Chambers, R. W., Kajwara, T., and Kearns, D. R.: *J. Phys. Chem.* **78**, 380 (1974).
4. (a) Barone, G., Crescenzi, V., Quadrifoglio, F., and Vitagliano, V.: *Ric. Sci.* **36**, 503 (1966).
   (b) Birshtein, T. M., Anufrieva, E. V., Nekrasova, T. N., Ptitsyn, O. B., and Sheveleva, T. V.: *Vysokomolekul. Soedin.* **7**, 372 (1965).
   (c) Stork, W. H. J., de Hasseth, P. L., Schippers, W. B., Kormeling, C. M., and Mandel, M., *J. Phys. Chem.* **77**, 1772 (1963).
   (d) Stork, W. H. J., van Boxsel, J. A. M., de Goeij, A. F. P. M., de Hasseth, P. L., and Mandel, M., *J. Phys. Chem.* **2**, 127 (1974).
   (e) See also Braud, C., Muller, G., Fenyo, I. C., and Sélégny, E., *J. Polym. Sci.* **12**, 2767 (1974); and this volume p. 28.
5. Liquori, A. M., Barone, G., Crescenzi, V., Quadrifoglio, F., and Vitagliano, V.: *J. Macromol. Chem.* **1**, 291 (1966).
6. Lerman, L. S.: *J. Mol. Biol.* **3**, 18 (1961).
7. LePecq, J.-B. and Paoletti, C.: *J. Mol. Biol.* **59**, 43 (1971).
8. Goldberg, I. H. and Friedman, P. A.: *Ann. Rev. Biochem.* **40**, 775 (1971).
9. Gale, E. F., Cundliffe, E., Reynolds, P. E., Richmond, M. H., Waring, M. J.: *The Molecular Basis of Antibiotic Action*, Wiley & Sons., London, 1972, p. 244.
10. Shiao, D. F. and Sturtevant, J. M.: *Biochemistry* **8**, 4910 (1969).
11. Crescenzi, V. and Quadrifoglio, F.: *Eur. Polymer J.* **10**, 329 (1974).
12. Quadrifoglio, F., Crescenzi, V., and Giancotti, V.: *Biophys. Chem.* **1**, 319 (1974).
13. Schwarz, G., Klose, S., and Balthasar, W.: *Eur. J. Biochem.* **12**, 454 (1970).
14. Rabinovitch, E. and Epstein, L. F.: *J. Am. Chem. Soc.* **63**, 69 (1941).
15. Rohatgi, K. K. and Singhal, G. S.: *J. Phys. Chem.* **70**, 1695 (1966).

16. Padday, J. F.: *J. Phys. Chem.* **72**, 1259 (1968).
17. Schwarz, G. and Balthasar, W.: *Eur. J. Biochem.* **12**, 461 (1970).
18. Bretz, R., Lusting, A., and Schwarz, G.: *Biophys. Chem.* **1**, 237 (1974).
19. Crothers, D. M., Sabol, S. L., Ratner, D. I., and Müller, W.: *Biochemistry* **7**, 1817 (1968).
20. Behme, M. T. A. and Cordes, E. H.: *Biochim. Biophys. Acta* **108**, 312 (1965).
21. Krug, T. R. and Neely, J. W.: *Biochemistry* **12**, 1775 (1973).
22. Sobell, H. M.: *Progr. Nucl. Acid. Res. Mol. Biol.* **13A** (1972).
23. Quadrifoglio, F. and Crescenzi, V.: *Biophys. Chem.* **2**, 64 (1974).
24. Müller, W. and Crothers, D. M.: *J. Mol. Biol.* **68**, 21 (1972).
25. Cariello, L., Crescenzi, S., Prota, G., Giordano, F., and Mazzarella, L.: *Chem. Comm.* **99** (1973).
26. Quadrifoglio, F., Crescenzi, V., Prota, G., Cariello, L., Di Marco, A., and Zunino, F.: *Chem. Biol. Interactions* (in press).
27. Armstrong, R. W. and Mazur Panzer, N.: *J. Am. Chem. Soc.* **94**, 7650 (1972).
28. Reinert, K. E.: *Biochim. Biophys. Acta* **319**, 135 (1973).
29. Zunino, F., Gambetta, R., Di Marco, A., and Zaccara, A.: *Biochim. Biophys. Acta* **277**, 489 (1972).

PART III

WATER PURIFICATION
PETROLEUM RECOVERY AND DRAG REDUCTION

ION EXCHANGE HOLLOW FIBERS

A. REMBAUM and S. P. S. YEN

Jet Propulsion Laboratory, California Institute of Technology, Pasadena, Calif., U.S.A.

and

E. KLEIN and J. K. SMITH

Gulf South Research Institute, New Orleans, La., U.S.A.

Abstract. Quaternized, crosslinked, insoluble polymers of unsubstituted and substituted vinyl pyridines and a dihalo organic compound are spontaneously formed at ambient temperature on mixing the two components in bulk, in solutions or in suspension. The amount of crosslinking may be varied according to the composition and reaction conditions. The polymer product exhibits a high ion exchange capacity and undergoes a reversible color change from black at a pH above 7 to yellow at a pH below 7. The polymer may be formed in the presence of preformed polymers, substrates such as porous or impervious particles or films to deposit an ion exchange resin in situ or on the surface of the substrate. The coated or resin impregnated substrate may be utilized for separation of anionic species from aqueous solution.

Porous walled hollow fibers prepared from polyacrylonitrile were used as scaffolds in which 4-vinyl pyridine dibromoethane ion exchange polymers were formed. The initial fiber hydraulic permeabilities of 6×10^{-5} cm/s atm were reduced to one fourth this value as the result of ion exchange resin formation in the pores of the walls. A concomitant decrease in permeability of non-electrolytes was observed, but very high rates of ion fluxes were noted when the fibers were used to separate ionic solutions. The Donnan pumping principle was demonstrated using strong Cl^- concentrations to pump $CrO_4^=$ against its concentration gradient. The thin fiber walls led to high transport rates, and the apparently high degree of semipermeability permitted an adequate driving potential to be maintained.

The need for a better theory of transport properties for ion exchange fibers is shown to be necessary. Several processes are operative simultaneously: the Donnan pumping process operates with constantly changing potential differences, and at the same time purely dialytic flows occur as the result of membrane imperfections. These flows are related through common concentration terms. The hollow fiber configuration allows the preparation of thin walled ion exchange devices with large surface areas; a recent cost calculation indicates such devices may make continuous water softening process costs (using the Donnan principle) economically competitive with present treatment methods.

1. Introduction

It was previously shown [1] that 4-vinyl pyridine (4-VP) polymerized spontaneously in presence of alkyl halides. The growing active species in this polymerization system exhibited high specificity. It was reacting only with monomeric vinyl pyridinium salts and not with other vinyl monomers including 4-VP. Therefore, in presence of excess of 4-VP, the polymerization was arrested after the alkyl halide was used up in the quaternization reaction. It was also observed that free radical inhibitors had no effect on the polymerization rate.

In view of these facts, Kabanov postulated a 'zwitterion' mechanism shown in Figure 1. The failure of addition of I to II (Figure 1) was explained by the relatively higher charge density and therefore more negative electrical character of the double bond in 4-VP than in the quaternary salt.

However, it was later found that in the quaternization of 4-VP with a weak nucleophile (e.g., *p*-toluene sulfonate) or with a dilute acid, a different polymerization mechanism is operative leading to a different structure of the polymer [2].

The alternative mechanism which does involve the addition of a neutral molecule of 4-VP to the quaternized salt is shown in Figure 2. This mechanism postulates a hydrogen transfer to the negatively charged carbon atom in III and formation of

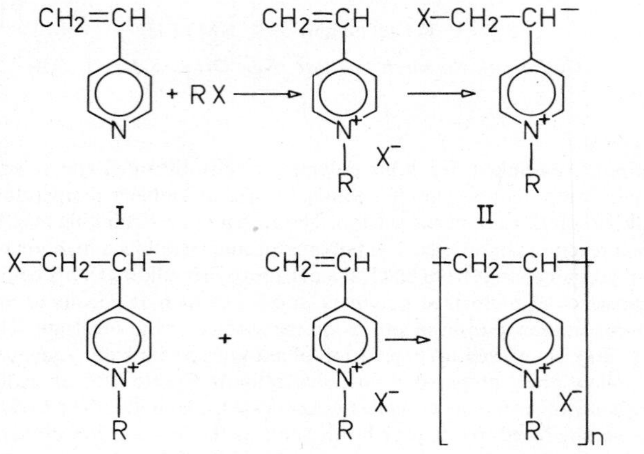

Fig. 1.   Mechanism of spontaneous polymerization of 4-vinyl pyridine.

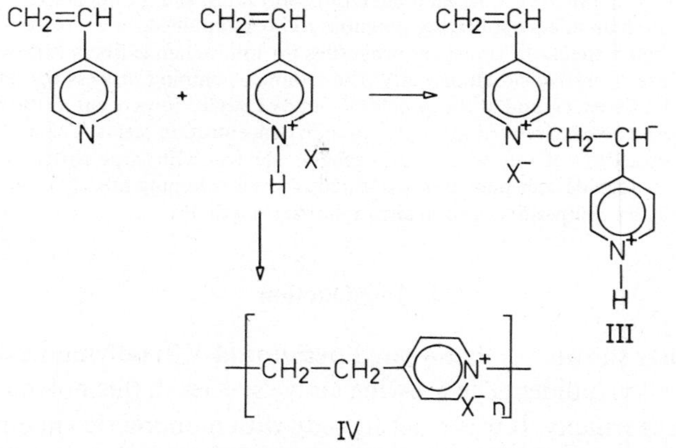

Fig. 2.   Mechanism of polymerization of 4-vinyl pyridine with weak nucleophiles or dilute acids.

ionene polymers (IV) where the positively charged nitrogen is located in the back-bone of the chain. Thus, the understanding of 4-VP polymerization can be sum-marized (see Figure 3).

In our investigations, 4-VP was reacted under a variety of conditions with α, ω dihalides yielding highly crosslinked polymers. Most of our findings can be accounted for on the basis of Kabanov's mechanism.

The use of dihalides involves a second quaternization reaction of the same mol-

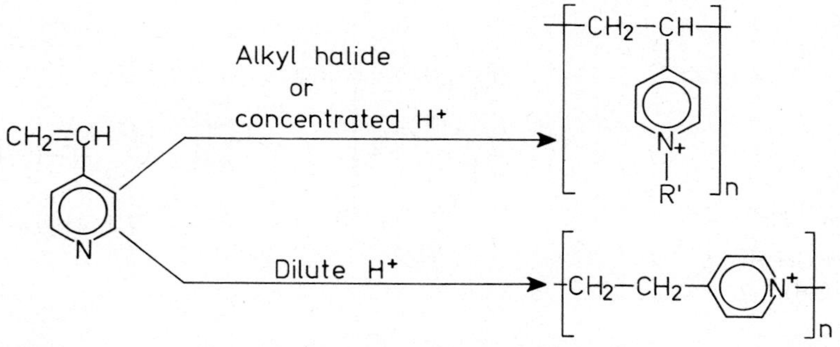

Fig. 3.   Postulated structures of Quaternized poly(4-vinyl pyridine). R' = alkyl group or H.

ecule and brings up the problem of the effect of the first positive charge on the second quaternization reaction.

It was previously shown [4] that a series of dicationic crosslinking agents (Figure 4) could be prepared and isolated.

All the dicationic crosslinking agents yielded crosslinked resins on addition of redox initiators.

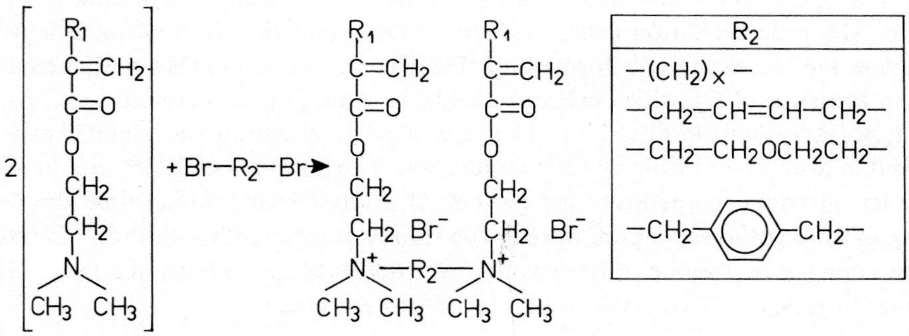

Fig. 4.   Preparation and isolation of dicationic crosslinking agents. X = 2 to 16. R_1 = H or CH_3.

Because of the spontaneous polymerization of quaternized 4-VP the isolation of a dicationic crosslinking agent was not successful with any α,ω dihalides investigated including dibromomethane. The latter, when reacted with two molecules of dimethyl-amino ethylmethacrylate (DEMA), yielded pure V indicating that the second quaternization was prevented by the proximity of the positive charge.

For similar reasons, we were unsuccessful to synthesize ionene polymers [5] of structure VI, and the reaction of tetramethylamino methane with dibromomethane yielded small molecules instead of VI. Furthermore, the crosslinked resins obtained by the reaction of 4-VP with dihalides contained unreacted bromine end groups. In order to determine the reason for this observation we have investigated the rate of quaternization of a model compound (structure VII) with pyridine.

The spontaneous formation of high ion exchange capacity resins at room temper-

$$
\begin{array}{c}
CH_3 \\
| \\
C{=}CH_2 \\
| \\
C{=}O \\
| \\
O \\
| \\
CH_2 \\
| \\
CH_2 \quad Br^- \\
| \\
N^+{-}CH_2Br \\
CH_3 \quad CH_3
\end{array}
$$

V

$$
\left[-\overset{\overset{\displaystyle CH_3}{|}}{\underset{\underset{\displaystyle CH_3}{|}}{N^+}} \overset{Br^-}{-} CH_2 - \right]_n
$$

VI

$$
\langle N^+ \rangle{-}(CH_2)_x{-}Br \qquad Br^-
$$

VII where x = 2,3,6

ature in presence of a variety of materials on the surface and in situ lead to the synthesis of these resins inside the walls of hollow fibers and to the examination of the transport properties of the latter. It was found that polyacrylonitrile hollow fibers can be conveniently impregnated with 4-VP crosslinked resins and used to remove dichromate from aqueous solutions as well as from industrial effluents.

2. Experimental

2.1. SYNTHESIS

Freshly distilled 4-VP was mixed with dihalides (molar ratio; 2:1) in bulk or in solution. The polymerization proceeded to virtual completion in bulk or in solvents (benzene, methanol, dimethylformamide (DMF) or mixtures of DMF with methanol) and in presence of e.g., silica gel, carbon black, sand, porous materials, e.g., paper, cloth, polyacrylonitrile fibers, etc. The rate of polymerization was significantly enhanced in absence of air or by $CO\gamma$ irradiation. The mixtures of 4-VP and dihalides were left at room temperature for periods of 5 days during which time the color changed from colorless to pink or red. The resin was isolated by addition of acetone and washing with acetone. After drying it was obtained in the form of a light yellow powder in yields of 70 to 100% of the theoretical amount.

2.2. EXCHANGE CAPACITY

Approximately 1 g of resin vacuum dried at 100 °C overnight and sieved to give a mesh size of 250–500, was placed in a burette. 3N NaOH (100 ml) was then added to the column and eluted with distilled water. The elutant was neutralized to pH 6 with N/10 HNO_3 and diluted to 250 ml.

    A 30 ml aliquot was analyzed for Br by the Mohr method. Exchange capacity = meq/g of dry resin.

2.3. RELATIVE SWELLING

Samples of dry resins (dried at 100 °C) were placed in containers at 100% humidity. The increase of weight was measured after 120 hours.

2.4. PREPARATION OF MODEL COMPOUNDS AND RATE STUDIES

Freshly distilled pyridine was mixed with α,ω dibromides (excess) and reacted at room temperature (7 days). Compounds VIII, IX and X were obtained in quantitative yields based on the amount of pyridine, and their structure was ascertained by NMR analysis.

VIII   IX   X

The rate of quaternization of VIII, IX and X was studied in excess of pyridine in DMF methanol (1:1 by volume) at 25° and 50 °C by following the increase of ionic bromine concentration with time.

2.5. IMPREGNATION OF HOLLOW FIBERS

Polyacrylonitrile hollow fibers fabricated at Gulf South Research Institute were used. Their hydraulic permeability was 9×10^{-5} cm/s atm, the wall thickness 50 μ, the inside diameter 200 microns and the wall micropore diameter about 100 Å. Hollow fibers (150) assembled in bundles with a total surface area of 140 cm^2 were washed first with water, and then with methanol and dried by passing nitrogen gas through them for one hour. They were immersed in a mixture of 4-VP and α,ω-dihaloalkane (2:1 molar). The reaction was permitted to proceed for 10 days in case of dibromo ethane and 2 days in the case of dibromohexane. A cross section of a typical fiber is shown in Figure 5.

2.6. ION-EXCHANGE

Removal of chromate from aqueous solution is achieved by using the Donnan pumping principle, e.g., chromate ions are pumped against their concentration gradient through the ion exchange hollow fibers by a second ion (of the same charge sign) present at a much higher concentration on the other side of the hollow fiber wall. In this manner dilute ions can be concentrated into the pumping ion solution and the cleaned up water can be reused or discharged.

3. Results

3.1. YIELDS, EXCHANGE CAPACITIES AND SWELLING

The yields (Table I) of resins varied with the method and temperature of preparation. The exchange capacities were found to be below the theoretical values. However, in most cases the exchange capacities were comparable to those of commercial strong base ion exchange resins determined by the already outlined procedure.

3.2. HALOGEN ANALYSIS

The total amount of halogen bound to 4-VP resins was found to be greater (Elek Analytical Lab, Los Angeles, CA) than the amount of ionic halogen (Table II) indicating presence of unreacted bromine end groups. After further reaction with trimethylamine the ionic bromine content could be increased (Table II).

TABLE I

Yields, exchange capacities and swelling properties of 4-VP resins
(room temperature)

Dihalide	Method	Yield (Wt. %)	Exchange capacity (meq/g)	Relative swelling (%)
1,2-dibromoethane	Solution	70	4.9	–
	Bulk	51	–	–
1,3-dibromopropane	Solution	97	–	–
	Bulk	70	4.52	67.1
1,4-dibromobutane	Solution	94	–	
	Bulk	80	4.59	58.0
1.6-dibromohexane	Solution	100	–	
	Bulk	79	3.11	
1,8-dibromooctane	Solution	100	–	
	Bulk	90	2.37	47.8

DMF, methanol (1:1 by volume).

TABLE II

Theoretical and observed ionic bromide content
of 4-VP resins

α, ω dihalide	% Br^- theoretical	% Br^- found	% Br^- After further reaction
1,2-dibromoethane	40.2	35.3	–
1,3-dibromopropane	38.44	31.1	36.5
1,4-dibromobutane	36.85	28.4	–
8,8-dibromooctane	31.58	19.4	26.5

3.3. RATE OF QUATERNIZATION OF MODEL COMPOUNDS

Examination of rates of quaternization of VIII, IX and X using a large excess of pyridine yielded straight lines when $\log a/(a-x)$ was plotted vs time (a and x=initial bromide concentration and concentration at time t, respectively). The first order rate constants are recorded in Figures 6 and 7 for reactions carried out at 25° and 50°C, respectively.

3.4. MASS TRANSFER MEASUREMENTS

The transfer rates of tritiated water, KCl, and $CrO_4^=$ ions were measured through the walls of the hollow fibers arranged as parallel tubes terminating in two headers made from silicone rubber (Dow Corning RT-11). To prepare the headers, 150 parallel fibers were drawn through a 1″ section of $\frac{3}{8}$″ diameter polyethylene tubing so that approximately 1″ of the fiber end protruded from the polyethylene placed at each end of the fiber bundle. Split single hole corks were inserted into the polyethylene tubing to retain the fiber bundle and to act as dams for the silicone rubber. The RTV-11 was introduced into the polyethylene tubes and forced to flow between the fibers, and then cured. The polyethylene tubes served as a base for subsequent connections between the fiber bundle and flow control devices.

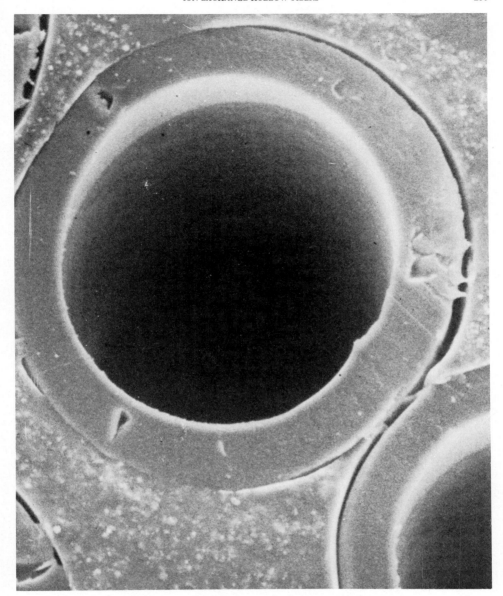

Fig. 5. Scanning electron microscope picture of a typical fiber cross-section.

To measure transport of tritiated water (THO), urea, creatinine, glucose, and vita-
min B-12 transport, one end of the fiber bundle was connected to the delivery end of
a Harvard syringe pump. The connections were made through bulkhead fittings in
the cover of a 1.0 liter jar, arranged so that the fiber exteriors could be bathed by a
dialyzing solution in the jar. The jar cover also contained bulkhead fittings through
which the dialysate could be recirculated from a 30 liter reservoir at 500 cc/min. These

high flow rates were used to reduce fluid boundary resistance at the fiber surface. Analyses of the solute concentrations in the feed syringe, in the fiber exit stream, together with flow measurements through the fiber bore allowed calculation of the fiber wall permeability.

Measurement of KCl leakage flow was carried out by connecting the two ends

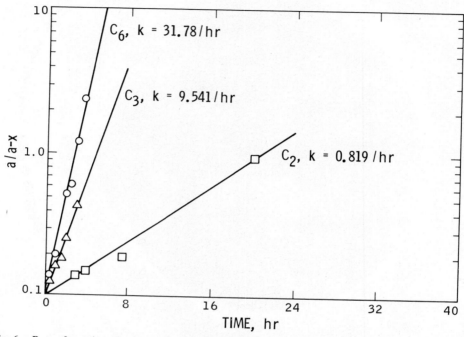

Fig. 6.   Rate of reaction of 1-pyridinium-2-bromoethane (C_2), 1-pyridinium-3-bromopropane (C_3), and 1-pyridinium-6-bromohexane (C_6) with pyridine at 25°C.

of the fiber bundle to a reservoir containing 1.0 M KCl via a peristaltic pump. The exterior of the fibers was bathed by a stirred volume of 200 cc of water. The appearance of KCl in the exterior volume was followed conductometrically to calculate the solute leakage rate.

Rates of ion exchange were determined in a configuration similar to that used for the THO transport measurements. However, the feed solutions were forced through the fiber bores from a large reservoir by a peristaltic pump. The external bath was not recirculated, but was stirred with magnetic stirring bar. It was initially equimolar in $CrO_4^=$ with the bore solution, and in addition was 0.25 molar in NaCl.

To measure hydraulic permeability of the fiber wall, one end of the fiber bundle was fitted to the delivery end of a 25 cc burette, and the fibers were flushed vigorously to remove entrapped air. The other end of the water filled burette was connected to a pressure source. When all the air had been displaced, the open end of the fiber bundle was capped, and the rate of water loss from the burette was timed at several pressures. At each pressure, steady state values were taken. From a knowledge of the number of fibers in the bundle, the unencumbered length of the bundle, and the diameters, the hydraulic permeability can be calculated.

3.5. TRANSPORT PROPERTIES OF BASIC FIBERS

The polymerization of the monomers described in this paper was carried out in fiber walls prepared from polyacrylonitrile. The fibers had a wall thickness of 0.0040 cm, and an inside diameter of 0.0250 cm. The scanning electron microscope picture of a typical fiber cross-section is shown in Figure 5. The structure appears to be uniform at this magnification, except for the presence of an occasional macrovoid; the macrovoid frequency and size have been related to extrusion and quenching conditions in the literature [6, 7]. Some evidence is available now that the walls of such macrovoids are not permeated readily.

The morphology of the fiber used for resin deposition determines the pore density and average pore size of the resulting ion exchange fibers. Characterization of membrane morphology is in itself a difficult study, so that phenomenological parameters have been used by many investigators to describe membrane properties; in this paper a similar approach is taken. The criteria found most useful are the water content of the membrane, the hydraulic resistance of the membrane, and the permeability rates of several solutes through the membrane walls. These values can sometimes be reconciled with a pore model of transport, when an independent measurement of either pore area or pore diameter can be achieved.

Pore diameters of the largest pores present can be measured by the pressure of air required to displace a fluid of known surface tension from the walls of the pores, if the fluid wets the pore wall. The relationships have been examined by Carman [8] for isotropic structures, and the relationship between air pressure (in psi) and surface tension γ (in dyn/cm) is given by:

$$P = \frac{0.415\,\gamma}{d}. \tag{1}$$

Attempts to measure displacement pressure, up to the pressure level that the fibers can contain, were unsuccessful. This indicates that the largest pore size is smaller than 0.2 microns.

Since the pressure at which the pore fluids are displaced from these fibers was greater than the rupture pressure of the fibers, another indirect method for determining pore size was investigated. Pappenheimer *et al.* [9], derived the relationship between the hydraulic flux of water through a membrane (which is dependent on the square of the equivalent pore radius) and the diffusion rate of tritiated water through the same membrane:

$$r = \frac{(8L_p n)^{1/2}}{(A/d)}, \tag{2}$$

where L_p is the hydraulic permeability coefficient determined by relating the volume flux J_w to the pressure gradient, P, applied to an area A:

$$J_w = L_p A \Delta P \tag{3}$$

and the term (A/d) is the effective cross-sectional area per unit length of pore derived from the measurement of THO permeability. The rate of tritiated water transport is related to the concentration gradient, membrane area, and diffusion coefficient of THO in H_2O:

$$J_{THO} = D_{THO}(A/d)\,\Delta c. \tag{4}$$

In Table III are shown the data required for calculation of the pore size of the hollow fibers.

TABLE III

H₂O and THO permeation of hollow fibers

Hydraulic permeability, L_p	6×10^{-5} cm s^{-1} atm^{-1}
	6×10^{-11} cm^3 dyn^{-1} s^{-1}
THO permeability, P_{THO}	1.36×10^{-4} cm s^{-1}
THO diffusion coefficient, D_{THO}	2.36×10^{-5} cm^2 s^{-1}
H₂O viscosity,	0.0089 dyn s cm^{-2}
Pore radius	43×10^{-8} cm

Another sensitive criterion of membrane structure is the permeability coefficient of specific solutes. The arrangement for measuring the permeability coefficient utilizes a fiber bundle through which the solution is pumped at a constant and controlled rate. The fibers are bathed in a large excess of solution in which the concentration of the solute being tested is maintained close to zero. The solute, therefore, diffuses out of the fiber pores as a result of the concentration gradient across the fiber wall.

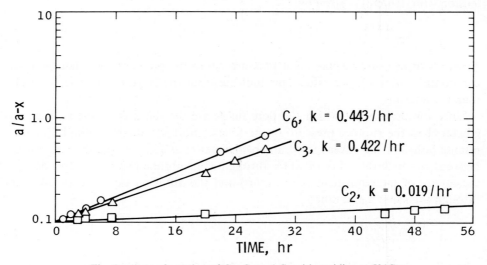

Fig. 7.   Rate of reaction of C_2, C_3 and C_6 with pyridine at 50 °C

Since the gradient is changing along the axis of the fiber, provision must be made for estimating the form of the gradient.

The material balance of the solute concentration can be written by:

$$-\dot{M} = Q_v(C_o - C_i),\tag{5}$$

where $\dot{M}$ is the rate of solute loss through the walls, Q_v is the axial flow of solution, C_i and C_o are the inlet and outlet concentrations respectively. Similarly, if the process is diffusion controlled, one can project that the rate of solute transport is directly proportional to the area available for permeation, and the average concentration gradient:

$$-\dot{M} = \bar{P}A(\Delta\bar{c}),\tag{6}$$

where $(\Delta\bar{c})$ is an average concentration gradient.

Using a log average depletion of solute, we get:

$$(c) = (C_o - C_i)/\ln\frac{C_i}{C_o}.\tag{7}$$

Equating Equations (2) and (3) leads to:

$$P = \frac{Q}{A}\ln\frac{C_i}{C_o}.\tag{8}$$

With Equation (8) it is relatively simple to estimate the permeability coefficient using steady state analyses. As in the case of the hydraulic permeability, the solute permeability can be conveniently expressed as a resistance; i.e., $1/\bar{P}$.

In Table IV are shown the initial properties of the fibers used in these experiments:

TABLE IV

Transport properties of unmodified fiber

Hydraulic permeability solute resistances	6×10^{-5} cm^3 cm^{-2} s^{-1} atm^{-1}
Creatinine	100 min/cm
Glucose	160 min/cm
Vitamin B-12	400 min/cm

3.6. TRANSPORT PROPERTIES OF ION EXCHANGE FIBER

After formation of the ion exchange resin in the fiber walls, significant changes in the transport properties of ionic species are expected. The end point of such transport is governed by the equilibrium expressions derived by Donnan [10], and the most significant property which governs this equilibrium is the charge density of the immobilized species. If the charge density is adequately high the membrane will be truly 'semi-permeable' and the Donnan assumptions apply. In these experiments the immobilized species was the quaternary nitrogen, and the gegenions were negative ions. The concentration of the fixed groups was determined by exchanging the gegenions to (OH^-) forms by thorough washing in 0.1 N NaOH. The fibers were then placed in a known excess of HCl, and after 24 hours the fibers were removed, rinsed, and the remaining HCl titrated.

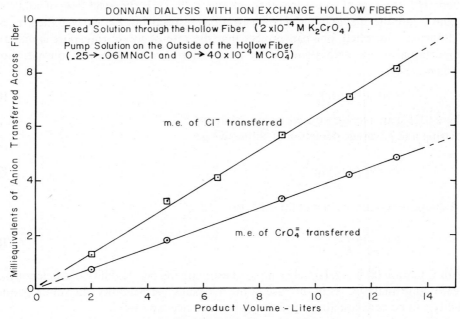

Fig. 8.   Donnan dialysis with ion exchange hollow fibers.

Titration of an aliquot of fiber indicated that 0.92 milliequivalents of (OH^-) were bound to 1.3 gms of fiber, to give 0.71 meq/gm. The sample represented 155 cm^2 so that the charge per unit area of membrane was 0.006 meq/cm^2.

The introduction of a significant weight of ion exchange polymer into the polyacrylonitrile fiber should change the permeability of the fiber wall to water and to non-ionic solutes, since the void volume is reduced. This reduction was found to be surprisingly small, as shown by the data in Table V.

TABLE V

Transport properties of ion exchange fiber

Hydraulic permeability solute resistances	1.6 cm^3 cm^{-2} s^{-1} atm^{-1}
Creatinine	550 min/cm
Glucose	1400 min/cm

The hydraulic permeability has been reduced by a factor of four, and the membrane resistance for creatinine has increased by a factor of five. The solute measurements were performed both initially, and after ion exchange deposition, with a 0.86% NaCl supporting electrolyte concentration.

The property that is most significant, of course, is the semipermeability of the fiber wall with respect to anions. To examine this function we used $CrO_4^=$ anions, because of the ease with which we could measure very small concentrations, and because of the current interest in removing such ions from water cooling system wastes. A concentration of 2.0×10^{-4} M $K_2CrO_4^=$ was pumped at 3.0 cc/min through a fiber bundle containing 150 fibers (153.1 cm^2). The fiber was bathed externally in a slowly stirred solution containing initially 0.25 M NaCl. The core solution leaving the fiber bundle contained less than 4×10^{-6} M $CrO_4^=$ (our lower limit of analysis), and 6.5×10^{-4} M Cl^-. The external bath increased continuously in $CrO_4^=$ concentration, and decreased in Cl^- concentration.

In a perfectly semipermeable system, the number of milliequivalents of $CrO_4^=$ permeating the wall would be exactly equal to the number of milliequivalents of Cl^- moving in the other direction. The actual situation is less than perfect, and in Figure 5 we show the number of milliequivalents of each ion that permeated, as a function of the total volume of fluid processed through the cores of the fibers. Considering the very large gradient of (Cl^-) available, and the small gradient of $(CrO_4^=)$, the lack of ideality is not surprising. At the end of the experiment, when 15 liters had been processed through the fibers, the external bath was 40×10^{-4} M in $CrO_4^=$, and the (Cl^-) was depleted to 0.06 M. The total volume in the external system had increased to 1000 cc as a result of osmotic pumping of water.

Considering the mass transfer of $CrO_4^=$ in the same way that non-ionic species were characterized, we can take the initial rate of transfer and calculate a membrane permeability:

$$P = \frac{3.0}{60 \times 153} \ln \frac{2 \times 10^{-4}}{4 \times 10^{-6}} = 1.3 \times 10^{-3} \text{ cm/s}$$

$$R = 13 \text{ min/cm}.$$

This is seen to be significantly higher than the rates attained for non-electrolytes. The comparison is, of course, not valid; the derivation assumes that only the concentration difference in $(CrO_4^=)$ provides the driving force. Instead, the work of Wallace [11], Smith [12] and Melsheimer et al. [13], indicates that the driving force is the Donnan potential, which drives the system toward the equilibrium:

$$N = \frac{(Cl^-)_1}{(Cl^-)_2} = \frac{(CrO_4^=)_1^{1/2}}{(CrO_4^=)_2^{1/2}}.$$

The semi-permeability of ion exchange membranes is conventionally determined by establishing the transference number of either K^+ or Cl^- in the membrane. In such experiments the potential difference is measured using reversible electrodes on either side of the membrane as it separates solutions of the same species held at different concentrations. In the case of hollow fibers this technique is difficult because the fine fiber inside diameters would require special electrode development. Instead we have relied on a transport method for estimating the semi-permeability.

The transport of co-ions is a function of the 'leakage' flow of ions through the fiber wall. No flow occurs through a perfectly semi-permeable fiber wall and the electromotive potential observed corresponds to the Nernst potential. However, when leakage flow occurs, as it does with real membranes, one can assume that the flow occurs through water filled pores. The cross-sectional area of pores allowing such flow, divided by the total available cross-sectional area is then a measure of the semi-permeability of the membrane. On this basis the rejection of ionic species is estimated by:

$$1 - \frac{(A/d)_{\text{KCl}}}{(A/d)_{\text{THO}}}.$$

The terms $(A/d)_i$ have been calculated from the measured permeability coefficients, as shown earlier in Equation (4). For these calculations the diffusion coefficient of KCl was taken as 1.89×10^{-5} cm^2 s^{-1}; the measured permeability of KCl was 5.7×12^{-7} cm s^{-1}, when the feed side concentration was 0.1 M. With these data the fractional rejection was found to be 0.995, indicating the very high efficiency with which the ion exchange fiber operates in dilute electrolyte solutions. When the ionic strength of the medium is higher, the semi-permeability would of course be somewhat lower, as shown by the $CrO_4^=$ data presented earlier.

In future work we hope to develop the methodology for distinguishing between the transport effected via the Donnan potential and the leakage flow of ion through uncharged pores, and how these variables are controlled by choice of fiber morphology and polymerization conditions. The recent publication by Dressner [14] pointing out the requirements for economical Donnan softening provides incentives for this line of investigation.

4. Conclusions

On the basis of the results obtained so far the following conclusions can be reached.

(1) The reaction of 4-VP with α,ω dihalides yields highly crosslinked anion exchange resins, the structure of which can be systematically varied by using a variety of dihalides.

(2) The first quaternization reaction activates the double bond of 4-VP leading to an addition polymerization of 4-VP monomers containing quaternary nitrogens. This conclusion is confirmed by the fact that the crosslinked resin is free of unquaternized pyridine rings.

(3) The rate of formation of crosslinked resin is determined by the reaction of the second bromine end group of the dibromide after the first end group has reacted. This rate determining step increases as the distance between positive charges, i.e., as the number of CH_2 groups in the dibromide increases and is probably due to the fact that a close approach of positive charges increases the coulombic repulsion and the free energy of formation of diquaternary salts (see Figures 6 and 7).

(4) The deposition of vinyl pyridine polymers initiated by the action of dihalides in the pores of hollow fibers having less than 100 Å radius leads to ion exchange surfaces which can readily transfer ions according to the Donnan formulations; i.e., the concentration gradient of a pump ion can be used to transfer another ion across the membrane wall against the latter's concentration gradient.

(5) The requirements for the level of polymer deposition needed to effect such transport is not well understood. The interaction between fiber properties, such as pore size, porosity, and wall thickness, and the ion exchange properties following resin deposition must still be investigated. However, it appears that the classical techniques of characterizing membrane transport can be utilized in such studies. The development of hollow fiber ion exchangers affords opportunities in ion exchange, ion separation and ion concentration not readily achieved with ion exchange resins in membranes.

Acknowledgements

This paper represents one phase of research carried out at the Jet Propulsion Laboratory, California Institute of Technology, under Contract No. NAS7-100, sponsored by the National Aeronautics and Space Administration. Authors Klein and Smith express their appreciation for financial support from the Louisiana State Science Foundation.

References

1. Kabanov, V. A., Aliev, K. V., Vargina, O. V., Patrikeeva, T. I., and Kargin, V. A.: *J. Polymer Sci., C.* **16**, 1079 (1967).
2. Salamone, J. C., Snider, B., and Fitch, S. L.: *J. Polymer Sci., A-1* **9**, 1943 (1971); Mielke, I. and Ringsdorf, H.: *Makrom. Chem.* **153**, 307 (1972).
3. Salamone, J. C., Ellis, E. J., Wilson, C. R., and Bardolivallan, D. F.: *Macromolecules* **6**, 475 (1973).
4. Rembaum, A., Singer, S., and Keyzer, H.: *J. Polym. Sci., B* **7**, 395 (1969).
5. Rembaum, A. and Noguchi, H.: *Macromolecules* **5**, 261 (1972).
6. Grobe, V., Mann, G., and Duwe, G.: *Faserforschung* **17**, 142 (1966).
7. Craig, J. P., Knudsen, J. P., and Holland, V. F.: *Textile Res. J.* **32**, 435 (1962).
8. Carman, P. C.: *Flow of Gases Through Porous Media*, Academic Press, N.Y., 1956.
9. Pappenheimer, J. R., Renkin, E. M., and Borrero, L. M.: *Am. J. Physiol.* **167**, 13 (1951).
10. Donnan, F. C.: *Chem. Rev.* **1**, 73 (1924).
11. Wallace, R. M.: *I&EC Process Res. Develop.* **6**, 423 (1967).
12. Smith, J. O.: *Exchange Diffusion as a Pretreatment to Desalination*, OSW R&D Report No. 655, U.S. Government Printing Office, May, 1971.
13. Melsheimer, S. S., Kelley, H. M., Landon, L. D., and Wallace, R. M.: *74th National Meeting of American Institute Chemical Engineers*, March, 1973, New Orleans, La., Paper #51b.
14. Dressner, L.: *I&EC Process Res. Develop.* **12**, 148 (1973).

RECENT STUDIES ON GRAFTING POLYELECTROLYTES TO MEMBRANES FOR REVERSE OSMOSIS

V. STANNETT, H. B. HOPFENBERG, F. KIMURA-YEH, and J. L. WILLIAMS*

Dept. of Chemial Engineering, North Carolina State University, Raleigh, N.C. 27607, U.S.A.

Abstract. The grafting of polyelectrolytes to polymer films leads to considerable increases in their water sorption and changes in their water permeability and salt rejection properties. Cellulose acetate polyvinyl chloride and polyoxetane films have been grafted with styrene-2 vinyl pyridine and with methyl methacrylate-2-vinyl pyridine mixtures. Grafting was readily carried out directly on the unmodified films by the mutual irradiation method. Tetrachlorethylene was used as the main swelling agent to accelerate grafting to cellulose acetate, and dimethyl formamide mixed with methanol was used to swell the polyoxetane films. After grafting, the films were quaternized or formed into the hydrochloride salt and their reverse osmosis properties studied. The cellulose acetate films showed different behavior from both the straight vinyl pyridine copolymer films and those grafted with either monomer above. Fluxes were found to steadily increase with increasing vinyl pyridine content of the grafted side chains. The salt rejections, on the other hand, behaved differently. For example, with a thirty percent total graft and increasing the proportion of vinyl pyridine in the graft composition the salt rejections remained constant, at greater than 99%, until the vinyl pyridine content exceeded 15% in the grafted copolymer after which they began to decline. Thus, high flux membranes with good salt rejection could be obtained. Preliminary results (on the corresponding polyoxetane and polyvinyl chloride films) indicate somewhat similar behavior to that found with cellulose acetate.

1. Introduction

A large number of different polymer films have been investigated as reverse osmosis membranes. Cellulose acetate, aromatic polyamides and various polyelectrolytes appear to be the most promising materials at this time [1]. Polyelectrolytes rely on the Donnan equilibrium principle for their salt rejecting properties. Since the Donnan effect diminishes with increasing salt concentration polyelectrolyte membranes are particularly suitable for the treatment of brackish rather than sea waters. Polyelectrolyte membranes tend to swell greatly in water and although this leads to high fluxes the salt rejections become less efficient with increasing swelling. Furthermore, the membranes often become soft, compressible and fragile when swollen. This problem can be overcome by various techniques such as depositing on, or in porous substrates, by minimizing the fixed ion content or by the use of grafted or block copolymers with hard hydrophobic polymeric second components. This present paper is concerned with studies of polyelectrolytes grafted to various polymeric substrates.

The earliest work of this kind appears to have been the grafting of polyacrylic acid to cellophane films [2] although many years earlier ion exchange membranes were prepared by grafting polystyrene to teflon and polyethylene followed by sulfonation [3]. Later a number of commercial ion exchange membranes were studied for the removal of sodium chloride and other salts from water by a reverse osmosis

* Present address: Becton Dickinson Company Inc., Research Triangle Park, N.C.27709.

Alan Rembaum and Eric Sélégny (eds.), Polyelectrolytes and Their Applications, 249–262. All Rights Reserved.
Copyright © 1975 by D. Reidel Publishing Company, Dordrecht-Holland.

procedure [4]. Other work concerned with polyethylene as the substrate has been recently reported by Yasuda and coworkers [5]. A rather extensive investigation of grafted membranes for reverse osmosis in general has been undertaken in these laboratories under the sponsorship of the Office of Saline Water. These have mainly centered around the grafting of styrene to cellulose acetate membranes to improve their compaction resistance [6, 7]. However, a number of studies have also been conducted which were directly concerned with the grafting of polyelectrolytes to various hydrophobic polymer films for possible applications as reverse osmosis membranes. The results of some of this work will be presented in this paper.

2. Materials

2.1. Films

Cellulose Acetate films (0.03 mm. thick) were machine cast from solutions of Eastman Cellulose Acetate #298-3 (39.8% acetyl content) and kindly provided by the Gulf General Atomic Corporation.

Polyvinyl Chloride films (0.055 mm thick) were laboratory cast onto glass plates from solutions of polyvinyl chloride (Geon 101.EP) in cyclohexanone.

Polyoxetane [3,3 bis (chloromethyl) oxetane, 'Penton', (Hercules Co.) films (0.04 mm) were cast from hot dimethyl formamide solution onto glass plates which were heated to 150°C on a thermostatted grill. All the monomers and solvents used were obtained from the Fisher Scientific Company. The monomers were freed from inhibitors by distillation under vacuum before use.

3. Procedures

3.1. Grafting

The grafting was carried out with the mutual radiation method. The films were rolled and immersed in the grafting solutions in glass ampoules and degassed by several freeze-thaw cycles under a vacuum of 10^{-5} torr. With polyoxetane the films were found to be damaged by this procedure and the degassing was carried out separately on the films and the grafting solution and then the solution added to the film. A simple constricted tube assembly was found to be suitable for this purpose with "he film initially in the upper half. After degassing and sealing under vacuum the tubes were inverted to allow the solution to immerse the films. In each case, after sealing under vacuum, the tubes were irradiated at 30°C in a Cobalt-60 source at the stated dose rate.

After grafting the films were extracted to constant weight with a suitable solvent for the homopolymer. In the case of the 2-vinyl pyridine grafts the quaternizations were carried out under 2 atmospheres of methyl bromide at room temperature for 24 hours at room temperature unless otherwise states. The quaternizations rarely exceeded seventy percent, presumably due to the poor accessibility of the grafted films to methyl bromide vapor. This problem is now being studied in more detail [10].

3.2. REVERSE OSMOSIS TESTING

Most of the reverse osmosis testing was carried out at 3.5% sodium chloride concentration and 1500 psi. pressure. A number of tests were also conducted at 0.5% and 800 psi.

The various films prepared were about 1–2 mils thick and were neither ultra thin nor prepared as asymmetric membranes. Consequently the product fluxes were low and could not easily be determined with conventional reverse osmosis test cells. Measurements were therefore carried out in a test cell first described by McKinney [8]. The test unit is uniquely suited for characterizing transport properties in dense, low flux membranes because of the small downstream dead volume and the simultaneous determination of flux and rejection in a capillary connected to the downstream receiving volume instrumented with a microelectrode to monitor down-stream salt concentration.

4. Results and Discussion

4.1. CELLULOSE ACETATE FILMS

2-vinyl pyridine was readily grafted to cellulose acetate film using a 20% solution of the monomer in methanol. The weight gain of the film after washing increased linearly with dose as shown in Figure 1.

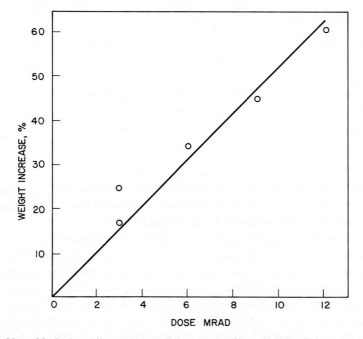

Fig. 1.   Grafting of 2-vinyl pyridine to dense cellulose acetate films at 30 °C. 20% monomer in methanol. Dose rate 0.13 Mrads per hr.

Styrene – 2-vinyl pyridine mixtures and pure 2-vinyl pyridine itself were grafted very efficiently in both carbon tetrachloride and tetrachlorethylene solutions. It was hoped that many shorter chains would result from the increased radical yields and the chain transfer properties of the chlorinated solvents. The results of these experiments are presented in Table I.

TABLE I

Grafting and reverse osmosis properties of styrene 2-vinylpyridine-grafted cellulose acetate

Composition of monomer solution	Dose (Mrad)	Weight increase (%)	Quaternized	Flux[b] (gfd-mil)	Salt[b] rej. (%)
Cellulose acetate (control)	–	–	–	0.040	98.0
Styrene: 2-Vinylpyridine: Tetrachloroethylene	7.5	42.2	CH_3Br	0.16	98.4
=15:5:80	7.5	42.0	HCl	0.08	94.4
10:10:80	7.5	33.4	CH_3Br	0.36	98.6
10:10:80	7.5	36.5	HCl	0.36	98.6
10:10:80	12.0	49.6	–	0.016	–
10:10:80	52.9	–	–	–	–
10:10:80	12.0	49.4	CH_3Br	0.30	87.0
10:10:80	12.0	51.1	HCl	0.49	87.0
10:10:80	12.0	51.1	HCl	0.43[a]	87.61
10:10:80	2.2	27.5	CH_3Br	0.24[b]	99.4[b]
10:10:80	6.0	42.8	CH_3Br	0.40[b]	99.0[b]
Styrene: 2-Vinylpyridine: Carbon tetrachloride	3.2	10.2	–	0.03	99.4
=20:20:40					
2-Vinylpyridine: Tetra-chlorethylene = 20:80	3	19.8	CH_3Br	0.20	98.0
2-Vinylpyridine: Tetra-chlorethylene = 20:80	7.2	30.9	CH_3Br	0.49	90.0
2-Vinylpyridine: Tetra-chlorethylene = 20:80	13.5	41.8	CH_3Br	–	–

[a] Dried and retested.
[b] 800 psi, 0.5% sodium chloride solution.

Methyl methacrylate, 2-vinyl pyridine and their mixtures were also successfully grafted in a 20% solution of the monomers in tetrachlorethylene. The results of these experiments are presented in Table II and Figure 2.

The reverse osmosis properties of the grafted films are presented in Tables I and II and in Figures 3 and 4. The results are most interesting in that the flux could be increased considerably, up to ten fold in the best case, while maintaining the salt rejection at a similar level to that of unmodified dense cellulose acetate. This was accomplished by the copolymerization of 2-vinyl pyridine and styrene or methyl methacrylate followed by quaternization. The full development of these findings would need the additional development of assymetric or ultra thin membranes from these materials.

TABLE II

Grafting and reverse osmosis properties of methyl methacrylate – 2-vinyl pyridine grafted cellulose acetate

Composition of monomer solution	Dose (Mrads)	Weight increase (%)	Quaternized	Flux (gfd-mil)	Salt rej. (%)
Cellulose acetate (control)	–	–	–	0.035	98.0
MMA-2. VP – Tetrachlorethylene					
15-5-80	0.7	19.8	CH_3Br	0.13	99.3
	2.4	38.1	CH_3Br	0.24	99.0
	3.2	44.6	–	0.03	–
10-1080	1.5	23.6	CH_3Br	0.23	98.4
	3.8	41.5	CH_3Br	0.34	97.6
	3.8[a]	41.5	CH_3Br	0.36	98.4
	5.3	48.9	–	0.015	95.4
5-15-80	5.0	40.6	CH_3Br	0.68	90.4
	12.8[b]	57.2	CH_3Br	0.62	94.6
	12.8	57.2	CH_3Br	0.05	94.0
0-20-80	6.0	34.0	CH_3Br	2.40	80.3
	9.0	44.5	CN_3Br	9.42	62.5

Tested at 0.5% sodium chloride and 800 psi.
[a] Tested, dried and retested.
[b] Tested, dried, quaternized and retested.

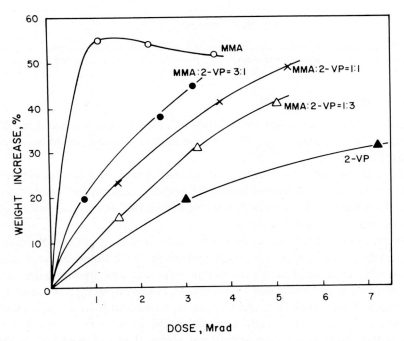

Fig. 2.   Grafting of 2-vinyl pyridine, methyl methacrylate and their mixtures to dense cellulose acetate films at 30 °C. 20% monomer(s) in tetrachlorethylene. Dose rate 0.13 Mrads per hr.

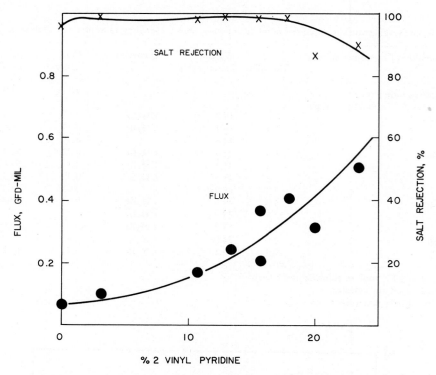

Fig. 3.   Reverse osmosis properties of styrene-2-vinyl pyridine grafted cellulose acetate membranes as a function of the 2-vinyl pyridine content. Approximately 30% graft. 0.5% NaCl – 800 psi.

4.2. POLYVINYL CHLORIDE FILMS

The grafting of 2-vinyl pyridine to polyvinyl chloride films proceeded smoothly and reproducibly in 20% monomer solution in methanol. Typical results are presented in Figure 5.

Acrylic acid was more difficult to graft because of the rapid homopolymerization and little or no grafting occurred before the grafting solution had completely polymerized. This was the case with both water and methanol as solvents and at different pH values, changed by adding alkali.

Cupric chloride and ferrous ammonium sulfate were then successfully used as inorganic inhibitors to prevent homopolymerization in the liquid phase. Films were sealed in 50% acrylic acid aqueous solution saturated with the inorganic inhibitor under nitrogen gas and irradiated with γ-rays at a dose rate of 0.13 Mrad/hr at room temperatures. After irradiation, the films were washed with water. The results are presented in Figure 6. The weight increase varied linearly with dose. The rate of grafting with ferrous ammonium sulfate was higher than that with cupric chloride.

The double grafting of acrylic acid and 2-vinyl pyridine to polyvinyl chloride film was accomplished by methods adapted from those used by Jendrychowska-Bona-

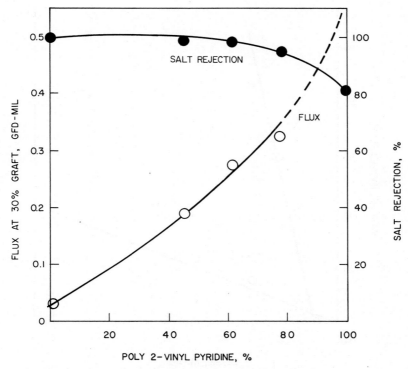

Fig. 4.   Reverse osmosis properties of methyl methacrylate-2-vinyl pyridine grafted cellulose acetate membranes as a function of the 2-vinyl pyridine content. Approximately 30% graft. 0.5 NaCl – 800 psi.

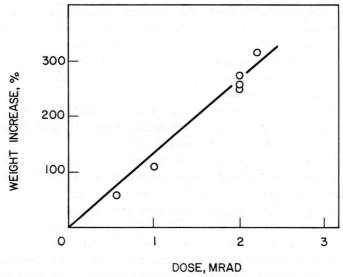

Fig. 5.   Grafting of 2-vinyl pyridine to dense polyvinyl chloride film under nitrogen gas at 30 °C. 20% monomer in methanol. Dose rate 0.13 Mrads per hr.

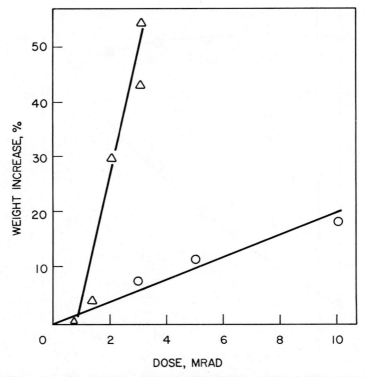

Fig. 6. Graftings of acrylic acid to dense polyvinyl chloride film in 50% acrylic acid aqueous solution saturated with cupric chloride (○) and ferrous ammonium sulfate (△). Dose rate 0.13 Mrads per hr.

mour [9] with polytetrafluorethylene film. The first method was to graft acrylic acid in the manner described above, neutralize with sodium hydroxide and then graft with 2-vinyl pyridine in the manner described earlier. The second method was to first graft with 2-vinyl pyridine, neutralize with dilute hydrochloric acid and then to graft with acrylic acid. The latter method was far more efficient, as can be seen from the results presented in Table III.

The polyvinyl chloride based membranes were mechanically strong and durable. The reverse osmosis properties are presented in Tables III and IV. Fluxes from 1.0–10.0 gfd-mil and respective salt rejections of 70–40% were found. These were well in the range of other polyelectrolyte type membranes reported in the literature. The double grafted films showed similar properties to the quaternized polyvinyl pyridine membranes. This was not, perhaps, surprising since no special efforts were made to change the polyacrylic acid component into the salt form and the aerated sodium chloride solutions were not of sufficiently high pH for adequate ionization. It is clear from Figure 8 that non-quaternized polyvinyl pyridine and polyacrylic acid grafts behave similarly to other non-ionic membranes with high fluxes but low salt rejections. The substantial increases on quaternization demonstrates the importance of the Donnan ion exclusion principle in this type of membrane.

TABLE III

Double grafting of acrylic acid and 2-vinyl pyridine to polyvinyl chloride film

First grafting			Second grafting			Composition (% Graft)		Flux	Salt rejection
Grafting solution	Dose (Mrads)	Weight increase %	Grafting solution	Dose (Mrads)	Weight increase (%)	(% 2-VP)	(%PAA)	(gfd-mil)	(%)
20:80 2-VP:CH$_3$OH	0.9	64.1	—	—	—	—	10.0	0	65.2
20:80 2-VP:CH$_3$OH	0.9	64.1	50:50 AA:H$_2$O (satd with Fe, AmSO$_4$	0.28	3.6	96.1	3.9	1.24	78.1
20:80 2-VP:CH$_3$OH	0.9	73.9	(satd with Fe, AmSO$_4$	0.28	8.5	83.3	16.7	1.55	68.0
20:80 2-VP:CH$_3$OH	2.4	184.2	(satd with Fe, AmSO$_4$	0.28	23.5	73.3	26.7	1.03	68.0
20:80 2-VP:CH$_3$OH	2.6	226.4	(satd with Fe, AmSO$_4$	0.3	46.9	59.6	40.4	—	—
20:80 2-VP:CH$_3$OH	2.6	226.4	(satd with Fe, AmSO$_4$	0.6	50.8	57.7	42.3	—	—
20:80 2-VP:CH$_3$OH	2.6	226.4	(satd with Fe, AmSO$_4$	1.0	156.5	44.2	55.8	—	—
50:50 AA:H$_2$O	0.9	9.8	20:80 2-V.P CH$_3$OH	0.8	0	0	100	—	—
(Sat'd. with FeAmSO$_4$)	0.9	9.8		2.4	9.5	50.8	49.2	—	—
	0.9	9.8		2.4	24.7	62.4	37.6	—	—

Tested at 3.5% sodium chloride and 1500 psi.

TABLE IV

Reverse osmosis properties of grafted
polyvinyl chloride films

Weight increase (%)	Quaterni- zation (%)	Film thickness (mils)	Average flux (gfd-mil)	Salt rejection (%)
12.4	0	1.77	~0	–
53.7	0	2.16	3.6	3.2
103.8	0	1.96	(43.3)	2.2
271	0	2.50	24.6	2.7
314.2	0	2.75	11.4	1.5
13.2	104.7	1.57	1.71	68.0
57.0	73.0	1.29	4.78	63.3
61.1	79.4	1.57	3.43	72.7
62.1	58.7	1.57	4.09	69.7
62.1	60.7	1.77	6.12	66.1
119.3	70.2	2.75	6.57	63.0
171.3	65.1	2.36	4.62	69.6
199.2	65.2	2.75	8.84	63.4
223.1	57.4	1.57	4.67	70.5
374.6	44.0	1.96	5.60	74.0

Tested at 3.5% sodium chloride and 1500 psi.

4.3. POLYOXETANE FILMS

Polyoxetane ('Penton', Hercules Co.) films have excellent mechanical strength and chemical stability and in principle should provide an excellent substrate for these grafting studies. An extensive investigation is presently being conducted therefore with this material. Briefly, it has been found initially that good grafting yields can be obtained using 20:30:50 Monomer:methanol:dimethyl formamide solutions. The results obtained with styrene, 2-vinyl pyridine and their mixtures are shown in Figure 7 and other results are included in Table V where the reverse osmosis properties are also presented.

TABLE V

Grafting and reverse osmosis properties of polyoxetane films

Monomer composition	Dose (Mrads)	Weight increase %	%Q	Flux	SR%
100:2-vinyl pyridine	3.36	40.0	52.5	0.14	92.8
	2.24	63.0[a]	41.3	1.87	33.2
	3.36	79.0[a]	44.8	2.64	47.7
80:20 2VP-styrene	1.68	52.8[a]	42.3	0.32	67.0
	1.68	62.2	31.8	0.69	21.0
	3.36	79.0	40.6	0.47	60.0
	3.36	82.5[a]	63.5	1.14	51.7
	5.04	100.0	45.6	0.56	64.0
50:50 2VP styrene	5.04	83.0[a]	46.0	0.44	92.0
	6.16	109.0[a]	43.8	0.78	38.6

[a] Non-frozen technique.

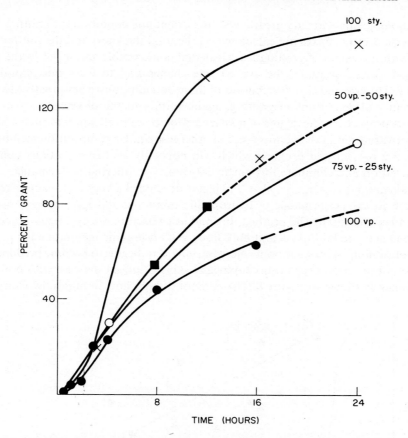

Fig. 7.   The mutual grafting of 2-vinyl pyridine-styrene mixtures to polyoxetane films at 20 °C. Dose rate 0.6 Mrads per hr. Thickness 0.01 cm. Solvent system: 20:30:50 monomer-methanol-dimethyl formamide by volume.

The grafted polyoxetane membranes showed an extremely high degree of variation. This is undoubtedly due, partly at least, to the fact that it is a semicrystalline substrate and the degree of crystallinity varies from film to film depending on the thickness and the conditions of drying. In addition the crystallinity will vary with the extent of grafting, the composition of the grafting solutions used, the dose rate and the other variables. These factors are under study and the results will be presented separately [10].

Substantial variations in the performance of grafted membranes with the same level and composition of grafting are frequent even with essentially noncrystalline substrates such as cellulose acetate and polyvinyl chloride. These are undoubtedly due to the variations in structure on at least two counts. Firstly, even with very thin films the grafting process is clearly diffusion controlled. The percent of grafting decreases through the half thickness of the membrane i.e. is greater at the two surfaces.

This inhomogenity will vary greatly with the extent and conditions of grafting. Since the reverse osmosis fluxes are calculated on the total thickness and the salt rejection varies with the extent of grafting it is clear that considerable variations in the properties will occur. Secondly the grafted side chains tend to form into mosaic-like domains. The size and size distribution of these domains must vary greatly also with the extent and conditions of grafting, again leading to the observed variations in transport properties. The degree of quaternization or neutralization is often a further cause for variations in the performance of otherwise similar composition membranes. Studies are presently in progress which will hopefully lead to essentially complete quaternization and eliminate this particular cause for differing performance.

Yasuda and his colleagues [11, 12, 13] have developed a very useful and ingenious approach to the performance of hydrophillic non-ionic and ionic reverse osmosis membranes. Essentially the product flux and salt rejection are explained in terms of combined activated diffusion and bulk flow. With non-ionic membrane [11], as the degree of swelling of the membrane increase, bulk flow begins to become predominant and the salt rejection drops catastrophically. This is indeed the case with non-ionic membranes as shown in Figure 8. The concept of bulk flow in highly swollen mem-

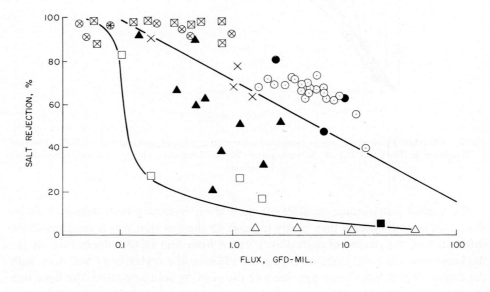

REGULAR CA. 398-3
S-VP -CA
MMA- VP - CA
QUARTERNIZED POLY-2-VP GRAFTED PVC
QUARTERNIZED POLY-2- VP GRAFTED CELL AC

POLYACRYLIC ACID GRAFTED PVC
POLY 2-VP GRAFTED PVC
NEUTRALIZED POLYACRYLIC ACID GRAFTED PVC
PVP- POLYOXETANE
DOUBLE GRAFTED PAA-VP-PVC

Fig. 8.   Salt Rejection vs log flux for various grafted membranes. At 3.5% NcCl and 100 psi except for the cellulose acetate grafts. Solid lines are generalized lines from Yasuda et al. (References 11, 12 and 13). cellulose acetate grafts. Solid lines are generalized lines from Yasuda et al. (References 11, 12 and 13)'

branes has been challenged by Paul [14, 15] however and it is not yet clear whether this approach is entirely valid. It does explain, however, the catastrophic drop in the salt rejection at high fluxes and provides a useful framework for correlating the behavior of various hydrophillic membranes.

In the case of ionic membranes [12] the behavior is different and the salt rejection was found empirically to be related by the simple relationship

$$\text{Flux} = A \exp(-B \cdot \text{SR}),$$

where A and B are constants.

The salt rejection vs product flux curves for nonionic and ionic membranes are shown in Figure 8 together with all the results found in this work. Except for the polyoxetane based membranes the results do roughly fit the relationships developed by Yasuda and coworkers. The superior salt rejections at high fluxes shown by the ionic membranes is clearly evident. This is due to the Donnan equilibria in the activated diffusion region and, according to Yasuda [12], to the effectively greater size of the salt ions in the charged environment of the 'pores' postulated to be present in the highly swollen membranes. This leads to a reduction in salt transport relative to the water flux. With both the nonionic and ionic membranes the salt rejection and flux are related by semi-empirical relationships linked by the degree of hydration of the membranes. Much further work is needed to show whether this treatment has validity. Grafted membranes have too many built-in variables in their structure to be ideal for such studies. Simple copolymers such as the styrene-4-vinyl pyridine copolymers studied by Williams and coworkers [16] are probably better suited and further studies along those lines should be highly rewarding.

Note added in proof: New work has shown that grafted polyoxetane membranes do show similar results to those obtained with polyvinyl chloride.

References

1. Podall, H. E.: in N. N. Li (ed.), *Recent Developments in Separation Science*, CRC. Press Cleveland, Ohio, 1972, Vol. 2, pp. 171–203.
2. Baldwin, W. H., Holcomb, D. L., and Johnson, J. S.: *J. Poly. Sci.* **A3** (1965).
3. Chen, W. K. W., Mesrobian, R. B., Ballantine, D. S., Metz, D. J., and Glines, A.: *J. Poly. Sci.* **23**, 903 (1957).
4. McKelvey, J. G., Spiegle, K. S., and Wyllie, M. R. T.: *Chem. Eng. Progress Symposium Series* **55**, 199 (1959).
5. Lamaze, C. E. and Yasuda, H.: *J. Appl. Poly. Sci.* **15**, 1665 (1971).
6. Kimura-Yeh, F., Hopfenberg, H. B., and Stannett, V.: in H. K. Lonsdale and H. E. Podall (eds.), *Reverse Osmosis Membrane Research*, Plenum Press, New York, 1972, pp. 177–203.
7. Bentvelzen, J. M., Kimura-Yeh, F., Hopfenberg, H. B., and Stannett, V.: *J. Appl. Poly. Sci.* **17**, 809 (1973).
8. McKinney, R.: *Anal. Chem.* **41**, 1513 (1969).
9. Jendrychowska-Bonamour, A. M.: *J. Chim. Phys.* **70**, 20 (1973).
10. Bittencourt, E., Williams, J. L., Hopfenberg, H. B., and Stannett, V.: to be published.
11. Yasuda, H. and Lamaze, C. E.: *J. Poly. Sci.* **A2 9**, 1537 (1971).
12. Yasuda, H., Lamaze, C. E., and Schindler, A.: *ibid.*, 1579 (1971).

13. Yasuda, H. and Schindler, A.: Reference 6, pp. 299–316.
14. Paul, D. R. and Ebra-lima, O. M.: *J. Appl. Poly. Sci.* **14**, 2201 (1970).
15. Paul, D. R.: *J. Poly. Sci. (Physics)* **11**, 289 (1973).
16. Williams, J. L., Schindler, A., and Peterlin, A.: *Makromol. Chemie* **147**, 175 (1971).

ON THE PICKUP OF HYDROCHLORIC AND
SULFURIC ACIDS ON A POLYAMINE RESIN

ROBERT E. ANDERSON

Diamond Shamrock Chemical Company, Redwood City, Calif., U.S.A.

Abstract. Weak-base ion-exchange resins are polymeric structures which contain primary, secondary and/ or tertiary amine groups. They are water-swollen and can form salts with a wide variety of acids. Their major use is for the removal of mineral acids from solution. These weak-base resins are an order-of-magnitude or so slower in picking up acid than are the strong-base resins having quaternary-ammonium functionality. This difference in rate has been shown to be due to the difference in internal ionization of the resins. Further, the relative rates of various weak-base resins can be correlated with their pK's as determined by titration curves. Thus, the rate of pick-up of a mineral acid from solution is primarily a function of conditions inside the resin. However, it has been known for many years that such resins show an appreciably higher capacity for sulfuric acid than for hydrochloric acid under many operating conditions, and that this difference is a function of kinetic factors. This puzzling difference can be explained in a way consistent with the well established rate-determining mechanism by the pick-up of bisulfate ion from these very dilute solutions. The bisulfate ion then appears to be converted to sulfate ion inside the resin.

Ion-exchange resins are highly crosslinked polyelectrolytes. When placed in water they take up water and swell until the osmotic forces are balanced by the elastic forces of the polymer network. In this swollen form they are best viewed as solid solutions.

1. Ion Exchange Resin Types

The four types of ion-exchange resins commonly used in water treatment are shown in Table I. The strongly ionized and weakly ionized resins have very different properties. The strongly ionized resins have a fixed number of exchange sites per unit of structure. The rate of neutralization is fast with half-times of the order of one to three minutes. Regeneration of these resins tends to be mass-action controlled. An excess of regenerant is required. The useful work that such a resin can do, its operating capacity, is largely dependent on equilibrium factors.

With weakly ionized resins the internal ionization, and therefore the number of actual exchange groups per unit of structure, is a function of the pH of the contacting solution. The capacity of the resin varies from very low to quite high, depending on the solution with which it is in equilibrium. The rates of neutralization are slow with half times reported from ten minutes to hours. Regeneration to the free acid or base form is pH dependent and may be essentially one hundred percent efficient.

The strong-ionized resins probably account for 90 to 95 percent of the volume of resin sold to the water conditioning industry. This same percentage is also approximately true for the technical literature on ion exchange. These resins with their fixed capacities and well defined properties have been fascinating playthings for many investigators. Even these resins have proved so complex that attempts at quantitative mathematical treatment have resulted in approximations. It is no wonder that only

Alan Rembaum and Eric Sélégny (eds.), Polyelectrolytes and Their Applications, 263–274. All Rights Reserved.
Copyright © 1975 by D. Reidel Publishing Company, Dordrecht-Holland.

a few brave souls have dared to tackle the weakly ionized resins which are much more complex in their behaviour.

The industrial chemist does not have the same latitude to pick and choose his subject for study as some. Neither is he required to develop the same kind of answer. He need only say what a resin will do, not why it works as it does. To get this answer,

TABLE I

Ion exchange resin types

Cation exchangers	Characteristics	Anion exchangers
	Strongly ionized	
RSO$_3$H	Fully ionized	RNR$_3$OH
Water treatment uses: Na cycle softening Decationization Mixed bed D.I.	Fixed number of exchange sites Rate of neutralization fast $T_{1/2} < 5$ minutes Regeneration mass-action controlled Excess of regenerant required Operating capacity dominated by equilibrium factors	Water treatment uses: Acid neutralization Silica removal Mixed bed D.I.
	Weakly ionized	
RCOOH	Ionization function of pH	RNH$_2$, R$_2$NH, R$_3$N
Water treatment uses: H$^+$ cycle dealkalization Base neutralization	Number of exchange sites varies from low to high Rate of neutralization slow $T_{1/2} > 10$ minutes Regeneration pH controlled, essentially 100% efficient Operating capacity dominated by kinetic factors	Water treatment uses: Acid neutralization

the resin is tested under simulated field conditions. A one-inch to six-inch diameter column is filled with the resin. A water of a typical composition is made up. This is passed through the column under set conditions and the effluent is monitored to determine how much work the resin will do. The water composition and column conditions are varied to give data from which a set of empirical curves is constructed covering the conditions under which the resin is likely to be used.

This does not demand any answers on why the resin acts as it does. However, often the empirical relationships which develop do not appear reasonable, and since chemists and engineers are involved, the why of the behaviour eventually does get some thought. The polyamine weak base resins are a case in point.

2. Polyamine Resins

The polyamine resins contain a structure of the form:

$$(-N-C-C-N-C-C-)$$

The number of nitrogens in the chain can vary from two on up. In many cases the nitrogen-carbon bond is an integral part of the polymer network. There are more than two carbons between nitrogens in some cases, and the carbons may carry other substitutions. The actual ionogenic functionality is a mixture of primary, secondary and tertiary amines. Polyamine resins made by condensing phenol, formaldehyde and polyamine were among the first commercial resins. Other methods of manufacture include the reaction of a polyamine with a chloromethylated styrene-divinyl-benzene copolymer, and condensation of epichlorhydrin with ammonia or poly-amines.

Outside of the technical sales literature, one of the few descriptions of the behaviour of these resins is an article by deJong, Brants and Otten [1]. They described a number of the characteristics of these resins. The three types of exchange capacity are of interest here. These are the theoretical exchange capacity, the equilibrium exchange capacity, and the operating exchange capacity.

The theoretical exchange capacity is the amount of acid that could be taken up by the resin if all of the amine groups were protonated. It is thus equal to the number of amine groups per unit volume or unit weight.

However, these amine groups have a wide range of basicities. This can be seen in Figure 1, which shows structures which could result from the condensation of epi-

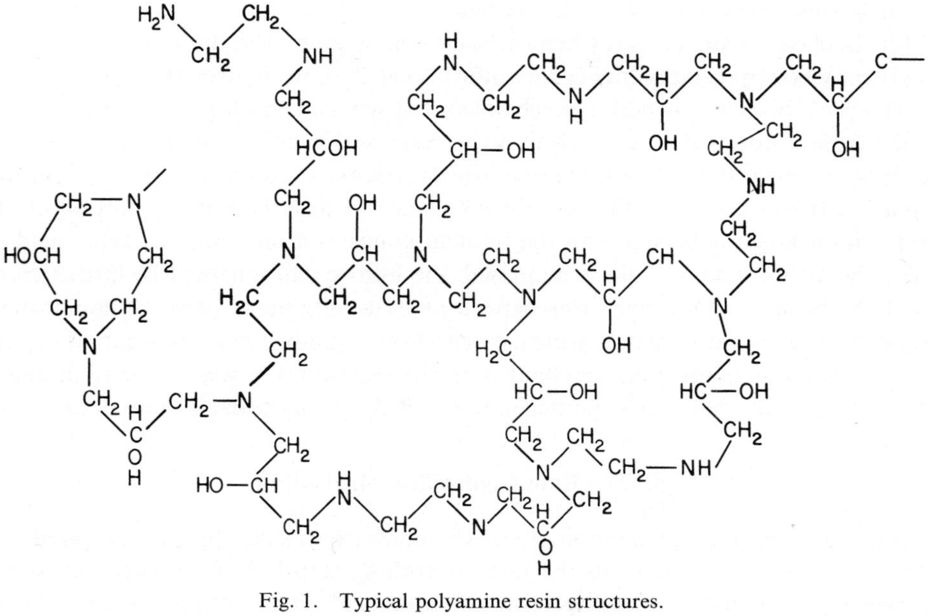

Fig. 1.   Typical polyamine resin structures.

chlorhydrin with the simplest polyamine, ethylene diamine. Not only is there a mixture of primary, secondary and tertiary amines present, there can be wide variations in their steric restraints. Since the orientation of the nitrogen-carbon bonds must adjust to accommodate the hydrogen of protonation, the effective basicity of each nitrogen will depend somewhat on the rigidity of the local polymer structure.

An even greater factor is the effect of partial protonation on the basicity of the adjoining nitrogens. When a nitrogen protonates, the positive center formed tends to withdraw electrons from the neighboring nitrogens in a polymer chain and reduces their basicity. Thus as the amine groups in the resin are converted to ion-exchange sites, the effective basicity of the remaining amines decreases, and a higher concentration of hydrogen ion is required to protonate additional sites. The result is an equilibrium capacity: the equivalents of acid the resin will hold per unit of volume or weight when in equilibrium with a given concentration of acid.

3. Operating Performance

In water treatment these resins are used to remove mineral acids from solution, usually as the second step in demineralization. The most important property to the engineer is their operating capacity. How much mineral acid can the resin remove under a given set of conditions? The operating capacity of a polyamine resin has been found to be a function of a number of operating conditions. The operating capacity:
  (1) decreases with increasing flow rate;
  (2) decreases with increasing resin particle size;
  (3) is higher with resins of higher basicity;
  (4) increases in some cases when carbonic acid is present in the water;
  (5) may be appreciably higher for sulfuric acid than for hydrochloric acid.
The problem is to propose a mechanism that will account for these observations.

If the flow rate is slow enough that the resin and solution are at equilibrium at each point down the column, the operating capacity will approach the equilibrium capacity. It will not equal the equilibrium capacity since the bottom portion of the bed cannot be equilibrated with the influent concentration without having acid in the effluent. As the flow rate is increased, the local conditions will be farther from the equilibrium and the operating capacity will deviate more from the equilibrium capacity. The variation of operating capacity from equilibrium is the result of kinetic factors. The rate controlling mechanism in the reaction of a weak base resin and a mineral acid will determine the dependency of operating capacity on column conditions.

4. The Rate Controlling Mechanism

The mechanism of neutralization of weakly ionized resins has been investigated and there is general agreement on the rate controlling step [2]. Two mechanisms are considered. The first is the diffusion of the total acid molecule into the resin where it reacts with an amine group.

$$RNH_2 + H^+Cl^- \rightarrow RNH_3^+Cl^-.$$

The rate of pickup in this case should be directly proportional to the acid concentration. There is some indication that such a situation may exist for a short time at the beginning of the neutralization. However, after the first few seconds of the exchange the rate is found to be independent of solution concentration. As soon as a small shell of resin has been converted to the highly ionized amine hydrochloride, further invasion of hydrochloric acid as such is inhibited by Donnan membrane forces.

The major part of the neutralization can be viewed as true ion exchange. For those amine groups that exist as positive ionic sites due to reaction with water, the neutralization can proceed by exchange of hydroxide ion for chloride ion.

$$RNH_2 + H_2O \rightleftarrows RNH_3^+OH^-$$
$$RNH_3^+OH^- + H^+Cl^- \rightarrow RNH_3^+Cl^- + H_2O.$$

For these ionized amine sites, the reaction is identical to the neutralization of a strong base resin.

$$R_4N^+OH^- + H^+Cl^- \rightarrow R_4N^+Cl^- + H_2O.$$

Except in very dilute solutions, the rate controlling step in such ion-exchange reactions is the diffusion of ions in the resin phase. It is a coupled diffusional process in that each time an ion enters the resin, an ion of the same charge must leave the resin in order to maintain electrical neutrality. The rate is dependent on the concentration of the leaving ion at the center of the resin and the concentration of the entering ion at the surface of the resin. These concentrations fix the gradients for ionic diffusion. In a strongly ionized resin the concentration of the leaving ion is high and the concentration gradient is sufficient to give a fast reaction rate. In a weakly ionized resin the concentration of the leaving ion is very low and the process is slow.

In summary, the rate determining step is ionic diffusion inside the resin. This is dependent on the internal concentration of the leaving ion, in this case the hydroxide ion. Its internal concentration is very low in a weak base resin because of the low ionization of the amine groups. As a result the pickup of a mineral acid on a weak base resin is inherently a slow process.

5. Rationalization of Column Performance

This rate mechanism explains most of the column characteristics of these resins.

(1) Acid pickup is a slow process, so sufficient time of contact between resin and acid must be allowed. Increasing the flow rate reduces the contact time and will tend to decrease the operating capacity.

(2) In processes that are controlled by diffusion within the spherical particle, the

rate is inversely proportional to the square of the radius. This relationship has been confirmed for acid pickup on weak base resins.

(3) Resins of higher basicity will have a higher internal hydroxide concentration, improved exchange rates, and will give higher operating capacities.

(4) The polyamine resins are strong enough bases to form bicarbonate salts. When a mixture of mineral acid and carbonic acid enters the column, both acids are picked up and the effluent is neutral. The bicarbonate is quantitatively displaced by the mineral acid and moves down the column in front of the mineral acid. This increases the rate at which the mineral acid can be picked up, since it now can exchange for a bicarbonate from the highly ionized resin. Although carbonic acid will appear in the effluent fairly early in the run, if the run is continued to a mineral acid breakthrough, a higher operating capacity will be obtained than if no carbonic acid is present.

These explanations are based on conditions inside the resin. Changing conditions outside the resin should have only minor effects over fairly wide ranges. The change from hydrochloric to sulfuric acid does not seem to fit this pattern. This difference in behaviour between hydrochloric acid and sulfuric acid is due to the step-wise ionization of the sulfuric acid molecule. Both equilibrium and kinetic components are involved. These can be seen in data taken on a recently introduced polyamine resin, Duolite® ES-340. This resin is made by an inverse suspension condensation of a polyamine and epichlorhydrin.

6. Equilibrium Capacity

The equilibrium capacity of Duolite ES-340 for hydrochloric acid and sulfuric acid was determined over a concentration range of 2.5×10^{-3} to 1.0×10^{-1} normal. Ten milliliters (measured by tapping to a constant volume in a graduate) of free-base-form resin was transferred to a 15 ml fritted glass funnel. A solution of acid of known concentration was allowed to drip through the column until the effluent concentration was equal to the influent concentration. The resin was dewatered by suction filtration, then eluted with an excess of sodium hydroxide. The amount of acid picked up by the resin was determined by the analysis of the eluate for chloride or sulfate. The results are shown in Figure 2.

Duolite ES-340 shows a 20–30% higher capacity in terms of hydrogen equivalents for sulfuric acid than for hydrochloric acid. This is due to the fact that a portion of the sulfuric acid is held as the bisulfate ion, thus in effect neutralizing two equivalents of hydrogen with one amine functionality.

The equilibrium between sulfate and bisulfate ions on a strong base resin has been studied as a function of sulfuric acid concentration [3]. As the acid concentration is increased, the equivalent fraction of the anion present in solution as bisulfate increases. Increasing the ionic concentration in solution also increases the apparent selectivity of a strong base resin for the monovalent bisulfate ions over the divalent sulfate ions. Thus, a strong base resin tends to convert to the bisulfate form in con-

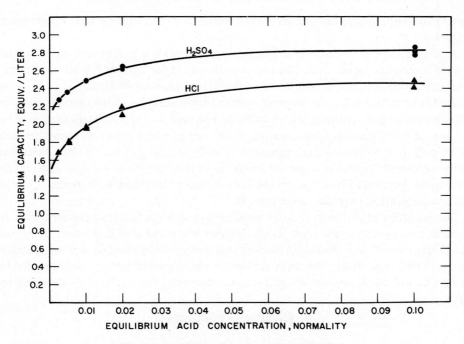

Fig. 2.   Equilibrium capacity of Duolite ES-340.

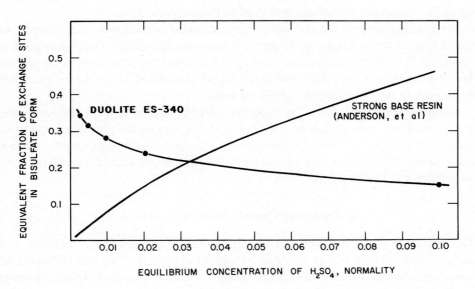

Fig. 3.   Resin composition vs acid concentration.

centrated sulfuric acid and to the sulfate form in dilute sulfuric acid as shown in Figure 3.

A similar type of behaviour might be expected with a polyamine resin, but the opposite seems to be the case. The composition curve for Duolite ES-340 in Figure 3 was calculated on the assumption that the actual number of exchange sites in the resin will be constant for a given hydrogen ion concentration in the external solution whether the acid is hydrochloric or sulfuric. The data of Figure 2 were corrected for the actual hydrogen ion concentration in the sulfuric acid solutions using the data of Sherrill and Noyes on the ionization of sulfuric acid [4] before the calculation of the equivalent fractions shown in Figure 3. In the case of the polyamine resin, the equivalent fraction of bisulfate on the resin actually decreases with increasing solution concentration over the range studied.

The tendency of the resin to hold bisulfate ion appears to vary inversely with the equilibrium capacity of the resin. As the number of ionized sites in the resin decreases, the proportion of such sites close enough together to share a sulfate ion will decrease sharply. This is probably the main reason for the apparent higher selectivity for bisulfate at low acid concentration, or more properly, for the lower selectivity for sulfate ion.

7. Equilibrium Composition

The electrostatic forces in the resin are quite appreciable. This can be seen in the relative volumes and compositions of three forms of Duolite ES-340 shown in Figure 4. The resin swells 35–40% over its free base form volume when equilibrated with 0.1 normal hydrochloric acid. When the free base resin is equilibrated with 0.1 normal sulfuric acid, there is essentially no change in its volume even though it picks up more equivalents of sulfuric acid than of hydrochloric acid.

The swelling of the hydrochloride form is a typical polyelectrolyte behaviour. As the number of ionized sites on the polymer increases, the electrostatic forces tend to straighten the polymer chains and increase the distance between neighboring chains. This movement is greatly restricted in the highly crosslinked ion exchange resin, but does result in the resin particle taking on water.

A similar behaviour would be expected with the sulfate form of the resin. The lack of swelling must be due to the decreased freedom of the polyamine chains to expand when a pair of amine groups are held adjacent to each sulfate ion by electrostatic forces. There is actually less water in the resin in the sulfate form than in the weakly ionized free base form.

8. Operating Capacity to H_2SO_4 vs HCl

The difference of 20–30% between the equilibrium capacities to hydrochloric and sulfuric acids seen in Figure 2 is not enough to explain the two-fold difference in operating capacity which occurs under some conditions. Figure 5 shows operating capacity data obtained on Duolite ES-340 with hydrochloric and sulfuric acid feeds

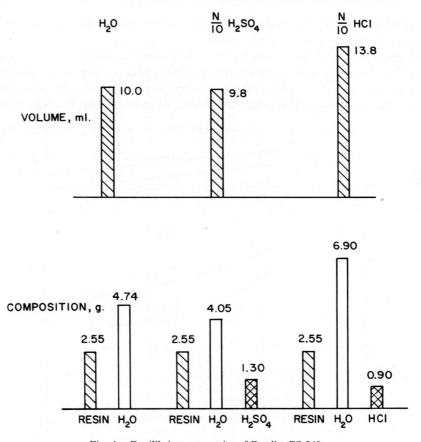

Fig. 4.   Equilibrium properties of Duolite ES-340.

at two concentrations and three flow rates. These data were taken in 2.5 cm diameter columns containing a 90 cm deep bed of resin. The feed waters were made up from deionized water and reagent-grade acid. The feed was pumped through the bed at a constant rate. The specific conductivity of the effluent was recorded. When the conductivity increased to 20 micromhos/cm, the run was stopped. The volume of effluent was measured. The operating capacity was obtained by multiplying the volume of effluent by the acid concentration of the feed and dividing by the volume of free-base form resin. All runs were made without interruption to minimize hysteresis effects. Some of them required as much as forty hours to complete. Each of the data points is an average of three or more runs.

The operating capacity is the measure of the useful work the resin can do. The work load is the equivalents of acid presented to a volume of resin per unit time. As this work load is increased, either by increasing the flow rate or increasing the acid concentration, the resin's relative performance in removing sulfuric acid versus hydrochloric acid changes. This is seen clearly in Figure 6, where the ratio of the two capacities is plotted against the flow rate. In 0.02 normal acid the ratio of operating capacities increases to approximately 2.

The removal of acid from solution requires a chloride, sulfate or bisulfate ion to diffuse into the resin. The rate controlling step for this reaction is the diffusion of hydroxide ion in the resin phase which is dependent on the concentration gradient of hydroxide ion inside. The rates of diffusion of chloride and bisulfate ions into

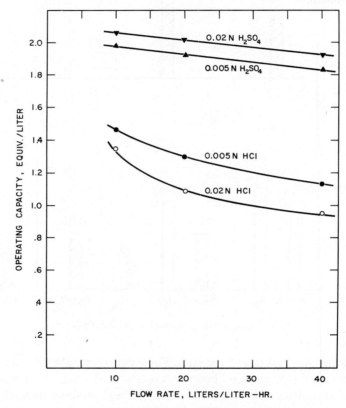

Fig. 5.   Operating capacity of Duolite ES-340.

the resin should be very similar and at least twice that of the divalent sulfate ion. However, when a bisulfate ion diffuses into the resin, effectively two hydrogen equivalents are removed from solution. The two-to-one ratio of sulfuric acid operating capacity to hydrochloric acid operating capacity observed in Figure 6 indicates that sulfuric pickup is occurring largely by the diffusion of bisulfate ion under these conditions of kinetic stress. This two-to-one advantage is enough to make the resin's operating capacity to sulfuric acid relatively insensitive to column conditions as compared to that to hydrochloric acid.

The diffusion of acid into the resin as bisulfate ion will also become more favorable as the acid concentration is increased, due to the higher relative concentration of bisulfate ions in the solution phase. In 0.005 normal sulfuric acid the ratio of bisulfate ion is roughly 1 to 4 while in 0.02 normal acid the ratio approaches 1 to 1. In either case, as fast as bisulfate ion is taken into the resin, the sulfate-bisulfate equilibrium will be reestablished in solution.

Once inside the resin, much of the bisulfate will be converted to sulfate since the

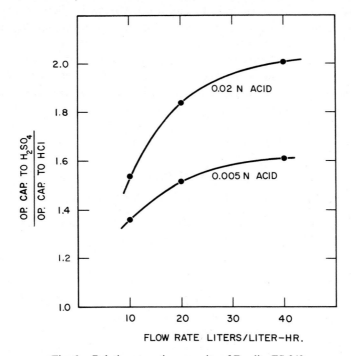

Fig. 6.   Relative operating capacity of Duolite ES-340.

equilibrium capacity of the resin must be established. This may occur by direct transfer of a proton from the bisulfate ion to a neighboring amine group. The pickup of sulfuric acid can be pictured as occurring in three steps.

$$RNH_2 + H_2O \rightleftarrows RNH_3^+ OH^-$$
$$RNH_3^- OH^- + H^+ HSO_4^- \rightarrow RNH_3^+ HSO_4^- + H_2O$$
$$RNH_3^+ HSO_4^- + RNH_2 \rightleftarrows (RNH_3^+)_2 SO_4^=.$$

9. Conclusions

Not all polyamine resins show these characteristics to the same degree. Duolite ES-340 has a relatively high basicity and good column performance. The basicity of some polyamine resins is low enough that their operating capacity to sulfuric acid is condition dependent. The polyamine resins based on chloromethylated styrene-divinylbenzene have the property of losing basicity, apparently by an autoxidation process. Their operating capacity thus becomes more dependent on conditions as they age.

The weakly ionized ion-exchange resins are much more complex in their behaviour than the fully ionized sulfonic acid and quaternary-ammonium types. The information on these resins will continue to be empirical. However, chemists and engineers in the water treatment industry must become familiar with the unique properties of the weakly ionized resins and the rationale behind their behaviour.

There are concurrent needs to reduce energy consumption, minimize waste discharges and reduce the salinity of certain water supplies. The application of conventional strongly ionized exchange resins to desalination processes is usually not attractive because of the excess of regenerant required and the added volume of waste which results. The imaginative use of weak acid and weak base ion exchange resins has potential in these areas that is very promising.

References

1. deJong, G. J., Brants, J. P., and Otten, G.: *Vom Wasser* **XXIX**, 242 (1962).
2. Helfferich, F.: *J. Phys. Chem.* **69**, 1178 (1965).
3. Anderson, R. E., Bauman, W. C., and Harrington, D. F.: *Ind. Eng. Chem.* **47**, 1620 (1955).
4. Sherrill, M. S. and Noyes, A. A.: *JACS* **48**, 1861 (1926).

SOME OBSERVATIONS ON DRAG REDUCTION
IN POLYACRYLIC ACID SOLUTIONS

O. K. KIM, R. C. LITTLE, and R. Y. TING*

Naval Research Laboratory, Washington, D.C. 20375, U.S.A.

Abstract. Polyacrylic acid was chosen as the model polyelectrolyte for drag reduction experiments designed to study selected variations in solution pH and salt concentration. Measurements were carried out in both a turbulent pipe system and a rotating disk apparatus. The observed drag reduction had a strong dependence on solution pH; it increased with pH up to 6 and then leveled off at higher pH values. Progressive addition of salt (sodium chloride) to the polymer solutions decreased the percent drag reduction sharply. The shear stability of polyacrylic acid was shown to be superior to that of polyethylene oxide.

1. Introduction

The drag reduction phenomenon observed in dilute polymer solutions has been the subject of great interest and activity for the past decade. It has often been demonstrated that a substantial reduction in frictional drag may be achieved by the addition of only a few parts per million of high molecular weight polymers to a given solvent system, usually water. While this phenomenon offers great practical potential in the design development of modern marine systems for higher speed and smaller size, a generally accepted explanation for the mechanisms responsible for drag reduction has not yet evolved. The state of the art seems to indicate that the important factors which contribute to the effectiveness of drag reducing polymers are: molecular linearity and flexibility [1], high molecular weight [2] and good solubility [3]. Most of the materials which were tested are commercial products. Because of possible questions concerning the purity of samples and the methods of solution preparation and characterization, and the sensitivity of drag reduction behavior to molecular properties, it is difficult to draw definite conclusions about various polymer types from the available literature. For example, in the case of polyacrylamide, most of the materials tested were either partially hydrolyzed or copolymerized polyacrylamide derivatives, and were somewhat loosely termed 'polyacrylamide'. This kind of complication greatly adds to the difficulty of analyzing and comparing information published by different investigators. Consequently, it was felt that a systematic study of the relationship between drag reduction and the relevant molecular parameters is not only desirable but necessary. A laboratory synthesis program was designed for this purpose with the hope that such an approach would shed further light on revealing the basic mechanisms of drag reduction.

In an earlier paper [4], the drag reduction properties of polyacrylamide modified by the incorporation of ionic groups in the main chain were reported. Such a modification enhanced the drag reduction of polyacrylamide to a significant extent. This

---

* To whom correspondence should be sent.

Alan Rembaum and Eric Sélégny (eds.), Polyelectrolytes and Their Applications, 275–285. All Rights Reserved.

enhancement in drag reduction reached a maximum at approximately 55% hydrolysis of amide to acid groups and showed a tendency to fall with further hydrolysis. The enhanced drag reduction was considered to be a consequence of molecular expansion resulting from the electrostatic repulsion between like charges and osmotic effects in the chain molecules. In order to better understand the effect of such electrostatic interactions, experiments were designed on the basis of selected variations in solution pH and salt concentration – factors known to substantially affect the extension of polyelectrolyte molecules [5]. Polyacrylic acid (PAA) was chosen as the model polyelectrolyte because of its close similarity to hydrolyzed polyacrylamide.

2. Experimental

2.1. MATERIAL

Acrylic acid (Eastman Organic Chem.) was polymerized after distillation under diminished pressure. Polymerization of the monomer (10% or 12% aqueous solution) was carried out with ammonium persulfate as catalyst under N_2 atmosphere at 35 °C overnight to produce a gel-like transparent polymer. This polymer was diluted with water and completely neutralized with sodium hydroxide. The polymer was then precipitated by adding acetone to the solution. The precipitation was repeated twice in essentially the same manner. Viscosity measurements of PAA sodium salt were made in 2N NaOH at 30 °C. The intrinsic viscosities of the prepared polymers were 5.72 and 6.27.

The PAA Na salt was dissolved in water to bring the concentration to ~ 0.05 N and then acidified to pH ~ 2 by addition of dilute hydrochloric acid. The PAA solution thus prepared was purified by dialysis using cellophane tubing in distilled water for 2 weeks until no Cl^- ion was indicated in the outer water.

The concentration of the PAA solution was determined by titration with N/10 NaOH. Partial neutralization of PAA was conducted by adding a certain amount of N/10 NaOH from a microburet. The partially neutralized PAA was diluted with ion-free distilled water in order to obtain a manageable viscosity.

2.2. TURBULENT PIPE FLOW SYSTEM

The drag reduction properties of the polymer solutions were characterized through the use of both a turbulent pipe flow system and a rotating disk apparatus. The pipe flow system was basically a metal syringe which, controlled by a D-C motor, drove the test fluid through a 0.62 cm diameter pipe. The flow rate was monitored by a small D-C generator coupled to the motor drive. Two pressure taps were placed at approximately 135 and 175 diameters from the upstream end of the flow. The pressure difference between taps was measured by a differential pressure transducer. The outputs from the D-C generator and the transducer were recorded continuously during the run on a dual-channel recorder. The flow rate and the pressure drop due to the resulting turbulent flow were then calculated by using the calibrated constants of the apparatus. The percent drag reduction (DR%) was computed by using the following relationship:

$$DR\% = \left[1 - \frac{f_{polymer}}{f_{water}}\right] \times 100,$$

where $f_{(\ )}$ is the friction factor, defined as the ratio of wall shear stress, τ_w, divided by the mean dynamic pressure head of turbulent flow, $\frac{1}{2}\varrho u^2$; ϱ is the liquid density and u is the mean velocity in the pipe. Because of the low concentrations involved in the experiments, the viscosity and density of water at the test temperature were used in all cases for calculation purposes.

2.3. ROTATING DISK SYSTEM

The rotating disk system consisted of a 450 ml Pyrex container, 5″ in diameter and 1.5″ in height, in which a 4.25″ diameter Teflon-coated stainless steel disk rotated. The disk was driven by a D-C electric motor equipped with a constant-speed control unit. The rates of rotation of the disk were such that turbulent flow existed over a major portion of the disk. Disk speed was measured using a strobe lamp. The output of a linear torque-sensing device built in the speed control unit was recorded continuously on a recorder. In this case, most of the measured torque was developed near the outer edge of the disk. The reduction in torque was proportional to the reduction in frictional drag on the disk surface.

3. Results and Discussion

3.1. EFFECT OF SOLUTION pH

The PAA homopolymer is a highly flexible chain molecule, the expansion of which is sensitive to pH changes in solution. A PAA sample ($[\eta] = 5.72$) was carefully purified to eliminate ionic contamination and partially neutralized prior to drag reduction evaluation. Figure 1 shows a plot of percent drag reduction versus the pH of the sample. It can be seen that drag reduction increases sharply with increasing ionization up to pH ~ 6 (corresponding to 45.5% neutralization), and then the effect levels off with further increases of pH. Similar drag reduction results for PAA polymers were reported recently by Parker and Hedley [6], who suggested that the large increase in effectiveness at higher pH is due to an extended molecular structure.

Since an increase in viscosity is a major consequence of molecular expansion one might attempt to correlate the observed drag reduction results for PAA with its viscosity data. The viscosity of aqueous PAA solutions was measured at 25 °C in a Cannon-Ubbelohde dilution viscometer having an efflux time of 108.5 ± 0.1 s for water. Figure 2 shows the reduced viscosity plotted against pH for different PAA concentrations. The reduced viscosity passes through a maximum at a different pH for different concentrations. The decrease in reduced viscosity at higher pH may result from an increasing shielding effect caused by a larger number of counter-ions in the domain of polymer molecule. This phenomenon seems to suggest that mo-

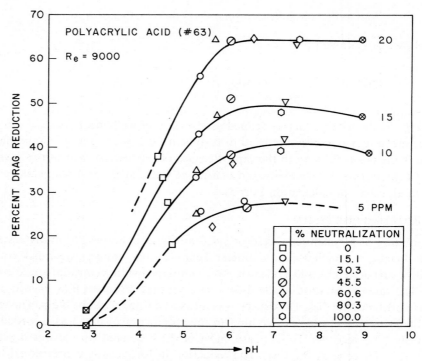

Fig. 1.   pH dependence of drag reduction in pipe flow.

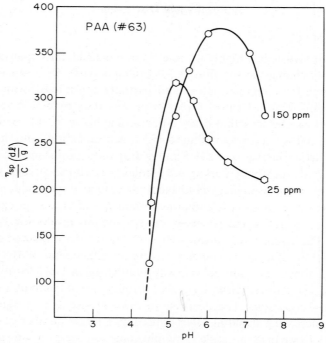

Fig. 2.   Reduced viscosity of PAA–Na as a function of solution pH.

lecular expansion of PAA diminishes at higher pH after the maximum value attained near pH $= 5 \sim 6$.

The data show that there is a correspondence between drag reduction and viscosity up to pH 5.5. The reason for the differing behavior at higher pH is not yet clear. It should be noted, however, that the viscosity data of PAA were obtained at fairly low shear rates, while the turbulent drag reduction took place at much higher shear rates. One explanation for this discrepancy at higher pH may be that the strong shear field caused by the turbulent flow conditions might have disturbed the shielding effect of counter-ions allowing repulsive forces to maintain the molecules in a more extended state than would have otherwise occurred.

3.2. EFFECT OF ADDED SALT

The solution properties of polyelectrolytes in general are markedly different from those of polyelectrolyte solutions with added salts. These differences are very striking-ly revealed in their viscometric behaviors. Viscosity, as pointed out in the previous section, is related to the size of polymer molecules and therefore is affected by mo-lecular expansion. When a small amount of a simple salt, such as sodium chloride, is added to a dilute polyelectrolyte solution, the ionic strength of the solution outside of the polymer coil is increased relative to the strength of the solution inside of the coil. Consequently, some of the mobile electrolyte diffuses into the polyion coil and the thickness of the ionic atmosphere around the polymer chain is reduced. This effect produces a significant contraction of the polyion coil and is reflected in de-creased values of the viscosity.

The viscosity of a PAA–Na salt ($[\eta] = 6.27$) was measured in a Cannon 50–S400 four bulb viscometer immersed in a water bath controlled at 25 °C. In the case of solutions of high molecular weight polyelectrolytes the macromolecules tend to es-tablish preferred orientations with respect to the direction of flow – especially when there is the tendency of these charged molecules to exist in extended configurations. As a result the viscosities usually vary with shear rate, and it is imperative that the viscosity data be taken and compared under the same shear condition. Figure 3 shows the reduced viscosities, η_{sp}/c, of 10 and 20 PPM solutions of this PAA–Na sample as a function of shear rate. The salt-free solutions exhibit a strong shear dependence of viscosity; the very high reduced viscosity observed at low shear rates decreases sharply with increasing shear rates. This shear dependence is gradually 'salted-out' with progressive additions of NaCl. In 0.1 M NaCl solution the shear dependence of reduced viscosity has almost disappeared. These results are cross-plotted in Figure 4 to demonstrate the effect of added salt on the reduced viscosity at three distinct shear rates. The rapid decline in the reduced viscosity with NaCl molar concentration indicates that large contractions in molecular size occur as a result of the changes in the surrounding ionic atmosphere.

The same PAA–Na sample was also tested in the turbulent pipe flow apparatus to study the effect of added salt on drag reduction. Figure 5 shows the percent drag reduction of 10 and 20 PPM solutions as a function of NaCl molar concentration

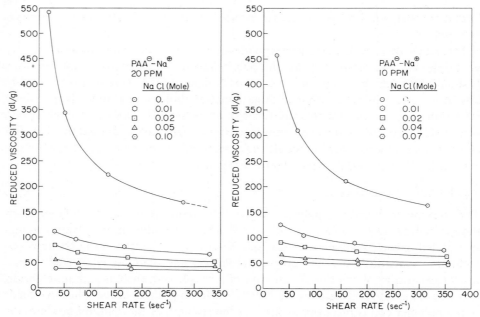

Fig. 3.   Reduced viscosity of PAA–Na solutions at 10 and 20 PPM as a function of shear rate.

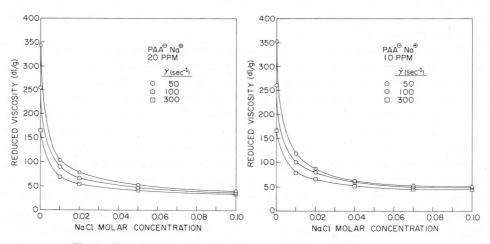

Fig. 4.   The effect of salt on the reduced viscosity of PAA–Na solutions.

at $Re=9000$. As the NaCl molar concentration increases the percent drag reduction decreases from approximately 55% in salt-free solution to about 18% in the 0.1 M solution. It should be noted that turbulent drag reduction has been established as a dilute solution phenomenon, i.e. above certain optimum concentration the drag reduction effect decreases with increasing concentration. The data at zero salt concentration in Figure 5 indicate this trend: 10 PPM solution is more effective than

20 PPM solution. However, the more concentrated polymer solutions seem to be less sensitive to the salt effect as NaCl concentration increases. While the observed difference between the percent drag reduction at different polymer concentrations is small, the concentration dependence of drag reduction as a function of salt concentration definitely requires further study.

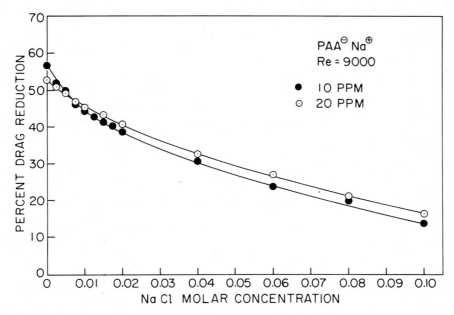

Fig. 5.   The effect of salt on the drag reduction of PAA–Na solutions in pipe flow.

A comparison of the viscosity with the drag reduction results seems to suggest that molecular expansion plays an important role in the turbulent drag reduction phenomenon. Since the reduced viscosity may be considered as directly related to molecular expansion, the viscosity data are correlated with the drag reduction data to clearly demonstrate the dependence of drag reduction on molecular expansion. This result is shown in Figure 6 for the 10 and 20 PPM solutions, respectively, at two selected shear rates. Now three remarks are in order:

First of all, Figure 6 suggests the existence of a cut-off 'molecular size', which may be estimated by extrapolating the correlation curve to zero drag reduction. For molecules contracted to dimensions smaller than the cut-off molecular size, there is no observable drag reduction. For the present case this cut-off molecular size is related to a reduced viscosity of approximately 40 dl/g. This concept of the cut-off point is consistent with the early observations by Hoyt and Soli [7] and Little [8], who showed the existence of a cut-off molecular weight for drag reduction of polyethylene oxide polymers.

Secondly, the percent drag reduction of the PAA–Na salt solutions seems to reach

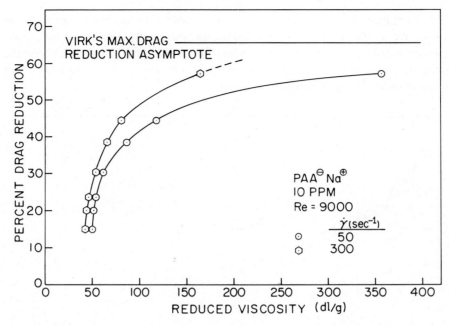

Fig. 6a.   Drag reduction vs reduced viscosity for 10 PPM PAA–Na solution.

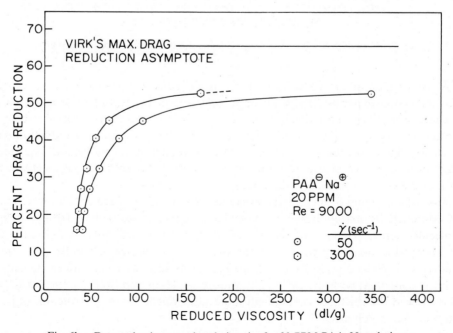

Fig. 6b.   Drag reduction vs reduced viscosity for 20 PPM PAA–Na solution.

a limiting value as the reduced viscosity, and therefore the molecular dimension increase to very high values. Virk and coworkers [9] have observed a maximum drag reduction asymptote in the turbulent pipe flow of dilute polymer solutions which is independent of polymer species, polymer concentration and pipe diameter. This asymptote, at $Re = 9000$, is approximately equal to 65.5% and is included in Figure 6 for purposes of comparison. The present result clearly supports Virk's concept as the percent drag reduction is approaching the maximum drag reduction asymptote at very high reduced viscosities.

Lastly, the rate of increase of percent drag reduction is large initially at small values of the reduced viscosity. This rate then decreases and gradually approaches a diminishingly small value. Such a fact seems to suggest that the effective increase in drag reduction is less as the molecular dimensions become greater and greater. Apparently when the polymer molecules become more expanded, the average distance between the polymer random coils decreases and intermolecular interaction becomes more important; this results in an adverse effect on the observed drag reduction. Little [8], in studying the concentration dependence of drag reduction, observed that drag reduction becomes less efficient on a unit concentration basis as the concentration increases. He attributed this effect to be a consequence of increased interactions among solute molecules as the solute concentration increases. Therefore, the present observation on the decreasing rate of change of drag reduction with increasing reduced viscosity or molecular expansion once again suggests that drag reduction is a dilute solution phenomenon and its efficiency decreases when molecular interactions begin to become important.

3.3. SHEAR STABILITY

The PAA–Na salt sample ($[\eta] = 6.27$) was also tested in the rotating disk system. Figure 7, as a typical drag reduction result, shows that the reduction in torque increases as the angular velocity increases. This is generally expected because a larger portion of the disk is exposed to turbulent flow as the angular velocity of the disk becomes higher. Figure 7 also shows similar salt effect on drag reduction as observed in the pipe flow. However, the effect of salt in reducing polymer drag reduction effectiveness seems to be less severe in the rotating disk configuration.

Dilute solutions of PAA–Na salt have been reported to possess remarkable shear stability in turbulent pipe flows [10]. This shear stability was attributed to the electrostatic interactions among adjacent charged groups and the counter ions. A polymer ladder-type structure, similar to that in linear associative colloids [11], was proposed, because such a structure with associative bonds has obvious advantages over linear structures such as those in polyethylene oxide or polyacrylamide. Generally associative bonds, if broken, could again be reformed in regions of low shear. The breakage of these associative bonds apparently removes a major portion of the turbulent shearing forces, which would otherwise have caused the scission of the molecular backbone and polymer degradation.

The shear stability of PAA–Na salt was also demonstrated in the rotating disk

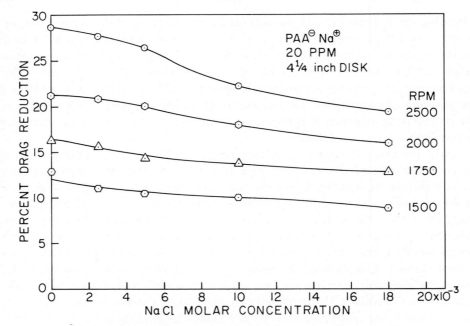

Fig. 7. The effect of salt on the drag reduction of 20 PPM PAA–Na solution in a
rotating disk system.

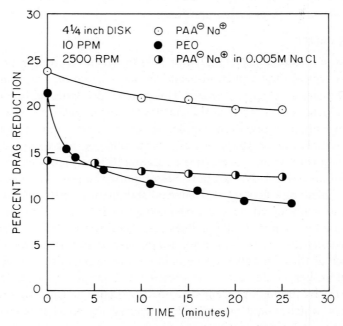

Fig. 8. Comparison of the shear stabilities of PAA–Na and PEO solutions in a
rotating disk system.

system, as can be seen in Figure 8. The results of PAA–Na solutions were compared with that obtained in polyethylene oxide solution (PEO, $M_w = 7.8 \times 10^5$). When the disk was rotated steadily at 2500 RPM over a period of 25 minutes, the percent drag reduction of the PAA solution decreased only 3%. However, the effectiveness of the PEO solution suffered a much greater and more rapid loss. In two minutes it dropped more than 25% in terms of the percent drag reduction. In five minutes the PEO solution became less effective than the PAA–Na salt in 0.005 M NaCl solution. The percent drag reduction of the PAA–Na salt in NaCl solution remained virtually unchanged over this testing period. This may be expected because the PAA molecules are more contracted in salt solution. As a result, the molecules, while less effective in reducing drag, are less vulnerable than those in salt-free solutions to chain scission caused by the intense interaction of polymer molecules with the turbulent shear field. It is felt that this remarkable shear stability of PAA–Na salt may be true in general for polyelectrolytes. If this turns out to be the case, then these high molecular weight materials indeed offer a significant potential in extending the application of the drag reduction phenomenon to various technological areas where closed circuit flow systems are involved, and thus deserve major research efforts in the future.

References

1. Hoyt, J. W. and Fabula, A. G.: *5th Symp. Naval Hydrodynamics*, ONR-ACR-112, 947 (1964).
2. Merrill, E. W., Smith, K. A., Shin, H., and Mickley, H. S.: *Trans. Soc. Rheol.* **10**, 335 (1966).
3. Peyser, P. and Little, R. C.: *J. Appl. Polym. Sci.* **15**, 2623 (1971).
4. Ting, R. Y. and Kim, O. K.: in N. M. Bikales (ed.), *Water Soluble Polymers*, Plenum Press, 1973, p.
5. Miller, M. L.: *The Structure of Polymers*, Reinhold, N.Y., 1966.
6. Parker, C. A. and Hedley, A. H.: *Nature-Physical Science* **236**, 61 (1972).
7. Hoyt, J. W. and Soli, G.: *Science* **149**, 1509 (1965).
8. Little, R. C.: *J. Coll. Interf. Sci.* **37**, 811 (1971).
9. Virk, P. S., Merrill, E. W., Mickley, H. S., Smith, K. A., and Mollo-Christensen, E. L.: *J. Fluid Mech.* **30**, 305 (1967).
10. Ting, R. Y. and Little, R. C.: *Nature-Physical Science* **241**, 42 (1973).
11. Honig, J. G. and Singleterry, C. R.: *J. Phy. Chem.* **60**, 1114 (1956).

SOME ASPECTS OF POLYMER RETENTION IN
POROUS MEDIA USING A C_{14} TAGGED POLYACRYLAMIDE

M. T. SZABO*

Calgon Corporation, Pittsburgh, Pa., U.S.A.

Abstract. Numerous single-phase flow and oil recovery tests were carried out in unconsolidated silica sands using C_{14} tagged hydrolyzed polyacrylamide solutions. The polymer retention data from these flow tests are compared to the adsorption data obtained from static adsorption tests.

The paper points out that the mechanism of polymer retention by mechanical entrapment has a predominant role in determining the total polymer retention in short columns of silica based materials with relatively low permeabilities.

Polymer concentrations in the produced water in polymerflooding tests were studied using various polymer concentrations, slug sizes, salt concentrations, and different permeability sands. In general, high polymer concentrations were noted in the breakthrough water.

Absolute polymer retention values showed an almost linear dependency upon polymer concentration. The effect of polymer slug size on absolute polymer retention is also discussed.

Distribution of retained polymer in sandpacks showed an exponential function of the distance. The 'dynamic polymer retention' values in short packs showed higher values than the 'static polymer adsorption' values due to mechanical entrapment.

A partly reversible polymer retention is observed during unsteady-state polymer flow. Mathematical equations describing the unsteady-state polymer flow are presented.

The mechanism of polymer retention in silica sands is described based on the observed phenomenon.

R_p	polymer flow resistance factor
$R_{p,or}$	polymer flow resistance factor at residual oil saturation
R_{bf}	brine flushed resistance factor
$R_{bf,or}$	brine flushed resistance factor at residual oil saturation
$\Delta P_{p,or}$	pressure drop during polymer flow at residual oil saturation
$\Delta P_{b,or}$	pressure drop during brine flow at residual oil saturation
$k_{b,or}$	permeability to brine at residual oil saturation
$k_{bf,or}$	permeability to brine at residual oil saturation after polymer treatment
PVI	pore volumes injected
C	concentration
C_i	injected polymer concentration, ppm
C_{ret}	polymer retention ($\mu g\ g^{-1}$)
$C_{ret,\infty}$	polymer retention at infinite distance from injection source ($\mu g\ g^{-1}$)
$C_{ret.\ mech.,\ \infty}$	polymer retention by mechanical entrapment at infinite distance ($\mu g\ g^{-1}$)
$C_{ad,\ equilibrium}$	polymer adsorption in equilibrium ($\mu g\ g^{-1}$)
$C_{ad,\ residual}$	residual polymer adsorption ($\mu g\ g^{-1}$)
C_{crit}	critical polymer concentration (ppm)
C_f	flowing polymer concentration (ppm)
C_{ef}	excess flowing polymer concentration (ppm)
C_{er}	fraction of excess flowing polymer concentration retained in 1 cm length (ppm)
C_{ert}	retained polymer in 1 cm^3 volume (μg)
F	cross-sectional area
F_1	constant
F_2	constant
F_3	constant
F_4	constant

* Presently with Gulf Research & Development Co., Pittsburgh.

Alan Rembaum and Eric Sélégny (eds.), Polyelectrolytes and Their Applications, 287–337. All Rights Reserved.
Copyright © 1975 by D. Reidel Publishing Company, Dordrecht-Holland.

F_5	constant
l	length (cm)
R_m	measured resistance factor
R_e	excess resistance factor
$R_{e,i}$	excess resistance factor at the injection face
R_i	initial resistance factor in unsteady-state flow
$R_{i,i}$	initial resistance factor at the injection face
$\bar{R}_i$	average initial resistance factor in $0-l$ length
$\bar{R}_m$	average measured resistance factor in $0-l$ length
R_o	original resistance factor during steady-state flow
$R_{o,\infty}$	original resistance factor at infinite distance from the injection face
$R_{i,\infty}$	initial resistance factor at infinite distance from the injection face
m_t	slope of the resistance factor vs time straight-line
m_1	slope of the resistance factor vs injected pore volumes straight-line in the first segment of a porous body
m_2	slope of the resistance factor vs injected pore volumes straight-line in the second segment of a porous body
t	time (s)
q	volumetric flow rate (cm^3 s^{-1})

1. Introduction

It is widely recognized that during polymer solution flow in porous media, a portion of the polymer is retained [1–11]. It is evident from the cited papers that both the physical adsorption of polymer onto solid surfaces and the polymer retention by mechanical entrapment play a role in the total polymer retention. The question "which of these mechanisms of polymer retention is more important?" has remained unanswered.

The effect of residual oil saturation on polymer retention and the polymer retention during the displacement of oil from porous media has not been reported. Although some phenomena [1–11] indicated that more polymer is retained in the first segment of a porous media, the literature lacks quantitative data on the distribution of retained polymer in porous media. The mechanism of polymer retention during unsteady-state flow [11] has not been adequately described.

The purpose of this study was to develop answers for the above questions. It will be shown that in low permeability silica sands the mechanical entrapment of polymer molecules is more dominant than physical adsorption. Physical adsorption is more important far from the injection face, particularly during the time when the porous rock is in contact with a polymer solution.

The absolute values of polymer retention were found to be not as great as was earlier reported [6, 10].

A mathematical description of the movement of a polymer slug in porous media was given earlier [12]. This mathematical model was based on the assumption that the porous media strips out a significant portion of polymer at the leading edge of a polymer slug. This process is in progress until all adsorption sites are satisfied; and, therefore, a connate water bank is always formed ahead of a polymer slug. In our laboratory studies, this phenomenon was not found to be the case. High polymer concentrations were observed in the breakthrough waters in oil recovery tests. However, a continuous decrease at the leading edge of the polymer slug was definitely noted. If these laboratory observations are extended to field dimensions, it is probable that the leading edge concentration of a polymer slug approaches low values at long distances from the injection well.

Polymer retention has an exponential distribution in porous media, whether the 'critical flow parameters' are exceeded [11] or not. This paper will point out that flow and retention measurements done on a short porous media do not yield adequate information to predict the flow and retention behavior for long distances. However, measurements taken at a minimum of three distances from the injection point could yield adequate information to predict the flow behaviors and retention values for a long distance using the mathematical models proposed in this paper.

2. Experimental

2.1. MATERIALS

Three different silica sands were used to make sandpacks of different permeabilities. The sands were prepared by an air and water flotation technique [11]. All sands were cleaned with hydrochloric acid, then washed in distilled water. Large amounts of dried sands were collected from each size, then each product was mixed in a large container. The average permeabilities of these sands were 1200, 670 and 173 md, and the porosities were 0.4384, 0.4284 and 0.4593, respectively.

An 18–20% hydrolyzed polyacrylamide was used in all tests. In all 300 ppm polymer solutions a radioactive C_{14} tagged polyacrylamide was used. At higher polymer concentrations (600 and 1200 ppm) a commercial product called Calgon Polymer 454 was added to the base 300 ppm radioactive solution. A special study was conducted to develop a radioactive polymer which has properties identical to the commercial product. Several experiments were run on both polymers to check these properties; such as, viscosity measurements, friction reduction flow tests, and flow tests in porous media. These special studies showed that performance of the radioactive product was equivalent to that of the commercial product. A typical result of these tests is shown in Figure 1. The small differences in the polymer flow resistance factors are due to small differences in the textures of different sandpacks, rather than to differences in the chemical structures of the polymers. Friction reduction, viscosity, and retention experiments showed even closer agreements between properties of the radioactive and commercial product.

During oil recovery tests a commercial refined oil was used, diluted with kerosene to 9 cp at room temperature.

The brine used in all tests was a 2% NaCl aqueous solution unless otherwise stated. The words fresh water or tap water refer to Pittsburgh municipal water.

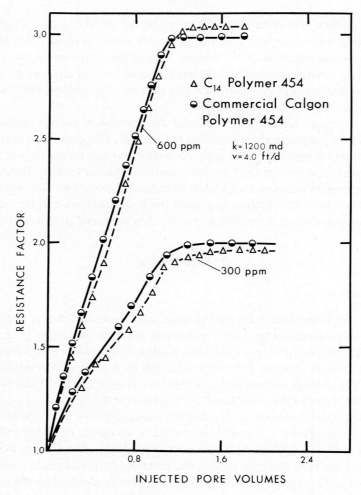

Fig. 1. A comparison of resistance factors during polymer flow developed by the radioactive and commercial polyacrylamides.

2.2. PROCEDURE

Known amounts (10 g) of dry sand were weighed into small glass vials. Thereafter, polymer solutions of different polymer concentrations and salinity were poured into the vials. The vials were mounted on horizontally rotating shafts and then rotated for 24 hours at 2 rpm.

Having settled the sand at the bottom of the vials after the rotation, liquid samples were taken for polymer concentration determinations from the bulk solution. Thereafter, the excess polymer solution was removed from the top of the sand. The sand was taken out and placed into a beaker. Brine or distilled water was poured on the sand, then the sand was gently stirred up. Having settled the sand at the bottom of the beaker, the aqueous phase above the sand was removed. This procedure was repeated three times. Since 125 ml of brine or water was used for one soaking, and the pore volume of the 10 gr sand represented 2.945 ml, during the four repeated soaking a 43.4^4 fold dilution of the polymer solution in the pore space took place.

For soaking, the same salinity of brine or water was used as had been used as solvent for the polymer. After soaking, sand samples were taken; and, the radio-activities thereof were determined in a Packard counter.

After the first desorption cycle, the sand samples were left in the vials with excess fresh brine for several days. The vials were gently shaken three times a day at 3 hours intervals to enhance the desorption. When a certain period of time had elasped, the sand samples were removed and soaked in fresh brine or distilled water as before. Sand samples were then taken for radioactive analysis. This procedure was repeated again with several days of standing and repeated soaking.

2.2.2. *Dynamic Retention*

In all flow tests the length of the sandpacks was 12 cm, unless otherwise noted.

2.2.2.1. Single-Phase Flow at 100% Water Saturation

The sands were packed into special sandholders using a wet packing technique. The sandholders and the packing technique were described earlier [11].

The injection of polymer solution was started after the brine permeability measurements. At a constant flow rate (6 ft/d) several pore volumes of polymer solution were injected through a sandpack. After the polymer flow cycle, the water lines before the sand face were flushed out with brine, and the injection of brine was initiated. During polymer flow and brine flush the injection pressures were recorded. Liquid samples were taken during both flow cycles. The polymer concentrations were determined from the radioactivities of the liquid samples.

After brine flush the sandholders were taken apart. Two sand samples were taken at 1 cm and 6 cm distances from the injection face. The retained polymer in the sand samples was determined via radioactivity.

2.2.2.2. Single-Phase Flow at Residual Oil Saturation

The packing technique was the same as before. After the permeability to brine measurement, oil was pumped through the pack at a 20 ft/day frontal advance rate. Approximately 40–50 pore volumes of oil were injected. Upon completing the oil saturation, a standard brineflood was performed on the sandpack. A 6 ft/day injection rate was applied. After 6 pore volumes of injected brine, negligible oil production was observed.

The pressure was continuously recorded until the abandonment of the brineflood. The pressure decline curve was then extrapolated to infinite injection. This extrapolated pressure value was used to calculate the permeability to brine at residual oil saturation.

Having attained the residual oil saturation, polymer solution was injected into the sandpack at the same 6 ft/day injection rate. During polymer injection, 1–2% additional change in oil saturation was noticed. The polymer injection continued until 2 to 3 pore volumes of injected solution had been reached.

After the polymer injection, a few pore volumes of brine were injected through the sandpack. The pressure during brine flush was again recorded.

Polymer flow resistance factors were computed from the equation below:

$$R_{p,or} = \frac{\Delta P_{p,or}}{\Delta P_{b,or}},$$

where $\Delta P_{b,or}$ is the computed pressure drop for brine using the extrapolated brine permeability.

The residual resistance factors at residual oil saturations were computed from the equation below:

$$R_{bf,or} = \frac{k_{b,or}}{k_{bf,or}}.$$

During both polymer flow and brineflush, effluent samples were taken for concentration determination. Sand samples were taken at 1 and 6 cm distances from the injection face for analysis.

2.2.2.3. Displacement of Oil with Polymer Solutions

2.2.2.3.1. *Oil recovery tests at irreducible water saturation.* Before each oil recovery test, the permeability of the sandpack to brine was determined. Next, oil was pumped through the sandpack until irreducible water saturation was reached. The total volume of displaced brine was determined volumetrically. A regular brineflood was carried out in the oil saturated sandpack. After the brineflood, the sandpacks were resaturated with oil. The polymerflood was carried out in the resaturated sandpack.

In some tests, no brineflood was performed preceeding the polymerflood. The polymerflood was carried out directly after the oil saturation cycle.

Regardless of whether a brineflood preceeded a polymerflood or not, before any polymerflood the irreducible aqueous phase was brine.

Effluent samples were frequently taken from the produced water. The polymer concentration of these samples was determined from the radioactivity thereof.

Four sand samples were taken; two at 1 and two at 6 cm distances from the injection face upon completion of a test.

2.2.2.3.2. *Polymer injection initiated at advanced stages of waterflood.* Two polymerfloods were performed in partly oil depleted sandpacks. The sequence of operations was as follows. Upon completing the brine permeability measurements, a sufficient amount of oil (20 PVI) was pumped through the sandpack. After saturation with oil, a regular brineflood was performed, applying a 6 ft/day frontal advance rate. Thereafter, the sandpacks were resaturated with oil.

Having completed the resaturation, a volume of brine (40% of the total pore volume) was injected into the sand. The brine was followed by a slug of polymer solution (20% of the total pore volume). After the polymer slug, brine was injected again until the end of the test (6 PV). During each cycle of the flooding operations, a constant, 6 ft/day frontal advance rate was applied.

Effluent samples were collected for polymer concentration determinations during both tests. Sand samples were gathered along the entire length of the packs upon completion of tests. At each location two sand samples were analyzed for radio-activity. The mean polymer retention for a given location is computed from the radio-activities of samples at the location in question.

In both tests the length of the sandpacks was 24 cm.

2.2.3. *Unsteady-State Flow Investigations*

Three sandholders were used during the experiment described below. Each sand-holder was packed with the same type of silica sand. The permeability of brine to each pack was determined independently.

After permeability tests, one pack was used as a filter. Six pore volumes of polymer solution were injected through this pack. Thereafter, the effluent of the filter was introduced to a second pack which was connected with the third pack in series.

First, the same flow rate was applied as during the filtration. The pressure was recorded before the second and the third pack. The injection of polymer solution was continued until pressure stabilization was obtained at both the second and the third pack.

Upon attaining pressure stabilization with the first flow rate (11.0 ft/d) a higher flow rate (18.3 ft/d) was applied at which no stabilized flow was expected.

The pressures and injected volumes were recorded as a function of time. When the resistance factor vs. time relationship was adequately established, the flow was stopped for a short period of time to study the effect of a no flow period. After the first no flow period, adequate injection was applied again to establish the flow param-eters. A second stoppage was applied later. After the second no flow period, the in-jection was again resumed and maintained until the end of the experiment.

3. Theory

3.1. STATIC ADSORPTION

Polymer adsorption from liquid phase onto solid surfaces has been investigated by numerous authors. A good summary of these investigations was given earlier [13].

It is not the primary objective of this paper to fully analyze the structure of the polymer layer on the solid surface. However, a few thoughts will be given about the role of salt and polymer concentration on polymer adsorption.

From the above cited reference [13] it is obvious that only one portion of the functional groups of the molecules is in contact with the solid surface. The rest of the polymer segments are in the bulk solution.

Inorganic salts have a bifunctional role, in our judgment. First, many rocks have a negatively charged surface (sandstones, quartz, silica [15], limestones [14]). Cations from the brine account for a primary adsorption layer on these surfaces. There certainly exists a salt concentration at which all negatively charged adsorption sites are satisfied by cations. It is also known that mono or divalent cations can reverse the originally negatively charged surface of quartz in aqueous solution [16]. Thus, if the polymer solution contains cations, an anionic polymer molecule can attach to the surface by a rock surface – inorganic cation – anionic group of polymer bridge. The marked effect of trace amounts of divalent cations on polymer adsorption in NaCl brines was previously reported [10].

The second role of salts is their direct effect on polymer molecules. The greater the salt concentration, the less the intra- and intermolecular repulsion between the anionic groups. As a result of this phenomenon, the hydrodynamic volume of the individual polymer polyions are smaller in brine. In other words, the density of polymer in unit liquid volume in the adsorption layer is high.

Since the increasing salt concentrations in NaCl brines have negligible effects on polyacrylamide solution viscosities between 2 and 10% salt content, the hydrodynamic volumes of molecules are about the same in that range of salt concentration. Since the density of polymer in the adsorption layer is certainly proportional to the hydrodynamic volume of molecules in the bulk liquid, above 2% NaCl the salt concentration should have little further affect on polymer adsorption.

In brine, because of low repulsion forces between the carboxylic groups of the hydrolyzed polyacrylamide molecules, the likelihood of multilayer polymer adsorption is high. Increasing polymer concentration will result in higher polymer adsorption.

In weak brine or distilled water the surface coverage of silica by cations is incomplete. In distilled water the COO— groups of polymer molecules can attach to the solid rock surface only by weak hydrogen bonding. Because of the greatly increased hydrodynamic volume of the polymer molecules, the density of polymer molecules in the adsorption layer is small. The result of this phenomenon is less polymer adsorption.

In very weak brine or in distilled water, the repulsion forces are very high between the COO— groups. As a result of this high repulsion, the adsorption isotherm levels off above a certain polymer concentration; a complete monolayer adsorption covers the sand surface.

Polymer adsorption greatly depends on the polymer concentration in strong brine; therefore, a significant desorption should occur by lowering the polymer concentration. In our experiments, removing the equilibrated polymer solution from the inter particle-space by many fold rinsing with brine resulted in a significant reduction in polymer adsorption. In distilled water, however, the structure of the adsorption layer is close to a monolayer form. Multifold rinsing of the polymer solution from the pore space, resulted in considerably less polymer desorption.

3.2. DYNAMIC RETENTION

The term 'dynamic retention' will be used throughout the paper to refer to the polymer retention in porous media during polymer solution flow.

An attempt will be made to clearly separate the two basic retention phenomena; mechanical entrapment of molecules and physical adsorption. Factors influencing these retention mechanisms will be discussed in detail. The resultant distribution of retained polymer in porous media will be given.

Figure 2 illustrates schematically the retention mechanism in porous media. As

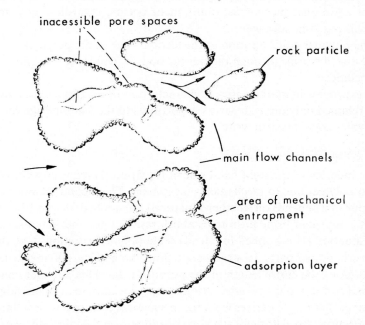

Fig. 2.   Schematic illustration of polymer retention mechanisms in steady-state polymer flow.

the polymer front reaches this segment of the porous media, physical adsorption takes place in the main flow channel. Also, polymer penetration occurs into small channels. There are pores which have large enough inlet mouths to let through the polymer molecules, but the outlet pore openings are small enough to trap the molecules. Having trapped the polymer molecule at the outlet end of this pore, the fluid motion in this pore is restricted. These polymer molecules cannot leave this pore. However, a slow flow of brine is allowed. Due to this slow motion of brine, a further compaction of polymer molecules takes place at the outlet end and in this pore itself. Since this process further reduces the permeability of this pore, the fluid motion further slows down. When steady-state flow is attained in the porous media, the polymer is retained by physical adsorption in the main flow channels, or it is entrapped in the small pores. When the porous media is flushed, a certain degree of desorption takes place in the main flow channels but little or no desorption occurs in the area of mechanical entrapment.

3.2.1. *Invasion of a Porous Body with Polymer Solution*

The number of mechanically trapped polymer molecules is linearly proportional to the polymer concentration that first appear in a segment of a porous body illustrated schematically in Figure 2. In other words, the amount of mechanical entrapment at a given location depends on the polymer concentration at the leading edge of the polymer front. Once this leading edge of the polymer zone passes a given location, only slight additional mechanical entrapment occurs, regardless how much more polymer will pass this location.

Since the polymer concentration at the leading edge of the polymer zone becomes less and less as it travels through the porous body, it will leave less and less retained polymer behind.

A similar process in mathematics is characterized by an exponential function. Because the retained polymer is linearly proportional to the leading edge concentration of the polymer zone, one can write:

$$C_{\text{ret}} = F_1 C_i e^{-F_2 l} + C_{\text{ret}, \infty}. \tag{1}$$

The first member of the right hand side of Equation (1) represents the distribution of retained polymer due to mechanical entrapment. The second member is constant, and the value of it depends on whether Equation (1) is applied for steady-state polymer flow or after sufficient brine flush to establish residual polymer retention.

When Equation (1) is applied for steady-state polymer solution flow, the value of $C_{\text{ret}, \infty}$ consists of two parts. In steady-state flow, the flowing polymer concentration is constant along the entire length of the porous media. This polymer concentration determines a constant physical adsorption in every location in the main flow channels. The second part of $C_{\text{ret}, \infty}$ derives from the mechanical entrapment at infinite distance from the injection face. Although at a great distance from the injection face the leading edge polymer concentration is practically zero, a certain degree of mechanical entrapment can also occur at this location. As the polymer concentration increases at this location, more and more polymer molecules can penetrate the small pore channels like the one in Figure 2. The first molecule that enters this pore does not necessarily 'plug' this pore. This and additional molecules can be adsorbed on the wall of this pore. Certainly, there is a minimum number of molecules necessary to effectively restrict the flow in this pore. When this minimum number is reached, further entrapment in this pore will be negligible. This means that when the major part of the polymer bank reaches this location, the mechanical entrapment of polymer molecules is already essentially completed.

Equation (1) can also be applied to describe the retained polymer after brine flush. The meaning of the first member on the right-hand side of Equation (1) is the same as during polymer flow. The $C_{\text{ret}, \infty}$ member, however, has a different meaning. It also consists of two parts. The first part originates from the physical adsorption in the main flow channels. Actually, this part represents a 'residual physical' adsorption on the wall of main flow channels, after brine flushing.

The second part of $C_{\text{ret}, \infty}$ during brine flush originates from the mechanical entrapment at infinite distance, and the value of it is the same as during polymer flow.

To summarize the thoughts above, values of $C_{\text{ret}, \infty}$ are given below for polymer flow and brine flush.

Polymer flow:

$$C_{\text{ret}, \infty} = C_{\text{ad, equilibrium}} + C_{\text{ret, mech.}, \infty}.$$

Brine flush:

$$C_{\text{ret}, \infty} = C_{\text{ad, residual}} + C_{\text{ret, mech.}, \infty}.$$

3.2.2. Factors Influencing the Distribution of Retained Polymer

There are two other constants in Equation (1); F_1 and F_2. Values of F_1 and F_2 depend on the pore structure, fluid saturations, type of polymer, salt environment, polymer slug size, and perhaps on the injection rate.

The role of pore structure can be generalized as follows: The smaller the pore, the greater the mechanical entrapment. Also, in smaller pores the specific surface area is greater. Therefore, the polymer concentration at the leading edge of a polymer front flowing through a low permeability medium decreases at a faster rate as a function of distance than in a high permeability medium. Since the mechanical entrapment of polymer molecules is related to the polymer concentration in the first portion of the polymer zone, the slope of the retained polymer distribution curve is very high close to the injection face. Consequently, the value of the product of F_1 and F_2 is higher in low permeability media than in high permeability materials.

The existance of non-wetting phase(s); oil or gas, in a water-wet system reduces the total cross-sectional area available for polymer solution flow. Therefore, the role of mechanical entrapment increases.

The role of polymer type can be given as follows. Both the physical adsorption [13] and the mechanical entrapment increases with increasing molecular weight. The role of functional groups of polymers is more important in determining the physical adsorption rather than the mechanical entrapment.

The role of inorganic salts is dual. The physical adsorption generally increases with increasing salt concentration. However, during brine flush, when desorption of polymer molecules occurs, the 'residual adsorption' does not depend as greatly on salt concentration. The mechanical entrapment also depends on the salt concentration. In stronger brine the hydrodynamic volumes of polymer molecules are smaller. The polymer molecules can, therefore, penetrate much smaller pores. However, polymer molecules in brine can also leave such pores; whereas, if the polymer was in fresh water, it might remain entrapped. Whether the resultant mechanical entrapment is more or less in strong brine certainly depends on the pore-size distribution.

The role of polymer slug size can be made clear with the following example. Consider two extreme slug sizes; first a 1% PV slug and second a continuous polymer injection. In both cases, the applied polymer concentration is the same. In the 1% PV slug, the amount of polymer is not adequate to satisfy the adsorption and mechanical entrapment sites, therefore, the slope of the retained polymer distribution curve shows a very high value close to the injection face. Very small amounts of polymer reach distances far away from the injection face. However, when continuous polymer injection is applied, the distribution of retained polymer will not show such a great decrease as a function of distance. Our experiments showed that below 20% slug size, the values of F_1 and F_2 greatly depended on slug size. However, the slug sizes between 20% PV and continuous injection had little effect on values of F_1 and F_2. Values of F_1 and F_2 using larger slug sizes than 20% PV are mainly determined by the pore structure, salt environment, saturation, and type of polymer. There is not enough data available yet to determine the role of injection rate in determining F_1 and F_2.

3.3. POLYMER RETENTION IN UNSTEADY-STATE FLOW

It was earlier reported [11] that using the same polymer and porous media, steady-state or unsteady-state flow can be obtained depending on the flow condition. Our further laboratory studies showed that this phenomenon is commonly observed with a wide variety of water soluble polymers. However, the critical flow parameters of different polymers can vary greatly.

The significance of this phenomenon necessitated the development of a molecular and mathematical description of this flow phenomenon.

Although the knowledge on unsteady-state polymer flow is incomplete, a working hypothesis can already be given based on observed phenomenon in the laboratory. This description of unsteady-state polymer flow must be in good agreement with well-established experimental facts, which are as follows:

(1) During unsteady-state flow the effluent polymer concentration is constant, and always less than the injected concentration. Generally the effluent polymer concentration is only a few percent less than the influent concentration.

(2) The resistance factor in most cases increases linearly with time or with injected pore volumes.

(3) A sudden break in polymer solution flow results in a greatly decreased resistance to flow upon resumption of polymer injection.

(4) The resistance factor increases at a faster rate close to the inlet face as a function of time than far away from the injection face.

(5) Unsteady-state flow can be brought about by:
(a) exceeding a critical flow velocity,
(b) exceeding a critical polymer concentration,
(c) using a salt concentration below a critical level,
(d) using porous media containing pores smaller than a critical size.

3.3.1. *Molecular Interpretation*

3.3.1.1. A Concept of Polymer Retention in Small Pores

Several assumptions can be made regarding the locations and the way of polymer entrapment during unsteady-state flow. A continuous polymer entrapment in small pore channels is precluded since the occurrence of this type of polymer buildup would not conform to the experimental finding #3 above. However, a certain degree of polymer penetration into smaller pores, which were already partly 'plugged', is not completely precluded. A higher flow velocity generally results in an increased flow resistance in the main flow channels, and as a result of this, a few polymer molecules can be dragged into a parallel connected smaller channel such as can be seen in Figure 2.

Stopping polymer flow results in a significantly increased permeability to polymer solution. This phenomenon cannot be explained by the polymer entrapment in small channels. In the no-flow period, the entrapped polymer molecules would have to leave these pores very easily, which could occur only by diffusion. But the motion of polymer molecules is highly restricted in these pores; therefore, a concentration equalization by diffusion is very unlikely.

3.3.1.2. A Multilayer Polymer Adsorption Concept

The decreasing permeability could be explained by the assumption of a multilayer polymer buildup in the main flow channel. It can be assumed that in the outer layer there is a permanent rate of disentanglement of polymer molecules. If the number of arriving molecules in unit time exceeds the number of molecules detached from the outer polymer layer, a multilayer polymer buildup could take place.

The assumption could explain the role of polymer concentration and flow velocity. Over a critical polymer concentration and flow velocity, the number of arriving molecules at a given location of the outer polymer layer exceeds the number of disentangled molecules.

However, this polymolecular adsorption concept is unable to give an answer for two basic experimental observations:

(1) The lower the salinity, the greater the likelihood of an unsteady-state flow.

(2) The smaller the pore dimensions, the greater the possibility of an unsteady-state flow.

In lower salinity solvents, particularly in distilled water, the electrostatic field in the immediate vicinity of the ionic groups on the polymer chains is only partially neutralized. A strong electrostatic repulsion exists not only between the ionic groups inside an individual molecule, but between the different polymer polyions. This strong electrostatic repulsion works against the buildup of a polymolecular layer. Also, because of the large number of electrostatically held dipole water molecules, a molecule in an outer adsorption zone would be easier dragged away.

The multilayer adsorption concept is also incapable to explain the role of pore dimensions. In a low permeability material, the shear rates are considerably greater at a given average frontal advance rate than in a high permeability material. Due to the higher shear in low permeability media, drag forces tearing off the polymer molecules from the outer adsorption layer are much higher than in high permeability media. But the experimental facts shows that the development of an unsteady-state flow can much easier occur in low permeability materials.

3.3.1.3. The Proposed Mechanism of Polymer Retention

Figure 3 illustrates the development of an unsteady-state polymer flow. In Figure 3a it is assumed that the polymer molecules have a coiled shape in brine at low linear velocity. The polymer molecules travel on the tracks of the stream lines. The molecules can easily adjust their own shape and their surroundings to the changing velocity field at any location.

At higher linear velocity, the shape of the molecules are elongated (Figure 3b). The time that elapsed during traveling through the pore is shorter and, therefore, the molecule can only partly adjust its shape and surroundings to the changing velocity field.

Figure 3c illustrates the state of a fully developed unsteady-state flow. The flow velocity is high, and the molecules assume a strongly elongated or rod-like shape. Because of the unwound shape of the molecules, the diffuse layer potential surrounding the individual molecules are high. The counterions of the ionic groups of the polymer molecules leave the immediate vicinity of the backbone of the molecules. The molecules become 'rigid' to a certain extent. They are unable to follow a streamline such as denoted on Figure 3c with '2'. Two flow zones develop inside a pore. In the center zone, the molecules follow the general direction of flow. Due to high double layer potential, both the mobility of the electrostatically held dipole water molecules and of the polymer molecules themselves are low; the flow is restricted. Because of the limited conductivity of the pore necks and of the electrostatic repulsion between the polymer molecules, there is a good opportunity for a polymer molecule at the location denoted with '1' in Figure 3c to become diverted into the top or bottom section of the pore. Certainly there is a slow circular fluid motion in these sections. As the molecule enters this section, it can gradually change its configuration, finally assuming the shape of a ball of thread because of the reduced flow velocity. As this process continues, the number of molecules not participating in the main flow increases. The polymer concentration in the areas of polymer retention increases. These molecules are in a dynamic trap. In order to leave this zone, the molecules must first assume an elongated or a rodlike shape and be accelerated to the velocity that prevails at the edge of the 'free flowing center zone'.

As the polymer concentration builds up in the retention areas, the repulsion forces between the dynamically held molecules increases. A final result of this process is an increasing force which tends to narrow the diameter of the 'free flowing center zone'. As the main flow channels contract, the flow resistance increases.

This picture of polymer retention allows us to understand why a short break in flow results in a greatly decreased flow resistance upon resuming the flow. When the flow is stopped, a fast equalization of polymer concentration takes place within the

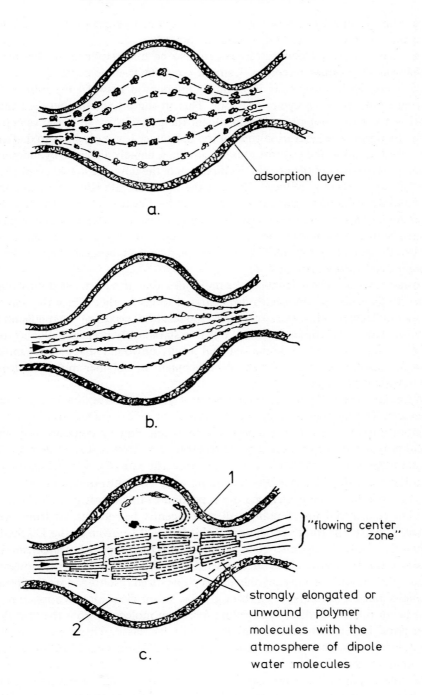

Fig. 3. Schematic illustration of polymer retention during unsteady-state flow.

pores. After resuming the polymer solution injection, a long period of time is needed to regain the previous high flow resistance factors.

The role of polymer concentration can also be understood from this proposed mechanism of polymer retention. Using a constant flow velocity, any increase in polymer concentration results in stronger interaction between the diffuse double layer surrounding the polymer polyions. This strong interaction between the molecules restricts the configuration adjustment the polyions make to the velocity field inside the pores. The relaxation time of the molecules is greatly increased. The molecules tend to follow the less tortuous stream lines, namely a coaxial cylinder around the longitudinal axis of the pore in the direction of the flow; the actual cross-sectional area for flow is thereby reduced. Because the injection rate is maintained constant, the reduced area of flow results in an increase in the linear velocity in the 'free flowing center zone'. This increased velocity can result in a certain degree of uncoiling of the molecules. In a stagnant polymer solution, the resultant vector of the repulsion forces between the polymer molecules has no definite direction, because it alternates every moment due to Brownian motion.

However, when shear is applied, the molecules assume a partly or a fully uncoiled shape. Statistically the molecules flow parallel to each other. Since the axis of the polymer molecules follow the streamlines, the resultant vector of the repulsion forces no longer has an indefinite direction, but it is statistically perpendicular to the axis of flow. It should also be pointed out that a fully unwound shape of the molecules is not an absolute necessity for the disattachment of polymer molecules from the 'free flowing center zone'.

As we have seen above, the development of unsteady-state flow is concentration dependent. The number of molecules 'expelled' into the 'dynamic trap' in unit time is proportional to the difference between the actual flowing concentration and the critical polymer concentration. This concentration difference, called excess polymer concentration is large at the inlet face, therefore, the rate of polymer buildup is also large. Far from the inlet face, this excess polymer concentration is greatly reduced since the rock has already stripped out a portion of this excess polymer concentration. The rate of polymer buildup will consequently be reduced far from the inlet surface. The experimental verification of this fact will be given in the Discussion.

The role of salinity can also be understood with the proposed mechanism. In less saline solvent, the shape of the molecules even in stagnant solution is elongated or rod-like. The anionic groups are only partly neutralized resulting in strong intra- and intermolecular repulsion between these groups. When shear force is applied on these solutions in porous media, the expelling forces on molecules from the 'free flowing center zone' into the 'dynamic trap' are greatly increased. Therefore, an unsteady-state flow will develop at a lower velocity or lower polymer concentration in distilled water than in brine.

The role of pore dimensions can also be explained by the proposed mechanism. The shear rate in finer pores is greater than in large pores at the same linear velocity. A higher degree of uncolling of molecules takes place in finer pores, thus the repulsion forces between the molecules are higher. At the same linear velocity, in small pores the molecules would have to adjust their shape and surroundings more frequently to the changing velocity field than in large pores in unit time. However, the relaxation time is given by the type of polymer and salt environment. Therefore, due to the finer pores an unsteady-state flow can develop easier in low permeability materials.

3.3.2. *Mathematical Evaluation*

3.3.2.1. Determination of Concentration Distribution in Linear System

Using a certain flow velocity at which the injected polymer concentration exceeds the critical polymer concentrations, the porous material will strip out all the excess polymer concentration above the critical concentration value. The accumulations of this excess polymer concentration take place not only at the inlet surface, but also in depth. If the injected concentration is equal to or less than the critical polymer concentration, no continuous polymer buildup will occur anywhere in the porous medium. The number of entrapped polymer molecules should be proportional to the polymer concentration above the critical polymer concentration. Since the excess polymer concentration is greatest at the injection face, the rate of polymer entrapment is also highest at this location.

The excess polymer concentration is defined as the difference of the free-flowing polymer concentration and of the critical polymer concentration.

$$C_{ef} = C_f - C_{crit}. \tag{2}$$

The fraction of excess flowing polymer concentration retained in unit length is proportional to the excess polymer concentration:

$$C_{er} = F_3 C_{ef}. \tag{3}$$

The retained polymer can be given as follows:

$$C_{ert} = C_{er} tq. \tag{4}$$

In the following analysis, the objective is to find an equation which describes the distribution of excess polymer concentration in a linear system.

Figure 4 can help to analyze this problem. If a linear system is cut at l distance from the injection face, the measured effluent polymer concentration will be equal to the flowing polymer concentration at that location.

The excess flowing polymer concentration at any l distance can be computed as the initial excess concentration less the excess polymer retention in the $0-l$ distance.

$$C_{ef}(l) = C_{efi} - \int_0^l F_3 C_{ef}(l) \, dl. \tag{5}$$

Differentiating both sides of Equations (3) and (4) to l;

$$\frac{\partial}{\partial l} C_{ef}(l) = -F_3 C_{ef}(l). \tag{6}$$

The general solution of the above differential equation is;

$$\ln C_{ef} = -F_3 l + \text{constant}.$$

The value of constant can be determined from the boundary conditions; at $l=0$,

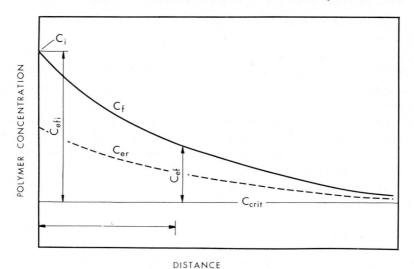

DISTANCE

Fig. 4.   Distribution of flowing polymer concentration in a linear system during unsteady-state flow.

the excess polymer concentration is equal to the difference in the injected and critical polymer concentration:

$$C_{efi} = C_i - C_{crit}.$$

If the excess polymer concentration is expressed as the difference of the flowing and critical polymer concentration, one can write:

$$C_f - C_{crit} = (C_i - C_{crit})\, e^{-F_3 l} \tag{7}$$

or

$$C_{ef} = C_{efi} e^{-F_3 l}. \tag{8}$$

Equations (7) and (8) postulate that knowing the polymer concentration in the injected solution and in the effluent, and the free-flowing polymer concentration at any particular location somewhere between the injection and outlet face, the value of C_{crit} and F_3 can be determined.

The technical problems related to solving Equation (8) for C_{crit} and F_3 are discussed in *Appendix A*.

3.3.2.2. Determination of Resistance Factor Distribution in Linear System

For determination of the resistance factor vs distance relationship, only one assumption has to be made; that is, the resistance factor measured at any location should be proportional to the dynamically retained polymer at the location in question.

It was earlier pointed out that after the first invasion of a porous body with a polymer solution, the retained polymer shows a characteristic distribution. If, for the first invasion the flow parameters are chosen in such a manner that finally a steady-state flow is attained, this characteristic distribution of retained polymer will not change with time. This function determines a distribution of resistance factors. Then, if the flow rate is increased to a value above the critical velocity, the initial distribution of the resistance factors at this increased velocity shows also a characteristic distribution. Further analysis of the initial resistance factor distribution during unsteady-state flow is given in *Appendix B*.

In a given moment after starting a higher flow velocity than the critical velocity, the measured resistance factor R_m, is always greater than the initial resistance factor R_i, at any location (Figure 5). The difference between the measured resistance factor and the initial resistance factor is called R_e, excess resistance factor.

$$R_e = R_m - R_i.$$

Applying the basic assumption that the excess resistance factor is proportional to the excess polymer retention:

$$R_e = F_4 C_{ert}$$

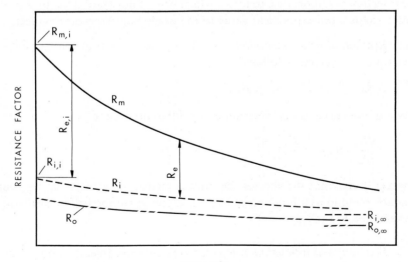

RESISTANCE FACTOR

DISTANCE

Fig. 5. Distribution of measured (R_m) and initial (R_i) resistance factors in a linear system during unsteady-state flow.

the equations below can be derived:

$$R_e = F_4 F_3 C_{ef} tq$$
$$R_e = F_4 F_3 C_{efi} e^{-F_3 l} tq.$$

The last equation can also be expressed in the following way:

$$R_e = F_4 F_3 (C_i - C_{crit}) e^{-F_3 l} tq. \tag{9}$$

At location $l = 0$, the excess resistance will be strictly proportional to the $(C_i - C_{crit})$ concentration difference and to the time.

$$R_{e,i} = F_4 F_3 (C_i - C_{crit}) tq. \tag{10}$$

The general form of the excess resistance factor distribution is given by Equation (11):

$$R_e = R_{e,i} e^{-F_3 l}. \tag{11}$$

Equation (11) suggests that knowing the resistance factor versus length curve at any moment, the F_3 factor can be determined from the slope of this curve.

However, the resistance factors can be determined experimentally only for finite distances. An average resistance factor measured between l_1 and l_2 coordinates does not belong to the $(l_1 + l_2)/2$ distance. Further analysis of this problem is given in *Appendix C*.

3.3.2.3. Resistance Factor as a Function of Time and Injected Pore Volumes

The resistance factors can be given as a function of time or injected pore volume. It has great technical importance to know which way of illustrating the increasing flow resistance yields a more practical value to the practicing reservoir engineer.

3.3.2.3.1. *Evaluation as a function of time.* The general form of the measured resistance factor can be given as follows:

$$R_m = R_{i,i} e^{-F_s l} + R_{i\infty} + R_{e,i} e^{-F_3 l}. \tag{12}$$

The slope of the curve can be determined by differentiating to t:

$$\frac{\partial R_m}{\partial t} = m_t = \frac{F_4 F_3 C_{efi} q}{e^{F_3 l}}. \tag{13}$$

Equation (13) says that the shorter the porous media, the greater the slope of resistance factor buildup as a function of time. The greatest slope will be at the injection face, where $m_t = F_4 F_3 C_{efi} q$.

3.3.2.3.2. *Evaluation as a function of injected pore volumes.* Injected pore volumes are linearly proportional to the duration of injection assuming constant injection rate. Consequently, the resistance factor versus injected pore volumes relationship is also linear. However, the slope of the curve not only depends on the rate of polymer retention for a given location, but it depends on the length, which serves as a base for computing the injected pore volumes.

The role of the length can be made clear with the following example. Consider a linear system with length l. Select an l_i distance from the inlet surface. The rate of polymer buildup as a function of time is given by Equation (13). Take an l_1' and l_1'' location before l_i, and an l_2' and l_2'' location after the location l_i. If l_1', l_1'', l_2' and l_2'' are selected so that the integral average value of the rate of polymer buildup is unchanged, and it belongs to l_i, we have two porous bodies with different lengths, but the rate of polymer buildup is the same for both bodies as a function of time. That is, within the same period of time, in both systems the ΔR change is the same. However, because the length of these two porous bodies are different, the slopes of the resistance factors versus injected pore volumes are proportional to the length thereof; $m'/m''=(l_2'-l_1')/(l_2''-l_1'')$. The shorter the length the greater the number of injected pore volumes during t time. As the length approaches zero, the injected pore volumes approach the infinite, consequently, the slope approaches zero.

In the further analysis, let us consider a porous body with length l_2. Furthermore, let us examine the resistance factor buildup vs injected pore volumes relationship for an l_1 and l_2-l_1 section, where $l_1<l_2$. During the same injection time, the number of the injected pore volumes in the $0-l_1$ section is $(l_2-l_1)/l_1$ times the injected pore volumes in the l_2-l_1 section. If $l_1<(l_2-l_1)$ the $l_1/(l_2-l_1)$ ratio would show how many times lower the slope of the resistance factor versus injected pore volumes relationship is in the $0-l_1$ segment. This is true provided that the rate of polymer buildup as a function of time is the same at any location. But it was earlier pointed out that the rate of polymer buildup as a function of time is the highest at the inlet face. Consequently, there should be an l_1 location, when for the $0-l_1$ distance the slope of R_p vs PVI is equal to the slope for l_2-l_1 distance.

The integral average resistance factors for the considered distances can be given as follows:

$$\bar{R}_{e1}=\frac{R_{e,i}}{l_1 F_3}\left[1-\frac{1}{e^{F_3 l_1}}\right] \tag{14}$$

and

$$\bar{R}_{e2}=\frac{R_{e,i}}{(l_2-l_1)F_3}\left[\frac{1}{e^{F_3 l_1}}-\frac{1}{e^{F_3 l_2}}\right]. \tag{15}$$

The injected pore volumes can be expressed as follows:

$$(\text{PVI})_1=\frac{Qt}{F\phi l_1}\quad\text{and}\quad(\text{PVI})_2=\frac{Qt}{F\phi(l_2-l_1)}. \tag{16}$$

Substituting the values of t from Equation (16) into Equations (14) and (15) and then dividing both sides of the obtained equations by the injected pore volumes, the slope of the R_p vs PVI can be determined. Thus, the m_1/m_2 ratio for the l_1 and l_2-l_1

section is as follows:

$$\frac{m_1}{m_2} = \frac{1 - \dfrac{1}{e^{F_3 l_1}}}{\dfrac{1}{e^{F_3 l_1}} - \dfrac{1}{e^{F_3 l_2}}}. \tag{17}$$

The value of F_3 is given by the pore structure. Equation (17) shows that within the $0 - l_2$ distance, only one l_1 exists where $m_1 = m_2$.

Equation (17) shows that depending on the selection of l_2 and l_1 the slope of R_p vs PVI function for the first segment of the system can be greater or smaller than in the second segment.

Equation (17) further shows that by measuring the pressure buildup at two locations in a linear system, the F_3 factor can be determined. An example for such a calculation will be presented in the Discussion.

3.3.2.4. Radial System

The mathematical treatment of the unsteady-state flow in a radial system is similar to what was given for the linear system. The equations describing the unsteady-state flow can be given for the analog of the Equations (2–17). The only new factor is the decreasing velocity as the function of the radius, which has to be taken into account in deriving the equations. The critical polymer concentration is no longer constant in the radial system. The greater the radius, the lower the velocity, and thus, the higher the critical concentration. In final analysis, it can be pointed out that a critical radius exists where the flowing polymer concentration is equal to the critical polymer concentration. Beyond this *critical radius*, steady-state polymer flow exists, no continuous polymer buildup occurs.

4. Discussion

4.1. STATIC ADSORPTION

Curves 1 and 2 in Figure 6 show the polymer adsorption as a function of equilibrium polymer concentration. The greater adsorption from strong brine (Curve 2) is striking compared to the adsorption from distilled water. The adsorption from 2 and 10% NaCl brine is the same. This result is certainly related to the hydrodynamic volumes of molecules. The hydrodynamic volume of polymer molecules does not change upon increasing the salt concentration from 2% to 10% NaCl. Therefore, during adsorption of polymer molecules onto the surface, the compaction of the layer should not be affected by salt concentration over 2% NaCl. This experimental finding is in disagreement with the data from Smith. Since the adsorption increases with polymer concentration, one has to assume a multilayer adsorption.

The smaller adsorption from distilled water is evident from Curve 1 in Figure 6. It is also important that over 140 mg/l equilibrium polymer concentration, the adsorption levels off. This fact suggests a monolayer type adsorption.

Curve 3 in Figure 6 shows the adsorption values in the first desorption cycle after 3 hours of soaking as described in the Procedure section. The results obtained were plotted against the first equilibrium concentration. The very striking fact is that the 'residual adsorption' values in this first cycle of desorption only slightly depended on the original polymer and salt concentration. This result suggests that after a short

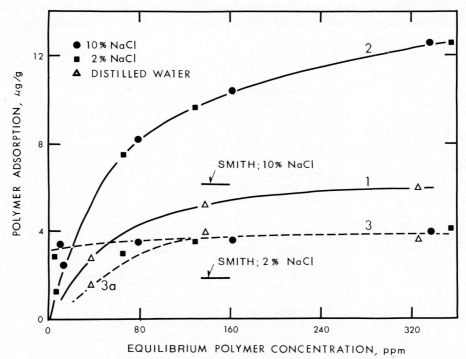

Fig. 6.   Static adsorption and desorption isotherms in a silica sand.

desorption time a monolayer coverage exists on the surface regardless of whether distilled water or strong brine was applied during the adsorption cycle. The difference between the values in the adsorption cycle (Curve 2) and desorption cycle (Curve 3) in strong brine is significant. These results show that as soon as the equilibrated polymer solution is replaced by brine, the outer polymer layers desorb at a relatively fast rate. The difference between the adsorption values in the adsorption cycle (Curve 1) and after a few hours of desorption in distilled water is small. This fact shows that the structure of the adsorption layer in an equilibrated polymer solution in distilled water should be close to a monolayer form. The desorption curve for distilled water (Curve 3a) shows a small decline from Curve 3 at lower polymer concentration values since the curve in the adsorption cycle also shows an imperfect monolayer coverage below 140 mg/l equilibrium polymer concentration.

Figure 6 also shows Smith's [10] values on silica. The surface area in this earlier study was 1 m^2/g sand determined by BET technique. In our study, the surface area was 0.123 m^2/g sand determined by BET technique. (The surface area calculated from permeability and porosity gave a 0.119 m^2/g value using equations from Pirson [14].) Therefore, Smith's mg/g values were divided by 8.13 in Figure 6.

In order to determine the effect of time on adsorption, the sand samples were submitted to further radioactive analysis after different periods of soaking time. The values from these desorption investigations are listed in Table I. Although the ex-

TABLE I

	Equilibrated polymer concentration in static adsorption test (ppm)				
10% NaCl	337.5	160.5	78.8	12.55	
2% NaCl	355.9	176.2	87.1	5.39	
Dist. water	594.9	326.7	138.4	37.09	

	Residual adsorption in μg/g after rinsing in 10% NaCl (sand analysis)				Average	Desorption time
#1 Run	3.959	3.568	2.494	3.300	3.747	3 hours
#2 Run	4.019	2.919	2.735	2.982	3.164	19 days
#3 Run	3.648	3.052	2.711	2.336	2.937	26 days
#4 Run	3.278	2.687	2.237	2.205	2.601	80 days

	Residual adsorption in μg/g after rinsing in 2% NaCl (sand analysis)				Average	Desorption time
#1 Run	4.176	3.420	2.910	2.803	3.327	3 hours
#2 Run	3.411	2.934	2.191	2.303	2.709	19 days
#3 Run	3.404	2.471	2.279	1.927	2.520	26 days
#4 Run	3.617	2.453	2.421	2.430	2.730	80 days

	Residual adsorption in μg/g after rinsing in distilled water (sand analysis)				Average	Desorption time
#1 Run	3.382	3.586	3.910	1.531	3.102	3 hours
#2 Run	1.698	1.592	2.265	1.322	1.719	10 days
#3 Run	1.965	1.442	1.718	1.180	1.576	17 days
#4 Run	1.770	1.484	1.737	1.079	1.517	64 days

perimental values showed a slight scattering, the average values showed a definite decrease in the amount of adsorbed quantity as a function of time. This relationship is illustrated in Figure 7. The amounts of 'residual adsorption' approach minimum values in each soaking fluid. It is anticipated that in a desorption cell fed by continuous injection of a fresh soaking fluid or in porous media in a desorption cycle, the rate of polymer desorption would be higher than it is seen in Figure 7.

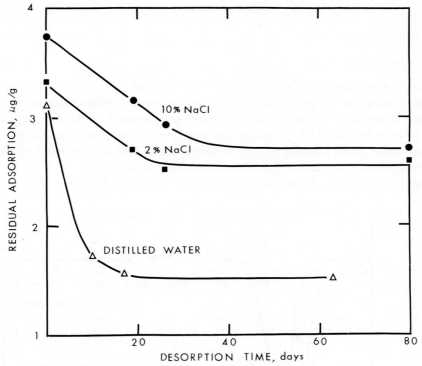

Fig. 7. The effect of desorption time on residual adsorption.

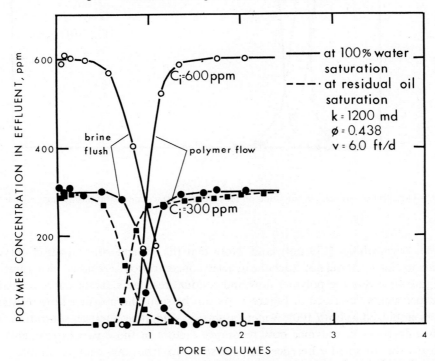

Fig. 8. The effect of residual oil saturation on the effluent polymer concentrations during polymer flow and brine flush.

4.2. DYNAMIC RETENTION

4.2.1. *Single-Phase Flow*

Figure 8 shows the effluent polymer concentration as a function of injected pore volumes in the highest permeability (1200 md) sand used. Both at 300 and 600 ppm polymer concentration, the effluent concentration attained the injected concentration level. When residual oil saturation was present, the characteristic S shaped effluent curve in the adsorption cycle is somewhat deformed. However, the effluent concentration did reach the injected concentration level.

Figure 9 shows the effluent curves at 300 ppm injected polymer concentration for

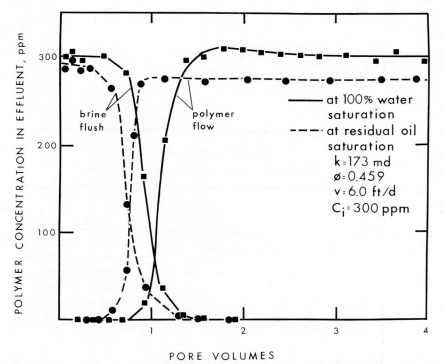

Fig. 9.   The effect of residual oil saturation on the effluent polymer concentrations during polymer flow and brine flush.

a lower permeability (173 md) sand. Note that the effluent concentration curve in the adsorption cycle did not reach the injected concentration at residual oil saturation.

Figure 10 shows the polymer flow and residual resistance factor curve related to the experiments described in Figure 8. As can be seen, the polymer flow resistance factors stabilized at both 100% water saturation, and at residual oil saturation. The small differences in polymer flow resistance factors in these two experiments are probably due to small differences in pore structure from one pack to another.

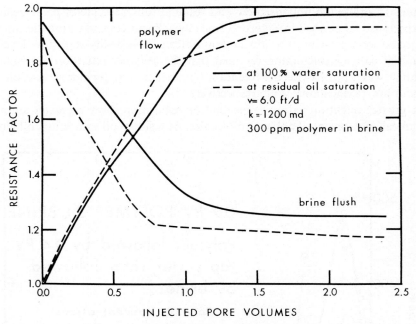

Fig. 10.   The effect of residual oil saturation on polymer flow and residual resistance factors.

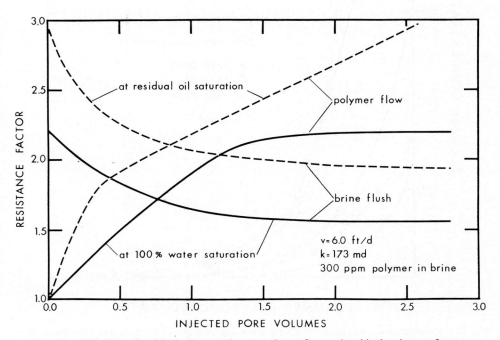

Fig. 11.   The effect of residual oil saturation on polymer flow and residual resistance factors.

Figure 11 shows the polymer flow and residual resistance factor curve related to the experiments described in Figure 9. Since the effluent concentration did not reach the injected concentration at residual oil saturation, an unsteady-state flow developed.

Material balance calculation revealed that the polymer retention at residual oil saturation was 1.5_4 or twice the retention obtained at 100% water saturation in the high and low permeability sand, respectively.

The greater retention values at residual oil saturation show a greater amount of mechanical entrapment of polymer molecules. At residual oil saturation, oil globules

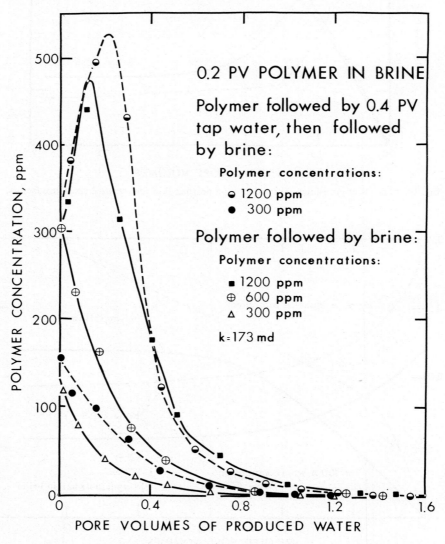

Fig. 12.   The effect of an intermittent 0.4 PV tap water slug on polymer concentrations in produced water during oil recovery.

occupy one portion of the total cross-sectional area for flow, mainly in the large pores. The polymer flow must occur in reduced size pores. This phenomenon results in a greater retention caused by mechanical entrapment. The reduced pore size at residual oil saturation results in increased linear velocity at residual oil saturation at the same injection rate. This increased linear velocity can lead to the development of an unsteady-state flow which is evident in Figure 11.

4.2.2. Two-Phase Flow

The polymer concentration in the produced water has been measured during many oil recovery tests using different permeability sand packs. Results from a 173 md sand appear in Figure 12. In all tests the polymer was dissolved in brine, and a 0.2 PV slug size was applied. It is to be noted that more polymer appeared in the produced water when a 0.4 PV tap water slug was injected after the polymer slug. The relatively high polymer concentration in the produced water at the water breakthrough show that no connate water bank was formed ahead of the polymer slug in any of these tests. It is also interesting to note that the leading edge polymer concentration of the slug drops to about one-half of the injected value. The originally injected 'square-shaped' slug spreads out, taking on a 'bell-shaped' form. The bulk of the polymer appears in the first 0.4 to 0.6 pore volumes of the produced water after breakthrough. This means that the original volume of the polymer slug has roughly doubled or tripled.

Figure 13 illustrates the same type of curves when a 0.4 PV polymer slug was applied in brine and then followed by continuous injection of brine. The peak concentration values are almost two-thirds the injected values. The polymer slugs have spread out to approximately twice the original injected volume.

Figures 14, 15 and 16 illustrate the effect of slug size on produced water polymer concentration at selected injection polymer concentrations in the 173 md permeability sand. When the polymer solution was continuously injected, the effluent concentration did not attain the injection concentration level.

Figure 17 compares the effluent polymer concentration curves for different salinity waters. In one test the polymer was dissolved in brine and followed by brine; while in the other test, the polymer was dissolved in tap water and then followed by fresh water. Although, the same polymer concentration and slug size were applied in both tests, considerably less polymer appeared in the produced water immediately after the water breakthrough when tap water was used. After about 0.3 PV of produced water, this situation was reversed.

Figure 18 shows the same types of curves for a 1200 md sand. It is noteworthy that a very sharp polymer breakthrough occurred at 0.4 PV's of produced water in the two tests when a 0.4 PV brine injection preceded the polymer slug. In these two tests the length of the sandpacks (24 cm) was twice as much as in all other tests.

Figure 19 illustrates the effect of injected concentration on the effluent polymer concentration curve when a 0.4 PV polymer slug size was applied.

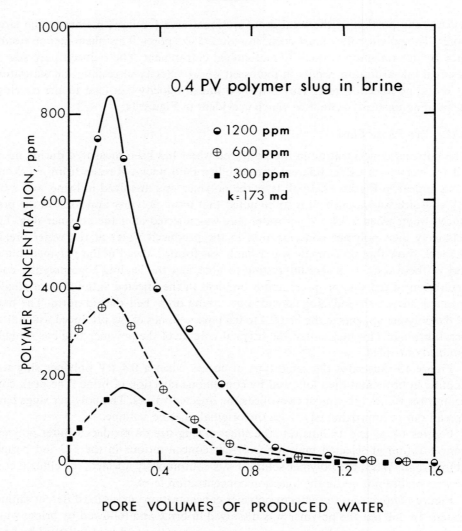

Fig. 13.   The effect of injected polymer concentration on produced water polymer concentration during oil recovery.

The effect of an intermediate 0.4 PV tap water slug on the effluent curves is shown in Figure 20.

Figure 21 shows the effect of salinity on recovered polymer in the produced water. When the polymer is dissolved in tap water, less polymer appears in the produced water immediately after the breakthrough than in polymer-in-brine applications. However, after 0.4 PV of produced water, more polymer is recovered when tap water is used.

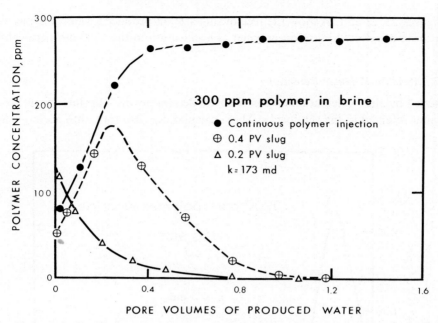

Fig. 14.   The effect of slug size on polymer concentration in produced water during oil recovery.

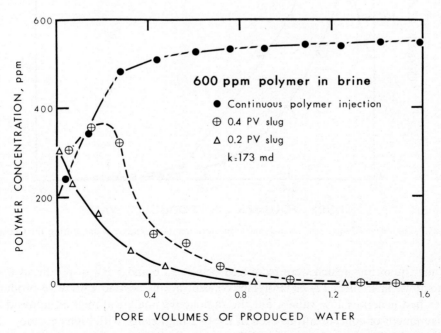

Fig. 15.   The effect of slug size on polymer concentration in produced water during oil recovery.

Figures 22, 23 and 24 show the role of slug size on produced water polymer concentration at selected injection polymer concentration in the 1200 md permeability sand.

4.2.3. *Absolute Polymer Retention*

Material balance calculations were used to determine the absolute quantities of polymer retained in the sandpacks. These calculations also took into account those

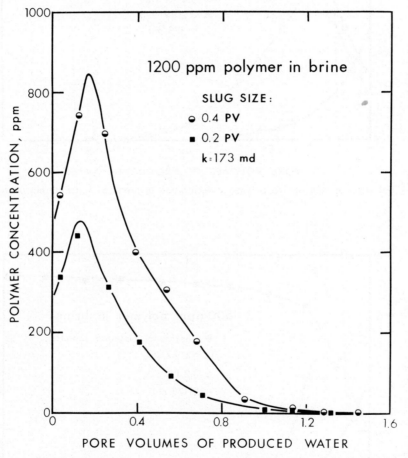

Fig. 16.   The effect of slug size on polymer concentration in produced water during oil recovery.

polymer quantities which were recovered between 1.6 and 5 PV of produced water (not shown in Figures 12–24). Polymer concentrations beyond 1.6 PV of produced water had generally low values, but the radioactive analytical method allowed the measurement of concentrations even in the 0–2 mg/l range with high accuracy.

Figure 25 illustrates the results of these material balance calculations for several oil recovery tests. The polymer retention is expressed as lbs/acre/foot values. In order to get μg/g dry sand values, multiply the lbs/acre/foot values by 0.257 and 0.247 factors in the 173 md and 1200 md sand, respectively.

The curves in Figure 25 demonstrate that in all oil recovery tests, mechanical

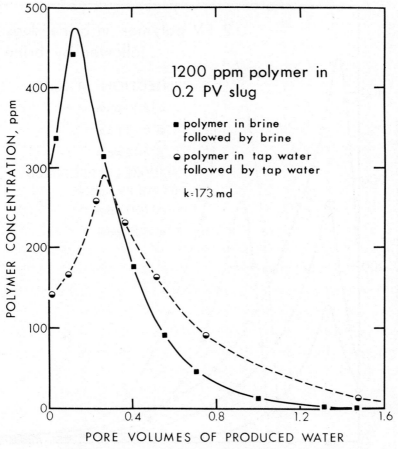

Fig. 17. The effect of salinity on polymer concentration in produced water.

entrapment was the main factor in determining total polymer retention. In Figures 6 and 7 we have seen that a 'residual polymer adsorption': on the 1200 md sand is in the range of 3 to 4 μg/g or less with prolonged desorption time. This value was almost independent of the polymer concentration used in the adsorption cycle. However, the polymer retention values in Figure 25 show a linear dependence upon the polymer concentration used. The absolute values of polymer retention in Figure 25 are one to five times the values determined for 'residual polymer retention' in static adsorption tests.

In the low permeability sand the absolute retention versus applied concentration curves show a leveling off section. This fact is related to the mechanism of polymer entrapment in small pores. In very small pores (173 md sand), less polymer concentration is needed to effectively restrict a continuous polymer buildup in these pores.

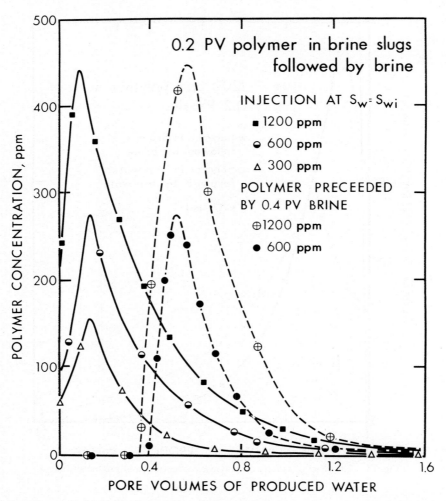

Fig. 18.   The effect of initial polymer concentration and initial water saturation on polymer recovery during displacement of oil.

When a larger (0.4 PV) slug size is applied, the flow resistance increases in the main flow channels whereby more polymer enters the small pores. Therefore, the leveling off section of the retention versus concentration curve begins at higher polymer concentrations.

The role of mechanical entrapment is well seen from Curves 5 and 6 in Figure 25.

Curve 5 shows the retention values in the 173 md sand when a 0.2 PV polymer-in-tap-water followed by brine injection was applied. The retention values are equivalent or higher than values obtained in polymer-in-brine followed by brine injection. The high retention values are explained by the larger size of the molecules in fresh water. Larger molecules can be mechanically entrapped easier in fine pores. When

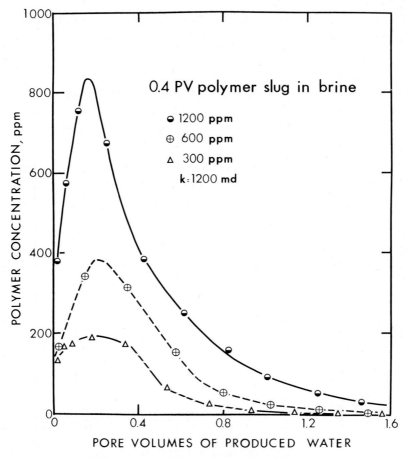

Fig. 19. The effect of initial polymer concentration on polymer concentration in produced water during oil recovery.

the pores are larger (in 1200 md sand) there is an adequate number of large pores which let the molecules through with low flow restriction. Curve 6 shows the polymer retention from oil recovery tests when polymer in tap water slug was followed by continuous injection of tap water. Although the hydrodynamic volumes of polymer molecules were greater than in brine, less mechanical entrapment occurred. In brine, the molecules are coiled up (Curves 1 and 2 in Figure 25). They can penetrate the smaller pores where they can be entrapped.

The role of slug size is evident from Curves 1, 2, 3, and 4 in Figure 25. These curves show that as the polymer slug passes a given location, the polymer retention becomes nearly completed. An increase in slug size from 0.2 PV to 0.4 PV results in only a slight change in the polymer retention.

4.2.4. *Percent Polymer Retention*

Figure 26 shows the percent polymer retention as a function of cummulative produced water in a few oil recovery tests in the low permeability sand. It is noteworthy

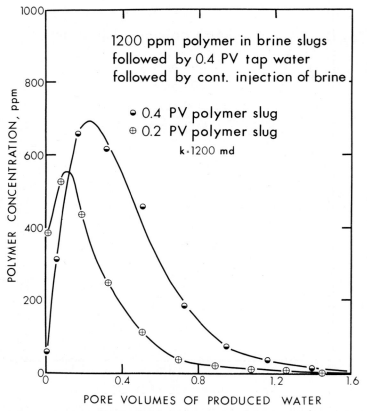

Fig. 20.   Polymer recovery in oil displacement tests when the polymer slugs are followed by a 0.4 PV intermittent tap water slug.

that whenever the polymer is applied in a tap water slug, the rate of polymer recovery is less than in brine application. When a polymer-in-fresh-water followed by brine injection is applied, the final polymer retention is higher than in the polymer-in-brine followed by brine application.

The polymer recovery during the polymer-in-fresh-water followed by fresh water injection test, reaches the same value as obtained in polymer-in-brine followed by brine test at higher pore volumes of produced water.

Similar relationships are shown in Figure 27 for the higher permeability sand. When the polymer is dissolved in fresh water, the rate of polymer recovery in the produced water is less than when it is dissolved in brine. However, the polymer recovery in brine application slows down as larger amounts of water are produced.

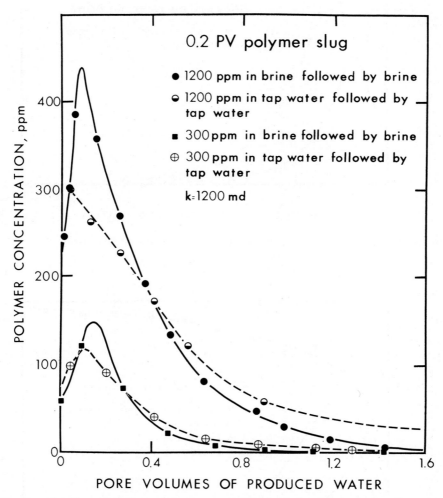

Fig. 21.   The effect of salinity on polymer concentration in produced water during oil recovery.

Finally, more polymer is recovered when fresh water is applied. The exact explanation of this phenomenon is not yet known. However, this phenomenon is certainly related to the low mobility of the individual polymer molecules in fresh water.

Figure 28 shows the precent polymer retention as a function of applied polymer concentration.

324

M. T. SZABO

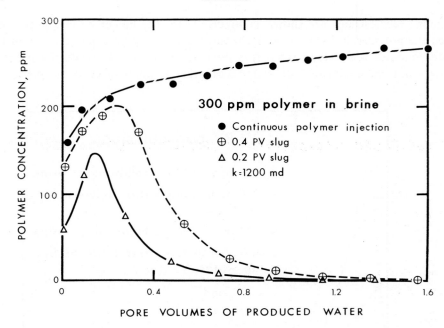

Fig. 22.   The effect of slug size on polymer concentration in produced water during oil recovery.

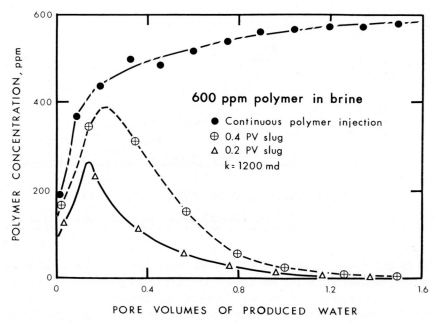

Fig. 23.   The effect of slug size on polymer concentration in produced water during oil recovery.

4.2.5. *Distribution of Retained Polymer*

Figure 29 demonstrates the role of pore structure on distribution of retained polymer. The exact shape of the curves in unknown since one has to know the retained polymer minimum at three locations to determine the constants in Equation (1). It should be noted that because of the $C_{ret, \infty}$ constant on the right-hand side of Equation (1), a

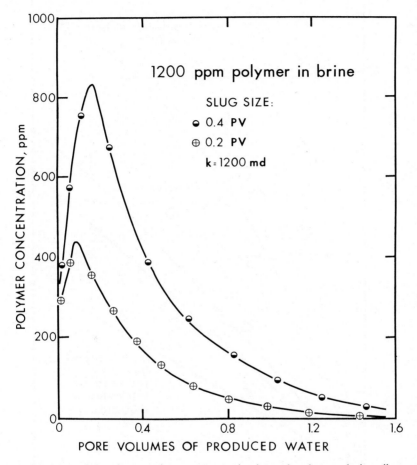

Fig. 24.   The effect of slug size on polymer concentration in produced water during oil recovery.

plot of retained polymer versus distance relationship in semi-log scale is only approximately linear. Therefore, the connection of the measurement points with a straight line serves only a demonstrative purpose.

The measurement points represent the polymer retention at one and six cm distances from the injection face after polymer flooding tests. The injected polymer solution volumes during oil recovery cycles are shown in Figure 29. After the oil recovery cycles, 5–6 pore volumes of brine were injected through the sandpacks.

From Figure 29, it is apparent that during these tests the polymer retention by mechanical entrapment was the main factor in the total polymer retention, especially in the first half of the sandpacks.

All retention values are considerably higher than those determined in the desorp-

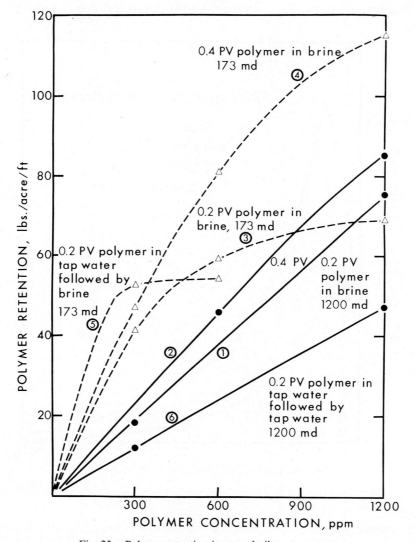

Fig. 25.   Polymer retention in several oil recovery tests.

tion cycles in the static adsorption tests. It is also evident that once the polymer retention by mechanical entrapment is completed, additional volumes of injected polymer solution or large volumes of injected brine do not influence the characteristic distribution of retained polymer by mechanical entrapment.

The effect of pore structure is evident. In lower permeability sand, the slope of the retained polymer vs length curve is great (Curves 1 and 2). In higher permeability materials the corresponding slopes have lower values (Curves 3 and 4).

Figure 29 can explain a few interesting peculiarities of Figure 25. If the physical adsorption were the dominant factor in determining the polymer retention, one

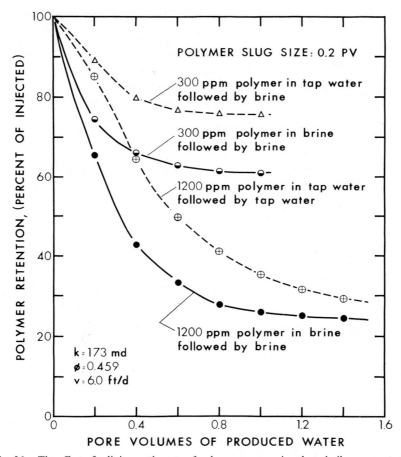

Fig. 26.   The effect of salinity on the rate of polymer recovery in selected oil recovery tests.

should expect a twofold retention in the 173 md sand over the retention in 1200 md because of a two times larger surface area. But Figure 29 shows that close to the inlet face, the polymer retention in the 173 md is about 5 times the value in the 1200 md sand. However, the retention rapidly decreased as a function of distance in the 173 md sand. At the center of the sandpacks, the polymer retention is approximately the same in both porous materials.

Another peculiarity of Figure 25, namely the leveling section of Curve 3, can also be correlated with Figure 29. The retention values obtained at 1 cm location in the

173 md sand shows that the mechanical entrapment at higher polymer concentration is not directly proportional to the polymer concentration. Further from the injection face (at 6 cm location) where the leading edge of the polymer was less concentrated in polymer, the polymer retention was almost doubled in the 600 ppm test compared to the retention in the 300 ppm test.

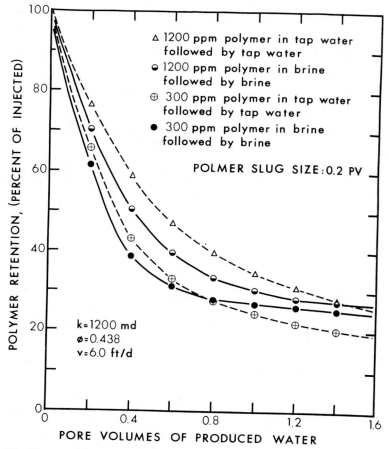

Fig. 27. The effect of salinity on the rate of polymer recovery in selected oil recovery tests.

Curve 5 in Figure 29 shows a more rapid decrease in retained polymer as a function of distance when a non-wetting phase was not present.

Figure 30 shows the distribution of retained polymer in two sandpacks. Preceding a 0.2 PV polymer slug in brine, a 0.4 PV brine slug was injected. After the polymer slug, five pore volumes of brine were injected.

The exponential shape of the curves is well seen. The F_1 and F_2 factors in Equation (1) have been determined from the slope of the curves. Regardless of which curve was used to determine the F_1 and F_2 factors, their values were the same. Using the actual values of F_1 and F_2, the general form of the equation describing the distribution of retained polymer in these tests is as follows:

$$C_{\text{ret}} = 0.035036 \, C_i e^{-0.080571} + C_{\text{ret}, \infty}.$$

The values of $C_{\text{ret}, \infty}$ can be determined by substituting any actual retention number

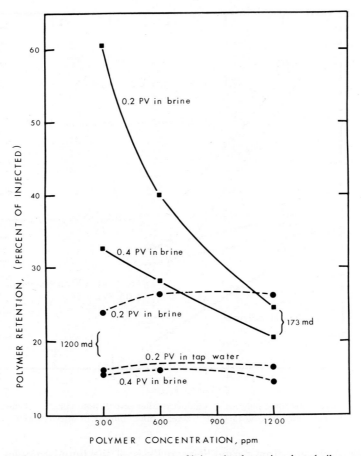

Fig. 28.   The polymer retention as a percent of injected polymer in selected oil recovery tests.

from the curves to Equation (1). So doing, values of $C_{\text{ret}, \infty}$ are equal to 6.63 and 3.31 μg g^{-1} for the 1200 ppm and 600 ppm test, respectively. The dependence of $C_{\text{ret}, \infty}$ on the polymer concentration shows that its value is related to the retention by mechanical entrapment phenomenon at infinite distance. It is also worthwhile to compare these values to the 'residual polymer adsorption' values from static adsorption tests (Table I).

4.3. Unsteady-state flow

Figure 31 illustrates the obtained resistance factors as a function of time in the experiments described in the Procedure section. In the first pack, the resistance factor buildup is much faster than in the second pack. The first stop in flow resulted in a significant recovery in permeability to polymer solution. After resuming the injection, several hours were required to regain the maximum polymer flow resistance

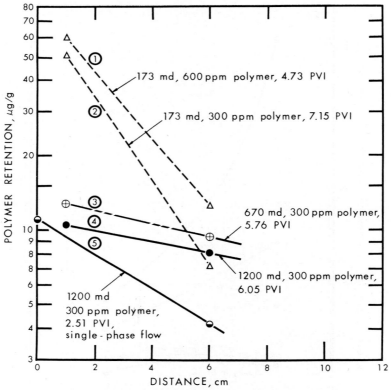

Fig. 29. Distribution of retained polymer in sandpacks after several pore volumes of injected polymer during oil recovery and thereafter, several pore volumes of injected brine.

factor of the preceding flow period. The analysis of the liquid samples produced shortly after the resumption of injection showed 30–40 mg/l higher polymer concentration than in the previous flow cycle. Later, the polymer concentration again decreased and reached a stable value. The linear dependence of resistance factor upon time is very definite. The effect of the second shutdown of the flow is also striking; an approximately 2.5 to 3 fold increase in permeability was obtained. The times elapsed in both no-flow periods (3 and 11 minutes respectively) are very short. The dramatic effect of these stoppages on flow behavior can only be explained by the 'proposed mechanism of polymer retention' during unsteady-state flow concept.

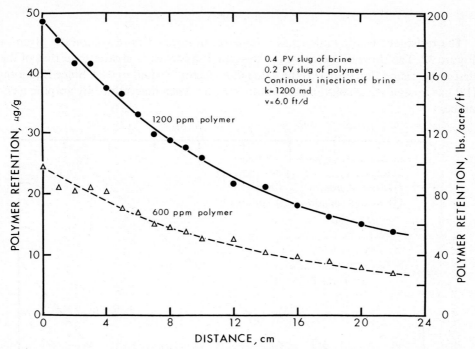

Fig. 30.   Distribution of retained polymer in sandpacks after completion of oil recovery tests.

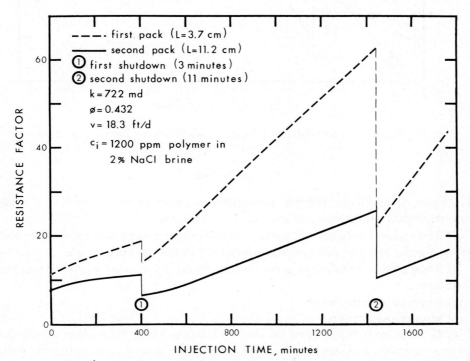

Fig. 31.   Resistance factor buildup in different segments of a linear system during unsteady-state flow as a function of time.

The resistance factor buildup as a function of injected pore volumes is given in Figure 32. The slope of the curve of the second sandpack is greater than that of the first pack. Using Equation (17) the value of F_3 factor is equal to 0.24. Since Equation (17) has no analytical solution, the value of F_3 has been determined by graphic technique.

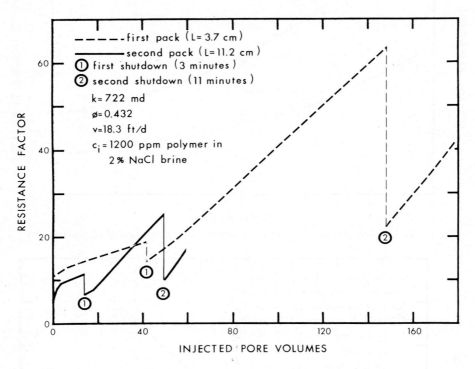

Fig. 32.   Resistance factor buildup as in Figure 31, as a function of injected pore volumes.

5. Conclusions

(1) Mono- or multilayer polymer adsorption can develop on the surface of silica sand depending on the polymer and salt concentration.

(2) The residual polymer adsorption depends slightly upon desorption time and on the salinity and polymer concentrations used in the adsorption cycle.

(3) The polymer retention in porous media during polymer solution flow has two basic forms:

(a) mechanical entrapment,

(b) physical adsorption.

(4) In laboratory experiments on short sandpacks made from clean sands, the polymer retention by mechanical entrapment exceeds by several times the retention caused by physical adsorption.

(5) The polymer retention far away from the injection face can be determined from laboratory experiments if the polymer retention is measured at a minimum of three locations. The data obtained from these tests can be extrapolated to infinite distances with the use of the equation developed in this paper (Equation 1).

(6) The factors determining the distribution of retained polymer in porous media are as follows:

(a) pore structure,
(b) type of polymer,
(c) salinity, type of salts,
(d) polymer concentration,
(e) fluid saturations.

The effect of velocity was not studied.

(7) No connate water bank has developed ahead of a polymer slug in all oil recovery tests.

(8) Unsteady-state flow develops when at least one critical flow parameter is exceeded. The resistance factor buildup is linear either as a function of time or injected pore volumes.

(9) The rate of polymer buildup in the first segment of a porous body is greater than in the rest of the body.

(10) An unsteady-state type flow is limited to the first segment of an oil reservoir. A critical distance exists in a radial system beyond which unsteady-state flow cannot develop.

(11) The unsteady-state polymer flow resistance factors for any moment and distance can be determined from laboratory experiments, provided that the pressures are measured at an adequate number of locations. Combining these experiments with the determination of the flowing polymer concentration at different locations, the value of C_{crit}, or the absolute quantities of retained polymer can be determined for any location and for any time.

(12) If an unsteady-state type flow develops in an injection well, and it causes injectivity problems, the injectivity of the well could be restored by periodically shutting down the injection.

Appendix A

The determination of critical polymer concentration and F_3 factor based on experimental data can involve some uncertainty using Equation (8).

If, for instance, the stabilized effluent concentration is 3% less than the injected concentration and the accuracy of the concentration determination is $\pm 1\%$, there is already a $\pm 33.3\%$ inaccuracy in the determination of the critical concentration.

Moreover, in order to compute the value of C_{crit} using Equation (8) from polymer concentration determinations, one would have to measure the flowing polymer concentration somewhere in the porous body. This task is rather difficult to implement technically, but it can be done if the porous system consists of at least two independent parts of the same porous structure connected to each other in series. Fluid sample can be taken at the connection between the independent parts. Since the polymer concentration at the connection is higher than at the end of the system, the uncertainty of the determination of critical concentration is further increased.

However, the value of C_{crit} can be determined even if the polymer concentration is measured only at the outlet face of the porous body. In that case, the value of F_3 factor must be known from other sources.

EXAMPLE

Suppose the $F_3 = 0.24$ from an independent study. The injected polymer concentration is 1200 ppm, and the length of the porous body is 14.9 cm. If the effluent concentration is 1100 ppm (representing a 8.33% polymer loss), the value of C_{crit} is 1097.1 ppm, applying Equation (8). Using a lower injection rate, the value of C_{crit} is higher. Lowering the injection rate, one can reach a flow velocity at which $C_{crit} = C_i$, that is, the unsteady-state flow converts to a steady-state flow.

Appendix B

The polymer flow resistance factor vs flow rate curve has a minimum in many cases [10]. It was also shown earlier [11] that over a critical flow rate an unsteady-state polymer flow can develop. In such a case, only the first value of the measured resistance factor will be on the resistance factor versus flow rate curve on its imaginary section (Figure 33).

The polymer flow resistance factor shows a characteristic distribution as a function of distance in steady-state flow. This uniform distribution is due to the first invasion

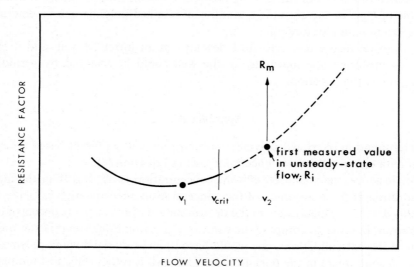

FLOW VELOCITY

Fig. 33. The real and imaginary section of the resistance factor vs flow velocity curve.

of the porous body with polymer solution. After this first invasion with polymer solution, the resultant polymer retention is an exponentially decreasing function of distance. Since the resistance factor involves the effect of permeability reduction and viscosity, the polymer flow resistance factor decreases with distance. This distribution of polymer flow resistance factor in steady-state flow is called 'original' polymer flow resistance factor distribution. This curve can be rather flat under certain conditions.

Figure 34 shows the resistance factors at selected distances from the injection face

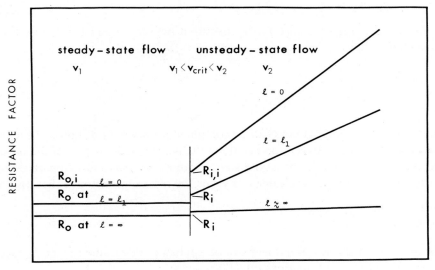

Fig. 34.   Steady-state and unsteady-state polymer flow resistance factors at different distances from the injection face as a function of time.

during steady-state and unsteady-state flow as a function of time. When a low velocity (V_1) is applied in a steady-state flow, values of polymer flow resistance factors (R_o) do not depend on time. When the flow velocity is increased over a critical value (V_2) a continuous polymer buildup occurs in any segment of the porous body. The fastest rate of polymer buildup occurs at the injection face. The first values of the polymer flow resistance factors in unsteady-state flow determine the R_i initial resistance factors at selected distances from the injection face. These values determine the distribution of the R_i initial resistance factors as a function of distance (Figure 5).

However, neither the 'original' nor the initial resistance factor can be measured at a given distance from the injection face. The pressure difference measured between two points in the system is determined by the integral average flow resistance between these two points. Scrupulously the integral average resistance factor calculated from pressures measured at two distances does not belong to the midpoint of the finite distance difference in question.

However, the exact values of the polymer flow resistance factors for any given distance coordinate can be determined if the pressure drops are measured in at least three segments of the porous body. The analytical solution of this problem is given in Appendix C.

Appendix C

Suppose that the initial resistance factor distribution during unsteady-state flow is given by the equation below:

$$R_i = R_{i,i} e^{-F_5 l}. \tag{17}$$

The integral average resistance factor measured between 0 and l distances can be expressed as follows:

$$\bar{R}_i = \frac{1}{l} \int_0^l R_{i,i} e^{-F_5 l}. \tag{18}$$

The average resistance factors can be computed for several l distances if the pressures are measured at different points along the porous body. A plot of the product of the average resistance factors and distances can be made as a function of distance. The slope of this curve at any value of l yields the resistance factor at that distance:

$$\frac{\partial \bar{R}_i l}{\partial l} = R_{i,i} e^{-F_5 l}. \tag{19}$$

The same mathematical treatment can be applied to determine the F_3, F_4 factors and the value of C_{crit} in Equations (9)–(12).

The average excess polymer flow resistance factor between $0-l$ distance can be expressed as follows:

$$\bar{R}_m - \bar{R}_i = -\frac{1}{l} \int_0^l F_4 F_3 (C_i - C_{\text{crit}}) e^{-F_{3l}} tq \, dl. \tag{20}$$

Applying the same principle as was used in connection with Equations (18) and (19), the relationship below can be derived:

$$\frac{\partial (\bar{R}_m - \bar{R}_i) l}{\partial l} = F_4 F_3 (C_i - C_{\text{crit}}) e^{-F_{3l}} tq. \tag{21}$$

The values of F_4, F_3 factors and C_{crit} can be calculated from the slopes of the $(R_m - R_i) l$ vs l curve determined at three different l coordinates.

References

1. Burcik, E. J.: 'A Note on the Flow Behavior of Polyacrylamide Solutions in Porous Media', *Prod. Monthly* **29**, 14 (June, 1965).

2. Sadowski, T. J.: Ph.D. Thesis, University of Wisconsin, Madison, Wis., 1963.
3. Gogarty, W. B.: 'Mobility Control with Polymer Solutions', *J. Pet. Tech.*, 161 (June, 1967).
4. Mungan, N., Smith, F. W., and Thompson, J. L.: 'Some Aspects of Polymer Floods', *J. Pet. Tech.*, 1143 (Sept., 1966).
5. Mungan, N.: 'Rheology and Adsorption of Aqueous Polymer Solutions', *J. Cdn. Pet. Tech.* **8**, 45 (1969).
6. Mungan, N.: 'Improved Waterflooding Through Mobility Control', *Cdn. J. of Chem. Engineering* **49**, 32 (February, 1971).
7. Harrington, R. E. and Zimm, B. H.: 'Anomalous Plugging of Sintered Glass Filters by High Molecular Weight Polymers', *J. Polymer Science*, Part A-2, **6**, 294 (1968).
8. Rowland, F. W.: 'Thickness and Structure of Layers of High Polymers Adsorbed from Solution onto Solid Surfaces', Thesis, Brooklyn Polytechnic Institute, Brooklyn, N.Y., 1963.
9. Jennings, R. R., Rogers, J. H., and West, T. J.: 'Factors Influencing Mobility Control by Polymer Solutions', *J. Pet. Tech.*, 391 (March, 1971).
10. Smith, F. W.: 'The Behavior of Partially Hydrolized Polyacrylamide Solutions in Porous Media', *J. Pet. Tech.* (February, 1970), 148.
11. Szabo, M. T.: 'Molecular and Microscopic Interpretation of the Flow of Hydrolyzed Polyacrylamide Solution Through Porous Media', SPE 4028 paper presented at the *47th Annual Fall Meeting of SPE of AIME*, San Antonio, Oct. 8–11, 1972.
12. Bondor, P. L., Hirasaki, G. J., and Tham, M. J.: 'Mathematical Simulation of Polymer Flooding in Complex Reservoirs', SPE 3524 paper presented at the *46th Annual Fall Meeting of SPE of AIME*, New Orleans, Oct. 3–6, 1971.
13. Eirich, F. R., Bulas, R., Rothstein, E., and Rowland, F.: 'Structure of Macromolecules at Liquid-Solid Interfaces', *Chemistry and Physics of Interfaces*, ACS publication, Washington, 1965.
14. Pirson, S. J.: *Oil Reservoir Engineering*, McGraw-Hill Book Co., 1958.
15. Ter-Minassian-Saraga, L.: 'Chemisorption and Dewetting of Glass and Silica', *Advances in Chemistry Series* **43**, ACS publication, 1964, p. 232.
16. O'Connor, D. J. and Buchanan, A. S.: 'Electrokinetic Properties and Surface Reactions of Quartz', *Faraday Society, Transactions, V.* **52**, 387 (1956).

INDEX OF SUBJECTS

Polyvinylimidazole, 94
Polyvinylimidazoline, 188
Polyvinylpyridine, 82
  N-oxide, 188
Potentiometric behaviour, 22, 37
  in hydro-organic mixtures, 22
  in urea solutions, 22, 24
  titration, 6, 7, 10, 17, 19, 21, 23, 25
Proflavine, 218
Propionylcholine, 178
Putrescine, 164
Pyrogenicity, 135

Quaternization, 250
  of polyvinylpyridine, 82, 90

Radioactivity, 289, 292, 293
Rate
  of copolymerization, 108
  of floculation, 56
Rausher leukemia virus, 135
Receptor, 175–184
Red blood cells, 66
Redox reactions, 59
Repulsive coulombic interactions, 17
Residual oil saturation, 288, 291, 310
Residual resistance factors, 292, 306, 310
Reticuloendothelial system, 133
Rotating disk, 275
Rotatory power, 22

Salt effect (primary), 71, 72, 73, 75, 78
Sandpacks, 291, 328, 332
Sheep red blood cells, 131
Short-range interactions, 6
Silane coupling agents, 119, 120
Silica sands, 289
Site binding, 42
Size distribution, 63
Slug sizes, 287, 298
Solvation, 108
Solvent, 5
  composition, 22
  effects, 108
  gradient column method, 110
Specific conductivity, 35, 36
Spermidine, 164
Spermine, 163
Staudinger (theory of), 97
Steric effects, 107

Styrafoam, 125, 126
Styragel, 113
Sulfated polysaccharides, 141
Surface
  charge densities, 51
  electrical potential, 56, 66
  groups, 59
Synapses
Synthetic water-soluble polymers, 3

Terpolymerization, 104
Tetraethylenpentamine, 75
Tetrahydrofuran, 108
Tetrahydronaphtoquinone, 109
Thermal transconformation, 197
Thermodynamic parameters, 48
Thrombocytopenia, 135
Titration curves, 26
Torpedo californica, 175, 176, 183
Toxicity factor, 113
Transition state of propagation, 103
Transmitter, 175, 179
2,4,6-Trinitrostyrene, 104
Tryptamin hydrochloride, 90
d-Tubocurarine, 175, 178
Turbulent pipe flow, 275

Ultrasonic absorption, 41
Ultraviolet, 107
Unsteady state flow, 287, 298–302, 311–316, 317
Urea, 23, 29, 81, 82
Uronic acid content, 148

Vinyl ethers, 3
Vinylidene cyanide, 104
4-Vinylpyridine, 104, 233
Viral infections, 99
Virus, 135, 137
Viscosimetric behaviour
  in hydro-organic mixtures, 22
  in urea solutions, 22
Viscosity, 17
  intrinsic, 5, 136, 228
  measurements, 26

Water treatment, 273–276

X-ray polymerization, 108

Yellow OB, 8